MOUNTAIN MONARCHS

WILDLIFE BEHAVIOR AND ECOLOGY
George B. Schaller, Editor

MOUNTAIN MONARCHS

Wild Sheep and Goats of the Himalaya

George B. Schaller

THE UNIVERSITY OF CHICAGO PRESS

Chicago and London

The University of Chicago Press, Chicago 60637
The University of Chicago Press, Ltd., London

Library of Congress Cataloging in Publication Data

Schaller, George B
  Mountain monarchs.

  (Wildlife behavior and ecology)
  Bibliography: p.
  Includes index.
  1. Bovidae. 2. Alpine fauna—Himalaya Moun-
tains. 3. Mammals—Himalaya Mountains. I. Title.
QL737.U53S3   599'.7358   77-1336
ISBN 0-226-73651-2

To William G. Conway

CONTENTS

PREFACE

While studying wildlife in peninsular India from 1963 to 1965, I occasionally glimpsed some of the high Himalayan peaks—Kangchenjunga, Nanda Devi, Trisul and others—that form the frozen barrier between the subcontinent and the Tibetan Plateau. On the forested slopes and alpine meadows of these mountains live bharal*, tahr, serow, and ibex, to name just a few of the large hoofed animals whose alien names conjure up visions of mysterious creatures in ice-bound valleys. Here was a whole assemblage of large mammals of which not one species had been studied in detail; here was a fauna about which virtually no information had become available since the 1930s. I yearned to study these animals, but it was not until 1968 that an opportunity first came to do so.

Over the years, the high Asian ranges were penetrated by many outsiders. During the 1800s and early 1900s explorers frequently roamed the uplands, some simply for the joy of discovery, others motivated by political considerations. Whatever the purpose, their books still provide much of the available information on Tibet and the various remote mountain ranges. Not being specifically interested in natural history, most explorers wrote little about wildlife. However, some, such as Przewalski (1884), brought back valuable descriptions and specimens. Sven Hedin (1898), the greatest of Tibetan travellers, mentions mainly those encounters with animals that interrupted his monotonous plodding across the gray wastes, and others, among them Younghusband (1896), were equally restrained in their wildlife observations. Fortunately Rawling (1905) was a perceptive traveller whose book contains valuable notes on animals. During the past eighty years, explorers of another type, mountain climbers, attempted or conquered many of the high Himalayan peaks. But the quest for sterile summits is

*Appendix A gives the scientific names of animals mentioned in the text.

so intense that little effort is expended on the living world, with the result that books about climbing are almost devoid of wildlife information, those by Shipton (1936, 1938) being happy exceptions. By virtue or rather folly of their training, scientists have limited objectives in the wilds, and those who roamed the Himalaya were most often archaeologists (Stein, 1912), botanists (Hooker, 1854; Ward, 1934) and ornithologists (Ali, 1949), to mention only a few whose notes on mammals were not extensive. However, much valuable information was collected by Schäfer (1933, 1937, 1950).

By tradition natural history consisted of watching birds, impaling butterflies on pins, and shooting large mammals, the last-named the domain of the trophy hunters. The virtues of previous generations are the foibles of this one, and so it is with hunting. The quest for trophies in an age when many species barely cling to their existence is anachronistic, yet the fact remains that most knowledge about the distribution and habits of large mammals in the Himalaya was collected by hunters. Starting in the middle of the last century, military officers, civil servants, and others penetrated valleys, crossed passes, and traversed plateaus, hoping for a trophy larger than all previous ones of that species. Some, like Darrah (1898) were mere butchers, but others, notably Kinloch (1892), Macintyre (1891), Stebbins (1912), Demidoff (1900), Wallace (1913), and Burrard (1925) were good observers. In a class by himself was Stockley (1928, 1936) whose books are the best ones on Himalayan wildlife. So many visitors penetrated the Indian Himalaya that by 1864 Kinloch (1892) grumbled that Kashmir had been spoiled by the incursions of tourists. Yet some regions remained untouched, and it was not until around 1890, for example, that Englishmen regularly began to visit the Hindu Kush. By the latter part of the century enough was known about wildlife on the subcontinent for Jerdon (1874), Sterndale (1884), and Blanford (1888–91) to publish compendia about mammals which are still of use today. Yet knowledge about the large mountain mammals has remained anecdotal and fragmentary, in contrast to the detailed studies that have been made in North America, Europe, and Russia (Geist, 1971a; Krämer, 1969; Nievergelt, 1966a; Heptner et al., 1966). The main purpose of this book is to present new information about several of the Himalayan mountain species, particularly about various sheep and goats, with the hope that these pages may stimulate research and promote conservation. I make no attempt to ask "a fine new question, a question that has never been asked before," as Rudyard Kipling's elephant child phrased it, nor is this volume intended as a summary of knowledge of Caprinae ecology and behavior. Although I discuss all species in the subfamily, the material from the literature is included primarily to place my data into perspective and to provide comparisons, and secondarily to present an overview of Caprinae natural history for those readers in Asia and elsewhere who lack access to scientific reports. The book is in effect a companion volume to *The Deer and the Tiger* in which I treated the ecology and behavior of some large mammals of peninsular India.

Taken in a broad geological sense, the Himalaya encompasses not only the Great Himalayan chain, but also the adjoining and connected ranges—the Karakoram, Hindu Kush, Sulaiman, and others—all of which were formed during upheavals in the Tertiary era. No one person can survey, much less study, the wildlife throughout this vast region, even if it were politically possible: a project has to have some limited objectives which might be attained within a few years. Geographically I restricted my work to the Great Himalaya of Nepal and India, and to the ranges in Pakistan. A basic decision was to concentrate on the hoofed animals. These presented some interesting biological problems which intrigued me, and the precarious status of several species demanded a conservation effort on their behalf. Two distinct assemblages of high-altitude hoofed animals exist in the Himalaya. One inhabits the Tibetan Plateau and includes those species which are adapted to the flat and rolling uplands such as the kiang, chiru, Tibetan argali, and yak; the other prefers rather rugged terrain, a habitat used by goats and their relatives, and by some deer, among them the hangul and muskdeer. With Tibet and most of Ladak inaccessible to foreigners, I selected the latter animals, and, to broaden the scope of the project, I included the sheep and goats living at low altitudes in the desert ranges as well.

Any scientific endeavor has to have some focal points beyond mere curiosity and the desire to gather knowledge. In perusing the literature, I had become fascinated by the bharal, commonly called blue sheep, a peculiar animal with physical traits so intermediate between true sheep and goats that taxonomists have had trouble in classifying it. Is the bharal behaviorally and ecologically closer to *Ovis* or *Capra*? To answer that question, I had to observe not only bharal but also various Asian sheep and goats to learn in just what way the bharal resembled or differed from the others. Of course for a species to be studied adequately, it has to be fairly accessible and locally abundant, a rare combination in the mountains, as I soon discovered, for vast tracts have been depleted of wildlife. Although it was not among my original intentions, I ultimately devoted much effort to making surveys with the hope of finding suitable study populations. While surveys took much time and produced few scientific data, they helped to ascertain the status and distribution of species. And I discovered areas which would make good national parks and sanctuaries. To preserve an animal and its habitat is now often more urgent than to study it. In retrospect I derive more satisfaction from having stimulated a government to establish or maintain a reserve than from having collected esoteric facts about a species.

Until recent years most field studies followed two diverging lines of research, one consisting of the ethological approach, mainly that of describing the relatively stereotyped signals used by interacting individuals, and the other ecological, an approach used especially by wildlife managers. Then, in the middle 1960s exciting insights into animal biology were gained by considering the relations between ecology, population dynamics, and social behavior (see Crook, 1965). By checking for possible correlations between habitat, body size, social organization, and food habits, Crook and Gartlan (1966) and Eisenberg et al. (1972) synthesized a large amount of information about primates into a coherent framework, and Geist (1971a), Estes (1974), and Jarman (1974) did the same for certain groups of hoofed animals. These authors elucidated evolutionary trends in primates, antelopes, and sheep, and they directed the thinking of field workers into new and promising lines of research. In presenting my data in this book I have used this socio-ecological approach, and my intellectual debt to my scientific precursors is therefore great.

I began work in October 1968 with a survey of hangul in Kashmir. October and November 1969 were devoted first to a study of Nilgiri tahr in southern India and then to another visit to hangul habitat. From October to December 1970, I was in Pakistan, and from January to April 1972 in Nepal (see table 1). Pakistan and Nepal were at that time more amenable to having an itinerant American zoologist in their mountains than India, and because the first-named country had the greatest variety of sheep and goats it was chosen for intensive work. I remained in Pakistan from June 1972 to May 1974, except for a visit to Nepal to study bharal from October to December 1973. Other trips to Pakistan were made from October to December 1974 and in April and May 1975. Iran and New Zealand were also visited briefly for a look at certain species. The total time devoted to the project abroad was thirty-seven months.

This report has a serious limitation in that my coverage of all species is superficial. Rather than concentrating on one kind of animal in one locality for years, in the accepted manner of current ungulate studies, I moved from area to area, spending one month with one species, then the next one with a different species (see fig. 1). There were several reasons for this diffuse and in many ways unsatisfactory mode of research. When I started the project there was no other zoologist working on high-altitude hoofed animals in that part of the world, and I felt that a widely ranging program would contribute more to conservation and provide more basic data than would one of more limited scope. All research results represent a mere passing phase of life in a changing world, and a baseline of knowledge different from the one established in the 1930s was needed. Furthermore, information on several species is essential for a comparative study, and this required work in several localities.

Some species received more attention than others, for I concentrated on those, such as the urial, wild goat, and bharal, about which little was known, at the expense of those which had been studied elsewhere, namely the ibex and Himalayan tahr. Goral, serow, and muskdeer were wholly slighted because I could not find good study areas, and others, such as the Tibetan argali and Marco Polo sheep, were too rare within my project area to warrant an extended effort. Indian and Bhutanese officials refused me access to takin habitat.

In the final analysis the political vagaries on the subcontinent determine where and for what length of time one can work. Most mountains lie on or near international borders. During the 1960s and 1970s China and India have fought one war, Pakistan and India two wars, and relations between Pakistan and Afghanistan

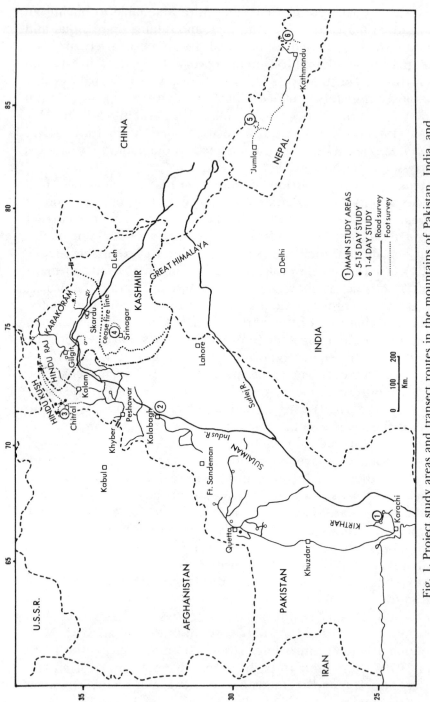

Fig. 1. Project study areas and transect routes in the mountains of Pakistan, India, and Nepal.

have been strained. Kashmir when under British control was one political entity, a mountain tract some 214,000 sq km in size. Following a dispute between India and Pakistan after the Second World War, Kashmir was divided along a cease-fire line with about 60% of the terrain going to India. China later took over about a fifth of India's portion. Still lacking formal borders, Kashmir is sensitive politically, and large tracts of it in India are or were prohibited to foreigners. Pakistan represents a unique situation, the consequence of its turbulent history in the past century. Waziristan and other Special Areas in the arid western mountains are largely under tribal control and closed to outsiders. The northern mountains, most of which constitute Pakistan-held Kashmir, were open intermittently until the Indo-Pakistan war of December 1971. Access was then barred, except for parts of Chitral, Swat, and Dir, until mid-1973. Then from mid-1974 onward the Indus and Karakoram highways were again closed, making access to most ranges difficult. Baluchistan had been peaceful for years and I had chosen some study areas there, but in 1973 political unrest made much of the province unsafe for field work. Given such problems, it seemed unwise, and in fact it was often impossible, to settle for long into a politically sensitive area. To work in the mountains of South Asia one must be flexible.

Two more problems need mention. Distance lends enchantment. To the uninitiated a project among the world's highest peaks sounds romantic. However, logistic difficulties may be so severe that more time is spent travelling to and from an area than in actual research. Because of torrential rains and snowcovered passes, it took five weeks of trekking just to reach one of my bharal study sites. I went to northern Hunza to census Marco Polo sheep, an endeavor that required special permission, only to find that all animals had moved to Sinkiang for the winter. Airplanes, if available, may be delayed for a week or more due to stormy weather: only three out of my nine flights into or out of Chitral left on the scheduled day. To move needed food and equipment into uninhabitated terrain requires porters or some form of animal transport whether it be yak, horse, camel, or donkey. My journals are filled with tirades against obstinate beasts of burden and rapacious porters. But to remember only the unreliable porters of Ringmo in Nepal, who abandoned us at 5000 m, far from the nearest village, without also thinking of the Tamang porters who on the same trip carried our belongings across a pass barefooted in deep snow is to deny man his altruism. Weather can be a problem. When watching wild goats among the barren crags with the temperature at 43° C, the sky and rock and sand a searing furnace, or when seeking bharal at 4900 m on an icy December day, one's mental energy sapped by altitude and by fierce blasts from the Tibetan Plateau, the search for shelter may take precedence over the search for facts. However, one tends to complain of discomfort and relish hardship: the more difficult a journey the greater the satisfaction of having accomplished a task. These comments do not excuse the shortcomings of this work, and I can do no better than to quote Richard Hakluyt, who in 1598 wrote:

"I ... presume to offer unto thy view this discourse. For the bringing of which into this homely and rough-hewn shape, which here thou seest, what restless nights, what painful days, what heat, what cold have I endured; how many long and chargeable journeys I have travelled; how many famous libraries I have searched; what variety of ancient and modern writers I have perused ...."

A few final words. To convey the impression that my research in the Himalaya was motivated solely by science would not be honest. Most men enjoy adventure, they want to conquer something, and in the mountains a biologist can become an explorer in the physical realm as well as the intellectual one. Mountains are symbols of the unknown, of the mysterious force that beckons us to discover what lies beyond, that tests our will and strength against the sublime indifference of the natural world. Research among the ranges affords the purest pleasure I know, one which goes beyond the collecting of facts to one that becomes a quest to appraise our values and look for our place in eternity. When at dusk the radiant peaks are deprived of the sun's fire, leaving them gloomy and desolate with cold prowling their slopes, one feels imprisoned by rock, but when later, white in the moon, the glaciers glow like veils of frozen light all difficulties vanish in the presence of such primordial beauty.

1 THE HIMALAYAN REGION

During the early Cretaceous, more than 100 million years ago, the land that is now Tibet and the Himalayan region was covered by the Tethys Sea which separated Eurasia from the southern continent, Gondwanaland. At that time the Indian Peninsula was still attached to Gondwanaland, but around the middle of the Cretaceous it separated and began an inexorable drift northward. When the two land masses came into contact around the late Cretaceous and early Tertiary, there began the most spectacular period of mountain building in the earth's history. The northern edge of India buckled during the collision, warped down, and slid beneath the Eurasian continent, forming a huge depression which over the eons has filled with sediments to create the Indo-Gangetic plain. Slowly the vertical uplifts and horizontal thrusts, the compressing and raising, moved the floor of the Tethys Sea upward to create the Tibetan highlands. The upheaval of the Himalayan system came in three major phases, the first one during the Eocene, a second after a period of quiescence during the middle Miocene—a period of intense activity some 23 million years ago which raised the Himalaya by intrusion of granites into the vast sedimentary deposits—and finally during the late Pliocene and Pleistocene (Wadia, 1966). Mountain building continued throughout the Pleistocene. In Nepal the Mahabharat Range, now part of the Middle Himalayan system, was at that time pushed up to a height of 3000 m,* and in India the Pir Panjal Range rose some 1800 m. The melting of glaciers from the end of the Pleistocene to the present is thought to cause a continuous slow increase in height as the icy burden is removed from the summits. In the beginning, before the meeting of the continents, only the Aravalli Hills in India and the Altai Mountains existed.

*To convert centimeters to inches multiply by .39, meters to feet by 3.38, kilometers to miles by .62, square kilometers to square miles by .385.

1

By the end of the Pleistocene the southern part of Eurasia had
been upthrust, crumpled, and folded into a huge and complex
mountain system (see fig. 2), its present height and conformation
determined by the nature of its formation. For example, Mount
Everest, 8848 m high, consists of about 4000 m of raised Tibetan
plateau, 4000 m of original Himalaya, and 2000 m of mountain
roots squeezed upward by the shifting of the continents along the
line of contacts, minus whatever erosion has occurred (Hagen,
1963). That the Himalaya were created after the raising of the
Tibetan Plateau is strikingly shown by the fact that the watershed
of many rivers lies not among the high peaks but on the plateau
beyond. Rivers, such as the Arun, being older than the mountains,
had to keep their channels open during the uplift, and in the

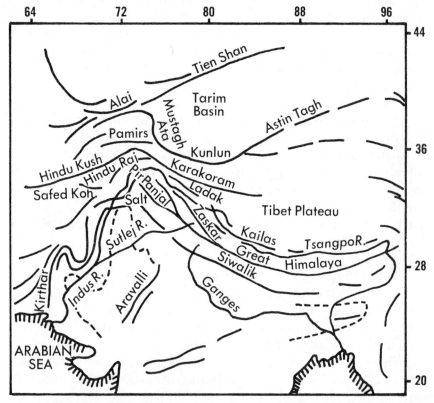

Fig. 2. The major mountain ranges in South and Central Asia. The stippled lines
indicate hidden extensions of the peninsular shield (adapted from Burrard and Hayden, 1907).

process they created the gigantic canyons that slice through the main Himalayan chain.

The parallel alignment of various Himalayan ranges, among them the Ladak and Kailas ranges and the series of some 20 great folds across Tibet which culminate in the Kunlun Range, point to a close structural relationship between these mountains (Wadia, 1966). When the small Indian land mass squeezed against the formidable barrier of Eurasia, sediments of the Tethys Sea were folded not only into the east-west-aligned Himalaya, but also into the north-south ranges in Burma and Baluchistan. In Pakistan this uplifting and horizontal thrusting created the Salt Range, and along the outer rim of the peninsular block several lateral ranges such as the Sulaiman and Kirthar. Wedges of ancient Gondwanaland penetrated as far as Quetta, probably explaining the cluster of high peaks around that city (Mani, 1974). Of course the broad features of all ranges have been modified by erosion working on rocks of different hardness to produce the maze of defiles, serrated peaks, and endless ridges of desolate rock. The Himalayan system in turn has connections with other Tertiary systems, including those of the Tien Shan, Zagros, Caucasus, and Alps, creating a network of peaks stretching from Europe across the Middle East to China and Russia.

The creation of new mountain systems from the sea during the Tertiary opened up new habitats and new pathways of dispersal to the flora and fauna. Movement into the subcontinent could proceed along the Himalaya from the east, from the Indochinese region, and, after the retreat of the Tethys in Iran and Rajasthan during the late Miocene, from Africa and Europe too. It was a period of constant change, with new ranges rising and climates modifying their usual patterns as a result of the upheavals. After having been covered by warm and humid forests for much of the Tertiary, the disappearance of the Tethys and the rise of the mountains changed the uniformity of the climate on the subcontinent (Cracraft, 1973). The Tibetan Plateau was once fairly lush in spite of its 4500 to 5000 m altitude, but the rain-bearing winds were intercepted by the rise of the Himalaya, leaving the land desolate, its many lakes turned into desiccated shallow basins. As Demidoff (1900) noted: "Most of the lakes in Asia, and especially in Tibet, have that well-defined line above their present level; they appear to be gradually drying away." The Zaidam, once one of the largest lakes in the world, consists now of marshes and dusty flats. Tibet's loss was India's gain, for the whole climate of the subcontinent is influenced by the Himalaya, the ramparts having in effect created the monsoon by turning back the rains toward the plains.

Tropical conditions persisted on the subcontinent well into the Pliocene as the startling variety of Siwalik mammals, most now extinct, shows. At least 10 species of elephants and 6 of rhinoceros roamed the forests and reedbeds, and present too were giraffe, hippopotamus, *Sivatherium, Hipparion,* sabre-toothed cats, various antelopes including some similar to kudu and eland, and many primates including the orang-utan. The Pleistocene ushered in a period of climatic turmoil during which glaciers advanced and retreated at least 4 times (Wadia, 1966). This period left the peninsula impoverished in species, many of which became extinct and had no ecological replacements. But most of the large Himalayan animals seem to have survived, adapted as they were to severe conditions. Compared to glaciations in Europe and North America, the extent of the ice was small, mainly because the mountains are arid and far removed from oceans. Still, glaciers were once more extensive than now. The Middle Himalayan region, seldom rising above 5000 m, has no glaciers today, whereas it did have them at one time, judging by morainal deposits on some summits. Snowlines in the Northwest Himalaya were previously some 800 m lower than now (Mani, 1968) and the same was true in the Kunlun (Trinkler, 1930). In the Pir Panjal, moraines are found as low as 1975 m though no ice sheets exist in that range today. Glaciers in the Pamirs once reached a length of 250 km (Mani, 1968). Now the largest glaciers outside of the polar regions occur in the Karakoram, with the Siachen glacier, 74 km long and 1180 sq km in size, being the most extensive. The advances and retreats of the glaciers no doubt forced the flora and fauna to shift up and down. During the last glaciation, for example, the flora of the Caucasus found refuge on the lower slopes of the hills (Gulisashvili, 1973), and similar survival for the Himalayan flora was postulated by Rau (1974). In spite of these vicissitudes the general type of vegetation at high altitudes seems to have remained fairly constant during the Pleistocene. Frenzel (1968) has shown that the vegetation of northern Eurasia alternated between vast steppes with cold dry climates and widespread forests. For example, during the Weichselian glaciation an arctic steppe vegetation covered much of the Himalaya and Tibet with a broad loess steppe and some deserts that extended northward. During the Holsteinian interglacial the same conditions prevailed in the uplands, but the steppe belt north of Tibet was narrow and soon was replaced by forest steppe and taiga of pine, spruce, and larch. Thus, the large Himalayan mammals seem to have had a relatively constant temperate to alpine environment in which to evolve.

"A hundred divine epochs would not suffice to describe all the marvels of the Himalaya," states a Sanscrit proverb. Having neither the space nor time to attempt this task, I will provide here only a brief sketch of the geography, geology, and vegetation of the Himalaya, basing my comments on books by Burrard and Hayden (1907), Gansser (1964), Wadia (1966), Mason (1955), Mani (1974), and Schweinfurth (1957). The main Himalayan system stretches for about 3000 km between Chitral in the northwest and western China in the southeast; in width it varies from as little as 80 km, with the mountains sweeping from the plains to the crest of the range abruptly as in Sikkim, to more than 300 km. The Great Himalaya dominates the system, arcing from Nanga Parbat (8126 m) and the great bend of the Indus eastward to Namcha Barwa (7715 m) on the Brahmaputra River, as the lower reaches of the Tsangpo are called. This expanse of peaks has been divided for convenience in various ways depending on whether the writer was a geographer, botanist, or political scientist. To a biologist several distinct vertical and horizontal zones are of relevance.

The main Himalayan system consists of three parallel mountain zones (fig. 4). Adjoining the Indo-Gangetic plain is the Outer Himalaya, or the Siwaliks, a low range seldom exceeding 1000 m in height and 50 km in width. Sometimes jammed against the main range, at other times separated from it by a flat valley, the Siwaliks trace the foot of the high mountains along much of their length. Of more recent origin than the other hills, the Siwaliks consist of consolidated river deposits that have been washed down from the mountains to the north. The Middle Himalaya, a complex series of ridges and valleys seldom more than 4000 to 5000 m high, stretch north of the Siwaliks for some 50 to 100 km. Several sections of the Middle Himalaya, among them the Pir Panjal and Mahabharat, branch from the Inner Himalaya, whereas others represent merely separate folds. And finally there is the Inner Himalaya, the home of the earth's giants—Everest, Makalu,

Dhaulagiri—whose massive crystalline cores of gneisses, granites, and quartzites tower into the realm of perpetual snow. Between the crest of the Himalaya and the Tibetan Plateau are other ranges lower than the main chain; Hagen (1970) has termed these the Tibetan marginal mountains. In Nepal, these mountains form the watershed between the Ganges and Tsangpo.

The relative humidity of the atmosphere is the most important factor in determining plant growth, and, based on the amount of precipitation, the vegetation of the main Himalayan system can be divided into several distinct horizontal zones (fig. 3). Situated at the edge of a huge and arid continental mass, the Himalaya receive precipitation only from the south. As the moisture-laden clouds collide with the peaks, the southern flanks receive most of the precipitation and the northern flanks and the mountains north of the crest line rather little (fig. 4). Pokhara, lying in the foothills of the Himalaya in Nepal, receives some 3500 mm of rain, whereas Jumla, located in the interior of the mountains, has only 750 mm (table 2). The monsoon rains are heaviest in the eastern half of the Himalaya, losing their force progressively toward the west. Darjeeling, for example, receives over 3000 mm of rain per year, but Simla some 1200 km to the west gets 1500 mm, and Srinagar, another 400 km westward, has only 650 mm, the last-named station receiving much of the total precipitation as winter snow and spring rain rather than summer monsoon. This gradation in the amount of rainfall greatly influences the kinds of forests that grow at low altitudes.

At the eastern end of the range, a tropical evergreen rain forest, to use Schweinfurth's (1957) terminology, covers the lowlands, but it gives way to a wet tropical deciduous forest which stretches from about the Manas River to Mount Dhaulagiri. This peak marks an important transition zone between the vegetation of the Eastern and Central Himalaya, between the tropical evergreen montane forest, in which *Castanopsis* is a distinctive tree genus, and the more arid conifer forests (see fig. 3). The Central Himalaya, its lowlands covered with a dry deciduous forest of *Shorea robusta* and *Terminalia*, extends westward to the Sutlej River. The gorge of the Sutlej is biogeographically important, for here the relatively moist forests of the east begin to be replaced by the vegetation of the northwestern mountains. The Western Himalaya, from the Sutlej to the Murree Hills in Pakistan is in

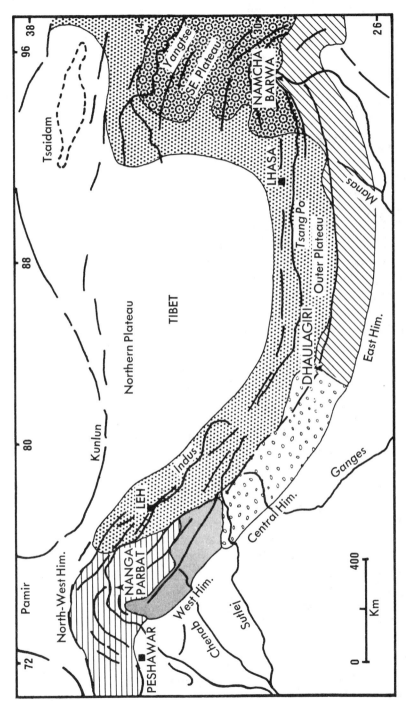

Fig.3. The horizontal climatic and vegetational zonation of the Himalaya (adapted from Troll, 1969, and Vaurie, 1972).

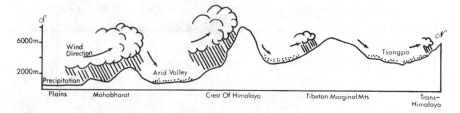

Fig. 4. The effect of the Himalayan mountain ranges on the amount of precipitation, showing the decrease from south to north (adapted from Hagen,
1970).

effect a transition area between the humid east and arid west. A
thorn steppe covers the plains, but, somewhat higher, pine forests
persist. The northwest, the Hindu Kush and Karakoram, is dry,
the valleys a semidesert, their lower slopes a sagebrush steppe
with stands of oak, birch, and conifers growing only in moist
protected places to give the land a harsh and desolate aura very
different from the green hills of the eastern Himalaya.

The mountains often tower above the surrounding lowlands,
with, for example, only 25 km separating the deserts bordering
the Indus River at 1200 m from the glaciers of Nanga Parbat at
8000 m. Such altitudinal differences, with their concurrent drastic
changes in temperature and precipitation, have created a striking
vertical zonation in the vegetation. This zonation has been intensively studied by German and Japanese investigators (Schweinfurth, 1957; von Wissmann, 1961; Troll, 1969; Numata, 1966),
and a general summary of their work is presented in figure 5. The
permanent snow line and the timberline are two limits that greatly
affect the distribution of plants and animals on mountain slopes.
Influenced mainly by the mean summer temperature and amount
of snowfall, the snow line varies considerably between ranges and
even on different exposures of the same range. On the south
slope of the Central and Western Himalaya the snow line may be
600 m lower than on the north slope. Mani (1968) noted that the
snow line on the south slopes of the Karakoram is 5650 m, the
Ladak Range 5790 m, the Zaskar Range 6090 m, Western
Himalaya 5180 m, Central Himalaya (Kumaon) 4720 m, and
Eastern Himalaya (Nepal) 4480 m, figures which differ somewhat
from those given by Troll (1969) in figure 5. Nevertheless, these
data show that arid mountains have a higher snow line than those
exposed to the monsoon. Some open ground, enough to support

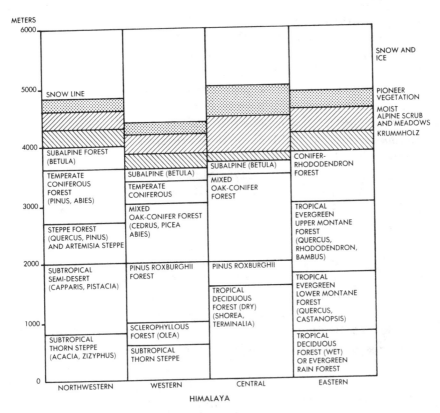

Fig. 5. Generalized vertical zonation of vegetation in different parts of the Himalaya (adapted from Troll, 1969).

an occasional plant, may exist above snow line because snow cannot settle on windblown precipices. Hingston (1925) collected *Delphinium glaciale* at 6260 m on Kangchenjunga, Holdsworth (1932) found a *Draba incompta* at 6200 m on Kamet, and Swan (1961) photographed a *Stellaria decumbens* at 6120 m on Makalu. The low precipitation at high altitudes coupled with the steep terrain also makes glaciers smaller and less common than expected. Snow has little opportunity to accumulate along the gorges, and those glaciers that do exist are often fed by avalanches rather than by a compacting of snow near their sources (Hagen, 1970).

Timberline coincides roughly with the 10–12° C isotherm at midday during May to September (Mani, 1968), a temperature which in the Himalaya is usually found between 3600 and 4000 m (plate 17). As a general rule, temperatures drop about 6.2° C for every 1000 m altitude, and table 3 gives some mean minimum temperatures for January, the coldest month in the mountains, for various stations. Above timberline are several vegetation zones, the first a patchy belt of *Krummholz,* low and twisted woody plants that have been savaged by wind and snow. Rhododendron dominates this zone in the Eastern Himalaya, but in the Central and Western parts juniper and birch are common too, *Rhododendron campanulatum* preferring northern exposures and juniper southern. Willow (*Salix hastata*) and some birch are the dominant *Krummholz* in the Northwestern Himalaya. Above this zone and reaching toward the snow line are alpine meadows, in some places large and lush, in others, particularly in the Northwestern Himalaya, mere tiny islands of green among barren scree and crags where rivulets of melted snow provide some moisture (plate 7). The dense turf of these meadows consists mainly of *Cobresia* and *Carex,* a splendid display of flowering forbs in season, and some procumbent woody plants such as *Salix* and *Potentilla.* As the winter snows recede between May and September—at the rate of about 10 m per day on the southern slopes of the Northwest Himalaya (Mani, 1968)—a succession of bright flowers appear, yellow daisylike *Doronicum,* delphiniums, gentians, poppies, roseroot, lousewort, anemones, asters, *Polygonum,* primulas, and *Mertensia* with minute blossoms of brilliant blue, to name a few of the showy plants. Higher up, the meadows give way to pioneering plants, cushions of *Saxifrage,* and lone *Draba* and *Gypsophila* huddling in the lee of boulders. Above this zone is another one, termed aeolian zone by Swan (1961), which extends to 6700 m and consists of spiders, mites, and insects feeding on windblown debris such as pollen, dead insects, and plant fragments.

Below an altitude of about 4000 m is a more or less continuous forest belt (fig. 5). In the Eastern Himalaya an extremely wet evergreen montane forest clothes the slopes between about 1000 to 3000 m. The lower parts of this forest remain free of frost and are tropical in nature with many laurel species and tree ferns, whereas the cool and misty upper parts are a cloud forest whose oak, rhododendron, magnolia, and other trees are festooned with *Usnea* lichens, mosses, orchids, and other epiphytes. Below the trees are thickets of spindly bamboo (*Arundinaria*). A coniferous forest of fir (*Abies*) and hemlock (*Tsuga*) and *Pinus excelsa* with a sparse undergrowth of bamboo and rhododendron marks the highest forest zone over much of the area. In some of the deep valleys of Bhutan and Sikkim are stands of dry boreal forest, an intrusion from the west, consisting of larch (*Larix*) and spruce (*Picea*). In the Central Himalaya, a mixture of sal (*Shorea*), *Terminalia, Erythrina, Anogeissus,* and other deciduous trees grows over the lower hills, and this is replaced at varying altitudes by monotypic expanses of tall *Pinus Roxburghii.* A mixed forest of oak, fir, and spruce reaches from an altitude of about 2000 m close to the timberline where a narrow belt of gnarled birch marks the transition to the alpine zone.

West of the Sutlej the lowland forests are typical of arid lands with thin stands of *Acacia* and *Zizyphus* giving way at about 600 m to brushy woodlands of *Olea, Punica, Oleander,* and others. Above these zones is one of *Pinus Roxburghii,* and at 2000 m the pines are replaced by a mixed boreal forest of evergreen oaks, maples, elms, fir, spruce, and in places stands of impressive deodar (*Cedrus deodari*). Fir mixed with birch and finally birch alone are the dominant trees at the timberline (plate 3).

The Northwestern Himalaya are semiarid to arid. The monsoon has here little effect and much of the precipitation falls in winter and spring (table 2). Temperatures in the valleys may be extreme—hot in summer and chilly in winter. The luxuriant forests are gone and the ones that persist have the stark look of those in the desert or arctic. Scattered *Pistacia, Capparis, Indigofera, Rosa, Ephedra,* and other stunted trees and shrubs grow on barren slopes; higher up are open stands of oak, *Pinus gerardiana,* spruce, and fir, which at 3600 m give way to thickets of birch (plate 4). To think of a forest belt in this region is misleading, for trees grow in isolated stands, their continuity broken by sterile expanses of scree and *Artemisia*-covered slopes. Much of the Northwestern Himalaya belongs to the steppes of Central Asia, and, in fact, all forests vanish in northern Chitral, northern Hunza, and in the valleys north of Skardu, except for occasional stands of poplar, birch, and willow along streambeds (plate 6).

A generalized outline of vegetation zones such as this provides little idea of variations in distribution as a result of exposure, degree of slope, local climate and so forth. In the western parts of the Himalaya, for instance, forests are often confined to the moist

and shady northern slopes of the mountains, whereas the southern dry sides, exposed as they are to intense solar radiation, support arid plant communities. Also in the western half, avalanches
modify the forest cover. Brittle conifers are snapped by the rush
of snow, but the colonizing alders, willows, and birches survive
later onslaughts, their supple stems bending under the white burden. Dry winds sucked into and funneling through some eastern
valleys, such as the Karnali and Kali Gandaki in Nepal, create such
arid conditions that the slopes remain treeless for 600 to 1000 m
above the valley floor. Table 2 shows the precipitation at Jomsom
in a dry valley as compared to Pokhara 60 km away.

Another major influence on the vegetation has been man, who
here as elsewhere has treated his environment with the ancient
principle "If it moves, kill it; if it doesn't, chop it down." With
cultivation practiced up to an altitude of 4500 m in northern
Nepal and 3500 in Northwestern Himalaya, and livestock driven
as high as 5000 m in search of forage, only the most inaccessible
terrain below the snowline has not been disturbed by man. The
forests below 3000 m in the Middle Himalaya have suffered most,
vast tracts having been cleared and the soil terraced for cultivation. The remaining forests are being cut for timber, plundered
for firewood, and in the case of some trees such as oak lopped for
livestock food. Even in those forests where cutting is regulated by
the government, disturbance in the form of livestock grazing and
wood collecting is so heavy that virtually no undergrowth remains.
In Nepal whole hillsides of juniper may be burned to create open
grazing lands, and the same is done elsewhere in the forests. In
the Northwestern Himalaya, where agriculture is dependent on
irrigation, virtually every alluvial fan, every level piece of ground
near water has been converted to stony fields (plate 10). In Nepal,
only about a third of the Mahabharat and Inner Himalaya are
now covered with forest, a third being man-induced rangelands
and cultivation, and the rest being wastelands (Numata, 1966).
Erosion has removed so much topsoil in some parts that up to 38 %
of the land area consists of fields whose productivity waned
until they were abandoned (Eckholm, 1975). Equally serious has
been the effect of man at the timberline and above, especially in
the vicinity of villages. Over the centuries most wood has been
burned for fuel, and people often have to walk for many kilometers to reach the last stands. Many treeless slopes, now useless

wastes of rock, as those around Chitral and Gilgit towns, were forested until recent times. About 22 million kilograms of fire-wood were used by the military in the Gilgit Agency during 1973. Herds of sheep and goats have denuded most uplands. With a predilection for certain grasses and forbs, livestock has eliminated much of the palatable forage, leaving behind primarily those plants which can somehow protect themselves. For example, mints and sagebrush contain aromatic phenols, and *Acantholimon lycopodioides* uses a hedgehog defense in that it presents a spiny cushion to voracious mouths. With the fragile soil trampled into a maze of paths and much of the plant cover removed, the alpine slopes are often eroded. To document the human devastation of the Himalaya adequately would take many pages (see Eckholm, 1975). Here I want simply to make the point that the Asian ranges consist not just of pristine peaks untrodden by man, an impres-sion left by the books of mountain climbers, but also of populated areas where man is an enduring presence whose activities have had a devastating effect.

Two areas adjoining the main Himalaya system need a brief description, for in one of them, the Baluchistan region, I did much work, and in the other, the Tibetan Plateau, I studied wildlife along the southern edge.

To the north of the Himalayan chain are the bleak uplands of Tibet, some 2,164,000 sq km of sedimentary deposits where once the Tethys Sea had been. The geographical boundaries of Tibet include the Kunlun and Astin Tagh ranges in the north, the Great Himalaya in the south, the Karakoram in the west, and roughly longitude 102° in the east. Vaurie (1972) divided the plateau into three natural units, as did Ward (1936), of which the Northern Plateau or Chang Tang covers about half of the region (fig. 4). The area is excessively bleak, with severe winds sweeping across the plains and eroded hills and with temperatures in winter drop-ping to -45° C. Rain is scanty in summer, though snow storms may hit during any month. Drainage is internal, the rivers ending in the brackish lakes that dot these wastes. Arid, cold, its soil often impregnated with salts, the Chang Tang supports only a sparse vegetation of grasses and forbs and a few woody shrubs, yet a distinct fauna of hoofed animals is adapted to this habitat. Hedin (1898) marveled that animals the size of a yak could live on "the miserable herbage which those highlands afford." But as Moorcroft (1816, quoted in Fairley, 1975) noted, "the bite of the yak is quicker and nearer the ground than any other species of neat cattle I am acquainted with, a peculiarity which fits them for the short and scanty herbage." While the variety of grazers is poor, numbers of some species may be great. Dalgleish saw a herd of 10,000 chiru (Bartz, 1935), and Wellby (1898) noted that "on one green hill we could see hundreds upon hundreds of yak grazing; there was I believe, more yak visible than hill."

The Outer Plateau includes the southern and western portions of Tibet, covering the enormous distance from Ladak to the headwaters of the Yangtse. The features of the area were to a large extent shaped by such rivers as the Indus and Tsangpo, which have outlets to the sea. Along the northern banks of the Tsangpo lies an extension of the Ladak and Kailas ranges, often referred to as the Transhimalaya, which catches whatever moisture escapes the Himalayan barrier. Consequently the Outer Plateau receives more precipitation than the Northern one and the weather is milder. Lhasa, for instance, has an annual rainfall of about 400 mm and temperatures seldom fall below -20° or rise above 27° C. Most of the large towns in the Tibet region—Leh, Lhasa, Shigatse—and much of the agriculture are located in the temperate Outer Plateau. Trees such as poplar and willow may grow in the valleys, often along irrigated fields, and forests of spruce, birch, and juniper occur locally in northeastern Tibet. Most of the vegetation in southern Tibet consists of what Schweinfurth (1957) calls "alpine steppe" with genera like *Caragana* and *Astragulus* dominant (plate 19). Toward the northeast the alpine steppe merges into grassy steppes (Vaurie, 1972), and toward the west, in western Tibet and Ladak, the country becomes a stony desert. Traveling near Pangong Lake, Macintyre (1891) commented on the "brown-colored, rounded and sterile hills ... the scanty tufts of herbage that existed on the sides of this huge natural water-basin ... were rendered almost invisible by distance." And Ali (1949) found near Mount Kailas a few narrow strips of grass along rivulets and commented that "this is the only green to break the drab and endless monotony of the stony landscape."

The third region, the Southeastern Plateau, differs from the popular conception of Tibet in being a rugged mountain land, deeply dissected by gorges the lower slopes of which are densely forested. The monsoon in this area hurdles the crest of the Himalaya to shed heavy rains on the hills, giving them a climate and plant composition similar to those of the East Himalaya.

The Baluchistan ranges, here taken to mean all the uplands west of the Indus in Pakistan from around the Khyber Pass and Safed Koh southward, comprise an arid plateau broken by barren ranges. Looking westward from the Indus plain, a plateau rises sharply to a height of some 300 to 1000 m and then stretches to Afghanistan and Iran. At the northern end the Safed Koh is the most striking feature, a rugged range which attains 4750 m, making it ecologically the most southernly extension of the Northwest Himalaya in Pakistan. South of the Safed Koh is a maze of eroded hills followed by two prominent mountain areas, the Toba-Kakar and Sulaiman, whose highest peak, Takht-i-Sulaiman, reaches an altitude of 3350 m. South of Quetta various ranges spread out sheaflike into a series of contorted and folded arcs, some like the Kirthar running north and south, others arching westward toward Iran, becoming lower as they extend west and south. The Makran Coast, Central Makran, and Siahan ranges remain below 1500 m, whereas peaks in the Raskoh and Chagai Hills rise above 2200 m. Between the ranges are several basins with internal drainage, the dry streambeds leading toward them filled with torrents only at long intervals. Baluchistan is a haggard land, a land from which protrude ridges lean and jagged, their skeletons laid bare (plate 12). Most hills are of limestone and sandstone and clay, but some peaks, such as Koh-i-Sultan and Raskoh, are volcanic in origin. Being outside the influence of the monsoon, much of the Baluchistan area is arid, with less than 25 cm of precipitation annually (table 2), some of it as snow. Temperatures are extreme, ranging at Quetta from -19 to 42° C, and these are made even more difficult to endure by the winds, scorching in summer and icy in winter. It is a somber plateau, sienna and gray in hue with folds of rusty red and a sky that often is earthen from dust.

A subtropical thorn steppe covers the hills from about 500 to 1500 m. Seldom exceeding 6 m in height, *Acacia modesta, Olea cuspidata* and *Pistacia* predominate, though such trees and shrubs as *Zizyphus, Cordia,* and *Ephedra* are also locally conspicuous. But trees are now rare in most areas, the last stands cut down long ago by pastoralists and agriculturalists. Forbs are scarce and grasses too, not so much due to the dry climate but to the intensive grazing pressure from livestock to which the terrain is subjected. Low bushes of *Artemisia, Caragana,* and *Astragulus* cover many slopes which once were forest, and the inedible *Dodenia* shrub—almost the ultimate sign in degradation—is common. In western and southern Baluchistan, where altitudes are low, *Tamarix* may grow along streambeds, and bushy euphorbias may dot the rocky slopes, their spiny branches providing acacia seedlings with the protection they need to grow the first few years. Above an altitude of 1500 m are, or were, forests resembling the steppe forest of the Himalaya. Termed dry temperate forests by Champion et al. (n.d.), this vegetation type includes two distinct associations, both found in areas with precipitation of at least 40 cm. Near Ziarat northeast of Quetta, on the lower slopes of the Safed Koh, and in some other hills at altitudes above 2500 m are almost monotypic stands of *Juniperus macropoda.* Some trees reach 10 to 20 m in height and 2 to 3 m in girth. Easily accessible to the human population, the juniper forests have been largely razed, only an occasional gnarled tree attesting to the fact that stands once shaded the slopes. Those that do survive are poorly managed, the forest so ravaged by livestock that regeneration of trees has virtually stopped. North and west of Fort Sandeman, in the high parts of the Sulaiman and Shingar ranges, as well as in the Shodar Hills of Waziristan among other places, are forests of low-branching *Pinus gerardiana* with a few *Pistacia* and *Acer.*

The distribution of plants and animals of the Indian subcontinent has been greatly influenced by the presence of the Himalaya. The mountains have prevented most of the Central Asian species from penetrating the forested lowlands on the peninsula and barred the Indian species from moving north. At the same time the Himalaya was a migration route for plants and animals and it became the crossroads of five major biogeographical subregions, the Mediterranean and Siberian subregions of the Palearctic Region and the West Chinese, Indochinese, and Indian subregions of the Oriental Region. The Tethys Sea vanished first in the northeast, bringing the subcontinent into contact with the ancient ecosystem of the western Chinese mountains (Rau, 1974). Plants from there migrated westward and covered most of the peninsula, which until the Miocene had a tropical climate (Cracraft, 1973). After the climate became more arid, the flora found refuge on the moist mountain slopes, some plants becoming isolated in the hills of southern India, others surviving on the slopes of the Himalaya. Today most species of humid Chinese mountain origin extend only to about 80° E, just beyond western Nepal, although a few plants, notably rhododendron, oak, fir, and bamboo, have penetrated far westward. The flora of the Western Himalaya is related to that of the Central Asian mountains.

The distribution of animals follows a similar pattern. The forest fauna at low to medium altitudes is mainly Oriental in origin and extends westward to Kashmir where it mingles with the Palearctic fauna. The biogeography of several animal groups in the Himalayan area has been studied, and I mention here just a few examples. Collecting insects above the timberline in the Northwestern Himalaya, Mani (1962) found that 3.5% of the species belonged to the Indochinese subregion and the rest to the Palearctic. However, endemism is more prevalent in insects than in other animals and in plants, over two-thirds of the Palearctic species having evolved in the Himalaya. Discussing the herpetological fauna, Swan and Leviton (1962) noted that the Indochinese species of lizards, frogs, and snakes become greatly reduced in number along the eastern frontier of Nepal, many forms there giving way to Indian plains species which have moved into the valleys and lower hills of Nepal. Several lizards and snakes from the Mediterranean subregion have a wide distribution, upward to 5450 m and horizontally from Kashmir to Nepal where they overlap with the Indochinese and Indian species. This, together with a certain amount of endemism, makes the herpetological fauna of Nepal "somewhat bewildering" and gives it "a labyrinthian quality," as Swan and Leviton (1962) phrased it. In the mountains of eastern Nepal, Gruber (1969) found different elements among the small mammals. For example, *Rattus eha* and the muskshrew are Oriental, the pika is typically Palearctic, and *Sorex cylindricauda* and *Pitymys sikimensis* are also Palearctic but with relatives in southwestern China. Kurup (1966) analyzed the mammal fauna of Assam, noting that most genera are of Indochinese origin but that western entrants, mainly Palearctic but also such Ethiopian elements as the elephant, comprise 21% of the fauna. Cooler and more humid conditions over much of India during some phases of the Pleistocene enabled many mammals, including some mountain forms, to spread over the peninsula and there become stranded in the hills during some drier climatic phase, a course of events illustrated well by tahr. Prakash (1963) found that 56% of the mammalian species in the Rajasthan desert are Palearctic in origin, 41% Indo-Malayan, and 3% endemic.

To delineate precise migration routes for the large Himalayan mammals is difficult because the Palearctic forms, to mention one example, did not just arrive from the west but also from the east and north. Such forked dispersal may explain the presence of one subspecies of red deer in Kashmir (the hangul) and another in southeastern Tibet (the shou), although the possibility exists that the two populations were once connected along the Tsangpo and upper Sutlej (Caughley, 1970a). I saw old shou antlers in the Dolpo District of Nepal, far west of the known range of this animal. The distribution of species today may have little relation to their range in the past, climatic changes having greatly altered habitats since the Miocene and Pliocene when many of the present-day mammals evolved and began their ascendancy. The tahr are now found in three widely separated areas, in the Himalaya, southern India, and Oman, but they once also occurred in Europe; and the Himalayan black bear, whose present range extends from Iran eastward along the Himalaya to China and Russia, was an inhabitant of Europe too (Kurtén, 1968). Such a distributional pattern indicates the presence of a continuous forest belt, probably during the Pliocene, which later became fragmented. The goral, serow, and takin are of Oriental origin, all three largely confined to forested slopes. The takin halts in Bhutan, but serow extend into the Western Himalaya, and goral reach the Afghanistan border, penetrating far into the realm of the Palearctic sheep and goats. Such species as the muskdeer and most large carnivores—brown bear, snow leopard, and wolf—are Palearctic animals. The leopard and tahr each have affinities to at least two faunal regions and this obscures their origin.

These few comments about the history of the region and its wildlife serve to emphasize that the distribution of animals, though influenced by temperature, vegetation cover, topography, and other factors, is also a product of the past, species having been buffeted over the continents by the earth's upheavals and the vagaries of climate. What we study now is but a fragment in time. Given the prevailing ecological conditions, takin might be able to extend their range westward, or black bear might once again occupy Europe. But the opportunities to do so have passed, for man is modifying the world so fast and so drastically that most animals cannot adapt to the new conditions. In the Himalaya as elsewhere there is a great dying, one infinitely sadder than the Pleistocene extinctions, for man now has the knowledge and the need to save these remnants of his past.

2 TAXONOMY

Ecological and behavioral analyses require sound systematics. Unfortunately taxonomy is in a state of flux and the problem of what constitutes a species remains unsettled, some investigators using the classical concept of morphological differences, others the biological one which stresses genetic distinctness. At the present state of knowledge it is not known which criteria—morphological, chromosomal, serological—are most useful in defining species. The taxonomy of *Ovis* and *Capra* and some of their relatives remains unsettled. No one has as yet undertaken the task of analyzing the morphology and biochemistry of sheep and goats throughout their range, using large samples to eliminate biases due to age, sex, and individual variation. Although I did little research on caprid taxonomy, I express here some opinions based on reading the literature and on looking at species either alive or in museums.

Simpson (1945) divided the subfamily Caprinae into 4 tribes and 14 genera. The number of species and subspecies remain in dispute but the divisions in the list on page 21 are accepted at least by some authorities.

SAIGINI
The Antilopinae, of which gazelle and blackbuck are examples, are difficult to distinguish from the Caprinae. Simpson (1945) and Thenius and Hofer (1960) included *Pantholops* and *Saiga* with the Caprinae, but others consider them Antilopinae (Kurtén, 1972). Bannikov et al. (1967) place the saiga into its own subfamily Saiginae. Being atypical and questionable members of the Caprinae, the two species receive only casual treatment in this report.

OVIBOVINI AND RUPICAPRINI
The takin, according to Pilgrim (1939), is nothing but a robust Rupicaprini, not an Ovibovini, and Pocock (1918) also felt that

20

Family Bovidae

Subfamily Caprinae	Common Name*	No. Species	No. Sub-species
Tribe Saigini			
Pantholops (Hodgson, 1834)	Chiru	1 (hodgsoni)	0
Saiga (Gray, 1834)	Saiga	1 (tatarica)	2
Tribe Rupicaprini			
Nemorhaedus (H. Smith, 1827)	Goral	1 (goral)	8
Capricornis (Ogilby, 1837)	Serow	2 (sumatraensis)	11
		(crispus)	2
Oreamnos (Rafinesque, 1817)	Rocky Mountain Goat	1 (americanus)	0
Rupicapra (de Blainville, 1816)	Chamois	1 (rupicapra)	9
Tribe Ovibovini			
Ovibos (de Blainville, 1816)	Muskox	1 (moschatus)	0
Budorcas (Hodgson, 1850)	Takin	1 (taxicolor)	3
Tribe Caprini			
Ammotragus (Blyth, 1840)	Aoudad†	1 (lervia)	4
Pseudois (Hodgson, 1846)	Bharal†	1 (nayaur)	2
Hemitragus (Hodgson, 1841)	Tahr	3 (jemlahicus)	0
		(hylocrius)	0
		(jayakeri)	0
Capra (L, 1758)	Goat	6 (aegagrus)	see text
		(ibex)	5
		(falconeri)	see text
		(pyrenaica)	4
		(cylindricornis)	0
		(hircus)	domestic
Ovis (L, 1758)	Sheep	6 (canadensis)	7
		(dalli)	3
		(nivicola)	3
		(ammon)	see text
		(orientalis)	see text
		(aries)	domestic

*See Appendix A for other common names in use

†The usual common names of "Barbary sheep" for *Ammotragus* and "blue sheep" for *Pseudois* are inappropriate, for these animals are not sheep. A widely used alternative name for Barbary sheep is "aoudad." The blue sheep is known as "ná" in Tibetan and "bharal" in Hindi. I prefer the latter.

muskox and takin lacked a close relationship. Yet other authors think that the two forms may have separated as recently as the Pliocene (see Thenius and Hofer, 1960). The relationship of muskox to the Caprinae rather than to the Bovinae was confirmed serologically by Moody (1958). *Nemorhaedus* and *Capricornis* are sometimes placed into separate tribes (see Benirschke et al., 1972). The rupicaprids have indeed evolved into a diverse assemblage, all being highly distinctive physically as well as variable in the presence of skin glands. All four genera have pedal glands. Chamois and mountain goat have postcornual glands. The serow is alone in having large preorbital glands. Pocock (1910) stated that goral lack preorbital glands, but Allen (1940) noted that they have small ones. Serow have inguinal glands according to Allen (1940) but not according to Pocock (1910). The takin, wherever its taxonomic home may be, has none of these glands (Pocock, 1910). Some taxonomists feel that goral belong in two species rather than one, but otherwise there is not much disagreement about the number of species in each genus.

With respect to the number of subspecies, I follow Dolan (1963) for the goral, serow, and chamois, Cowan and McCrory (1970) for mountain goat, Tener (1965) for muskox, and Ellerman and Morrison-Scott (1951) for takin.

CAPRINI

The members of the Caprini are so alike in their skeletal structure that fragmentary fossils are notoriously difficult to place into the correct genus. However, the gross morphological features of living animals, particularly those of males, are quite distinctive at least on the genus level. *Ovis* can be distinguished from *Capra* by the presence of preorbital glands, inguinal glands, and pedal glands on all feet; and by the absence of a callus on the knee, a potent body odor, and a beard in males. The enamel folds on the molars of sheep bifurcate whereas those of goats do not (Epstein, 1972). Goats have flat and rather long tails, bare underneath, in contrast to sheep, which have somewhat round, hairy tails. The horns of female goats tend to be proportionately larger than those of female sheep, occasional ewe populations having lost their horns entirely. These and some other morphological characters separate sheep from goats, but the fact that the domestic forms of the two genera may on rare occasions produce living hybrids (Schmitt, 1963) points to a basic similarity. In addition the tribe contains three genera—*Hemitragus, Ammotragus,* and *Pseudois*—which differ in important respects from the typical sheep and goats.

Capra

Horns have been the main taxonomic character used to classify goats. New species and subspecies have sometimes been designated on the basis of a minor variation in the shape of one set of horns, clearly an undesirable tactic with a structure that is highly malleable in evolution. The basic body pattern is similar in all goats, as shown by the fact that the females of various forms are sometimes difficult to distinguish. Distinctiveness has been achieved by embellishing the males with striking horns and with variations in coat color and hair length. Unless a character remains under strong selection pressure, it may become variable. In any herd of domestic goats in Pakistan there often are animals with scimitar-shaped and spiralling horns, and horns whose tips bend inward or flare outward. To place the markhor into a separate subgenus, as has been done by some authors, mainly on the basis of their spiralling horns seems unwarranted. With all goats possessing a 2n chromosome number of 60 (Nadler et al., 1974) and all interbreeding freely in captivity (Gray, 1954), morphological criteria remain at present the best means of classifying goats. In effect this requires arbitrary decisions on just what differences are of sufficient magnitude to warrant species status. A main problem is to find isolated groups which show consistent phenotypic variation and then to decide if these differences warrant subspecific or specific status. Many subspecies in sheep and goats lack reality in that better sampling methods would simply show a continuous variation of traits: the distinguishing criteria are just points in a cline. Ideally, subspecies should be confined to isolated populations which differ from neighboring populations in several characters.

Haltenorth and Trense (1956) recognized 4 goat species, Lydekker (1913) named 9, and Heptner et al. (1966), who also discussed taxonomy at length, named 8. Herre and Röhrs (1955) felt that only one species of goat exists, but then they divided this species into 3 groups (*aegagrus, ibex, falconeri*) with the ibex group containing such diverse animals as the Spanish goat, the Siberian ibex and the Dagestan tur. While it is true that taxonomic formalization tends to obscure relationships, forms being considered qualitatively distinct as soon as a scientific name has been attached

Fig. 6. Heads and horns of some Rupicaprini, the Ovibovini, several *Capra* goats, and the Himalayan tahr, illustrating different horn types. Chamois (top left), Rocky Mountain goat (top center), serow (top left, below chamois), takin (top right), muskox (top middle, below mountain goat), wild goat (center left), Spanish goat (bottom left), Kashmir markhor (bottom center), Himalayan tahr (center right), Alpine ibex (bottom right). (Sketches by Richard Keane.)

Fig. 7. Heads and horns of some sheep, of the round-horned caprids—aoudad, bharal, and
   Dagestan tur—and of the Saigini. Marco Polo sheep (top center), Rocky Mountain
   bighorn sheep (top left), mouflon (top right), Punjab urial (middle center), bharal
   (middle left), Dagestan tur (bottom left), aoudad (bottom center), chiru (middle
   right), saiga (bottom right). (Sketches by Richard Keane.)

to them, the system proposed by these authors does little to clarify matters.

As noted by Ellerman and Morrison-Scott (1951), the horn shapes of goats fall into several types (see figs. 6 and 7), and in the absence of better criteria these may be designated as species.

1. The horns are scimitar-shaped and have a sharp anterior keel, their sweep broken by occasional knobs on the anterior surface. This horn type is found in the wild goat (*Capra aegagrus*). Several subspecies have been designated. *C. a. cretica* is found on Crete and on several other islands, but only the animals on Theodorou Island are not somewhat mixed with domestic stock (Papageorgiou, 1974). Schultze-Westrum (1963) recognized another form, *C. a. pictus,* from Erimomilos Island and several other islands near the Greek mainland. However, these animals may be nothing more than domestic goat x wild-goat hybrids. *C. a. aegagrus* is found from Turkey to Iran and *C. a. blythi* from Iran and Turkmenia to Pakistan. The last-named subspecies differs from *aegagrus* mainly in having more slender horns with fewer knobs and in having a pelage which is shorter and of a lighter color, at least in the hot areas. These minor differences probably do not justify a subspecific distinction, a point also made by Roberts (1967), and in fact no wild goat subspecies may survive close scrutiny.

2. The horns are scimitar-shaped as in the wild goat but the anterior surface is relatively flat and broken by prominent transverse ridges. Such horns are found in the ibexes (*Capra ibex*), including Alpine ibex (*C. i. ibex*), Asiatic or Siberian ibex (*C. i. sibirica*), Nubian ibex (*C. i. nubiana*), and Walia ibex, (*C. i. walia*). The horns of these subspecies are similar except for minor differences such as the slight dorsal keel in *walia*. Ibexes differ somewhat in their facial features, the Alpine and Asiatic ones having short broad faces in contrast to most others, which have narrow features resembling the wild goat. Pelage color differs too, with, for instance, Asiatic and Nubian ibexes having prominent markings on their forelegs whereas Alpine ibex do not. Heptner et al. (1966) placed all ibexes, except the Walia, into separate species. Lydekker (1913) also recognized the Asiatic ibex as a species and divided it into 13 subspecies.

In the western Caucasus lives a goat, called the West Caucasian or Kuban tur, which has variously been designated as *Capra caucasica* and *Capra ibex severtzovi.* Its horns resemble those of Alpine ibex but are shorter and more massive. Heptner et al. (1966) recognized the resemblance of this animal to the Alpine ibex but preferred to place it into a separate species. I leave this goat among the ibexes, calling it the Kuban ibex, *C. i. caucasica,* a name which has priority over *severtzovi.*

3. The sharp-keeled horns are twisted into an open or tight spiral. Such horns are characteristic of markhor (*C. falconeri*). Seven subspecies are often recognized but, as discussed below, not all are valid.

4. The horns curve out and up and then back, inward, and up again, and they have a sharp posterior keel. Only the so-called Spanish ibex (*C. pyrenaica*) has this type of horn. This animal should be called the Spanish goat, leaving the name ibex to *C. ibex,* although the two are probably closely related. Other goats, among them the wild goat, markhor, and Nilgiri tahr are also called ibex locally, creating much confusion. Four subspecies of Spanish goat are accepted, the different forms supposedly recognizable by the amount of black in the pelage (de Beaux, 1956).

5. The eastern Caucasus harbors a goat, the East Caucasian or Dagestan tur, with heavy, almost round horns which curve out, up, back, inward, and up somewhat like those of the Spanish goat. Its beard is short, much shorter than that of the neighboring Kuban ibex, and its face is blunt. Ellerman and Morrison-Scott (1951) call it *C. caucasica,* a name first applied to the Kuban ibex, but I follow here Heptner et al. (1966) in terming it *C. cylindricornis.* The superficial resemblance between aoudad, bharal, and Dagestan tur has tempted some authors (see Geist, 1971a) to suggest an evolutionary cline. It should however be emphasized that the Dagestan tur seems to be a true goat, not a morphological link between sheep and goats as are the other two forms.

6. In keeping with the apparent custom of giving different scientific names to a domestic form and its wild progenitors, the domestic goat should be termed *C. hircus.* There has been much speculation about the wild ancestor of the domestic goat, the German literature on this subject being especially large though not necessarily illuminating. Epstein (1972) presents a useful summary of this topic. Herre and Röhrs (1955) and most other authors agree that only *C. aegagrus* was the ancestor of *C. hircus,* and some suggest that *C. falconeri* also contributed to the domestic stock.

However, the horns of all domestic breeds, except the Circassian one, spiral in the opposite direction to those of markhor (see Koch, 1931). On the basis of 3000-year-old seals that picture goats with horns of various shapes, Petzsch (1957) proposed that wild goat, Dagestan tur, and markhor contributed to the origin of domestic goats. Actually, domestic animals with twisted horns did not begin to replace those with scimitar-shaped ones in the eastern Mediterranean until about 4000 BC (Zeuner, 1963). While a polyphyletic origin of the domestic goat cannot as yet be disproved, there is no good evidence in favor of it.

Morphological divergence occurs and persists if two species are unable to interbreed. Differences between two such populations may be small, but it is not the magnitude of the differences that determines a species but the fact that the animals are unable to exchange characteristics. Interbreeding may of course be possible, but the species persist if hybridization is rare and if selection is against the hybrid. When two forms of goat occupy the same terrain the extent of interbreeding can provide clues about behavioral or morphological reproductive barriers that may have evolved, and this in turn can help to define the species. Wild goat and markhor in Pakistan apparently occupy the same two small ranges without evidence of interbreeding, and, farther north, Asiatic ibex and markhor may be found on the same ground, there being again no known hybridization. Wild goat and Nubian ibex may have existed together in Palestine during the Pleistocene (Zeuner, 1963). In the Caucasus, the Kuban ibex and Dagestan tur do interbreed along their zone of contact. However, Heptner et al. (1966) emphasize that hybrids are rare, suggesting some reproductive isolating mechanism. In a situation such as this, where hybridization probably results in reproductive waste, natural selection would tend to strengthen barriers to gene flow. The wild goat is found in about half of the range of the Dagestan tur, but no hybrids are known. Yet when seven Kuban ibex and Dagestan tur were introduced into an area inhabited by wild goat, the animals readily hybridized (Heptner et al., 1966). It seems that when two kinds of goat colonize the same range naturally, they evolve at least a partial reproductive barrier—and this, one could argue, represents a valid criterion for a species.

Taxonomy of Capra falconeri. The markhor are a good example of how the hasty naming of new subspecies, based sometimes on only one skull, can create taxonomic confusion. Seven subspecies are recognized by Ellerman and Morrison-Scott (1951) and Haltenorth and Trense (1956). Two of these, *C. f. heptneri* and *ognevi*, are confined to northern Afghanistan and southern Russia. They are not considered further here except to note that subspecific distinctions may not be justified, that only one form may be valid. The other five subspecies occur mainly in Pakistan (see fig. 17).

THE CASE OF THE CHILTAN GOAT. In 1882 a local hunter gave Colonel H. Appleton a set of horns from the "Chialtan Range" near Quetta, and these were later presented to the British Museum. There Lydekker (1913) described the animal as a new subspecies of markhor, *C. f. chialtanensis.* The horns formed a complete spiral and in general resembled those of markhor living on the Takatu and Zarghun ranges near Quetta—except that the Chiltan horns had a sharp keel in front in contrast to typical markhor horns, which have it in back. As it turned out, this first set of horns was unusual in that most Chiltan horns do not form a complete spiral but simply have the terminal halves twisted inward. Only one out of 27 sets of horns I examined had a complete spiral. While accepting Lydekker's taxonomic dictate, some authors found the Chiltan goat anomalous and speculated that it is a hybrid between markhor and wild goat (Roberts, 1969; Epstein, 1972) or markhor and domestic goat (Burrard, 1925; Dollman and Burlace, 1935). Lying, as it does, at the northern edge of wild goat distribution and the southern edge of markhor range, the Chiltan massif would be a logical site for a hybrid goat, especially since local people say that an occasional wild-goat male with scimitar-shaped horns visits the range, and, according to Stockley (1928), some straight-horned markhor live there.*

I visited the Chiltan Range in 1970 and three years later plotted goat distribution in the hills to the south. The horns of Chiltan goat resemble those of wild goat. Not only are they essentially scimitar-shaped, except for the terminal twist, but they also have a sharp keel in front with an occasional prominent knob or two. The horn core cross-section of Chiltan goat is like that of wild

*In this context the appearance of a Kashmir markhor × domestic goat hybrid is of interest. I saw this animal, a female, in Chitral. Her father was a tame markhor. Except for her longer hair and ears and a white blaze on her muzzle, the animal closely resembled a markhor. However, her horns were scimitar-shaped without any indication of a twist.

goat, not markhor (fig. 8). However, Chiltan goat horns are more massive and shorter than wild-goat horns. Average circumference at the base of 5 Chiltan goat horns, five or more years old, was 22 cm whereas that of 13 wild-goat horns of similar age was 18 cm (17–21 cm); average horn length of four 5-year-old Chiltan goats was 65 cm and that of 8 wild goats from southern Sind 78 cm; the world's record for a Chiltan horn is 91 cm and the record wild-goat horn is 144 cm. Another difference between the two types of horn is the frequency with which tips break. Checking trophy heads from various parts of Pakistan, I found that only 3% of 66 wild-goat horn tips were broken, whereas the figure for the Chiltan Range was 54% of 46 tips. The horns of the Chiltan goat are either fragile or the design is such that the tips come into violent contact when males clash. That the former may be the case is shown by the fact that scimitar-shaped wild-goat horns from an area where both types occur also break easily. Of 14 tips, 35% were broken. The pelage of male Chiltan goat during the rut, with its light color and black markings, resembles that of wild goat, not markhor, and the animals also lack a ruff. These features lead to the conclusion that the Chiltan goat is neither a markhor nor a hybrid, but a wild goat with distinctly shaped horns (Schaller and Khan, 1975).

This view was reinforced by a visit to the hills south of the Chiltan Range (fig. 9). Although the animals there have been almost exterminated, villagers told me that goats with both types of horns may occur on the same slopes and in the same herds. On the roof of one house were six sets of horns, all from the same area: two were wild-goat types, one was of the Chiltan variety, two were intermediate with the tips twisted in slightly, and one had both a Chiltan-type horn and a wild-goat-type horn. This small sample showed a complete gradation of types. From this evidence it is tempting to speculate that some wild goats became isolated on the cliffs of the small Chiltan Range (plate 11), contact with other populations possibly severed by the growth of dense juniper forests on the low-lying hills to the south, and there some minor mutation resulted in the Chiltan-type horn. After contact with neighbors to the south was established again due to destruction of the forests, the trait for twisted horns appeared there too. It is, of course, possible that markhor contributed some genes to the Chiltan population, although no evidence for this exists. Those vis-

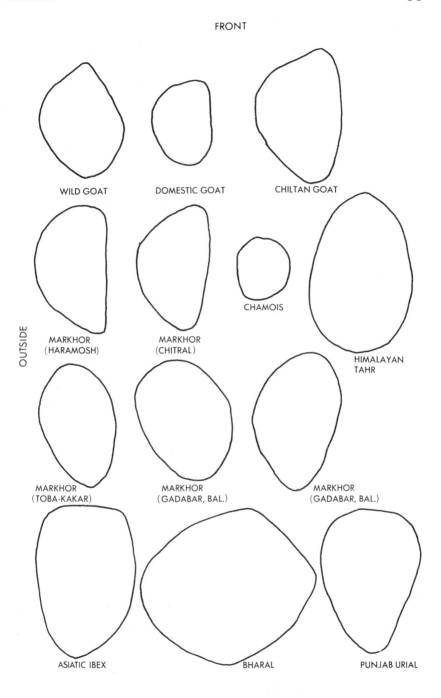

Fig. 8. Horn core cross-sections of various Caprini and the chamois, illustrating the shape and width of the frontal clash surfaces of the horns (natural size).

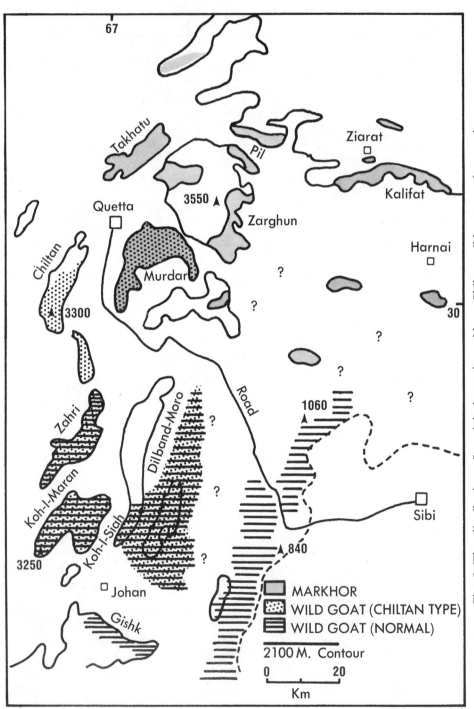

Fig. 9. The distribution of straight-horned markhor, Chiltan wild goat, and normal-horned wild goat on the various mountain massifs around Quetta, Baluchistan.

itors who claimed to have seen markhor on the Chiltan Range (Burrard, 1925; Stockley, 1928) may have been misled by the spiralling horns of some Chiltan goats. In conclusion, the so-called Chiltan markhor is a *Capra aegagrus* which does not merit subspecific status because intermediate forms exist between it and typical wild goats.

SUBSPECIES OF MARKHOR IN PAKISTAN. The spiralling horns of markhor have two basic shapes: straight and flaring. The Kabul and Sulaiman markhor have straight horns making up to 3 spirals, the former with an open twist and the latter with an open one as well as a tight one resembling a corkscrew. The Kashmir markhor has slight to moderately flaring horns, large heads having 2 to 3 twists to the spiral, and Astor markhor horns flare widely near the base and usually have no more than 1.5 twists (fig. 10). Each of these four markhor has been given subspecific status. In discussing the validity of these forms I continue to use the common names for convenience without implying that these represent taxonomic categories.

Stockley (1936) made an important point when he wrote that the various horn types "are by no means constant and must only be taken as the more usual form of the horns of the particular local race after which they are named." For instance, animals with Kashmir-type horns have been noted within the range of the Astor markhor (Cumberland, 1895; Stockley, 1936). Lydekker (1913) recorded specimens of Kabul markhor from Quetta far within the supposed range of the Sulaiman markhor, and he wrote an article, illustrated with a painting (1902), about a Kabul markhor from Chitral—which lies wholly within the range of the Kashmir markhor. Adding to the confusion are the observations of Burrard (1925), who asserted that most markhor within the range of the Sulaiman markhor resemble the Astor and Kashmir types and that the Chiltan markhor is found in "almost all the hill systems situated between Quetta and Chilas," the latter town being within the range of the Astor markhor.

In spite of these bewildering statements a quantitative geographical division between straight-horned and flare-horned animals can be made. Only 2 (3%) of 65 Kabul and Sulaiman markhor examined in Pakistan had slightly flaring horns, and 2 (2.4%) out of 83 Kashmir and Astor markhor horns were almost

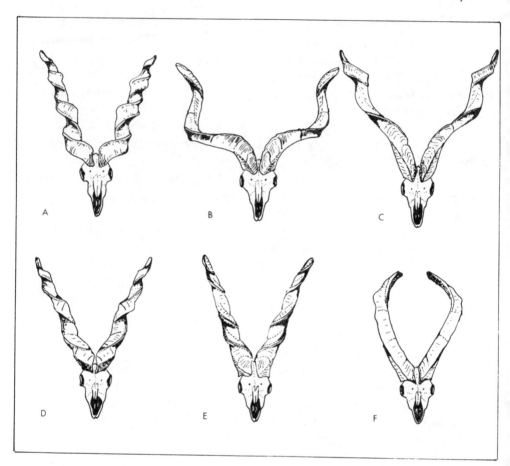

Fig. 10. Horn shapes of markhor. (A) a common shape among markhor from Russia; (B and C) flare-horned markhor, the former usually referred to as Astor markhor and the latter as Kashmir markhor; (D and E) straight-horned markhor, the former with an open spiral typical of the so-called Kabul markhor and most of the so-called Sulaiman markhor, and the latter with a tight spiral found only in some Sulaiman markhor; (F) the Chiltan wild goat, formerly designated as Chiltan markhor.

straight, both exceptions coming from Chitral. For the straight-horned markhor, horns of two types—open and tightly spiralled—may be found in the same population with, however, differing frequencies. Of 25 Kabul markhor examined from the Sakra, Safed Koh, and Marwat ranges, 100% had open spirals; of 10 Sulaiman markhor from the Gadabar Ghar 100% had open spirals, of 9 from the Toba-Kakar 89% had open spirals, and of 20 from around Quetta 65% had open spirals. The Sulaiman markhor was first given subspecific status on the basis of tight-spiralling horns, but such horns are in a minority at least in the populations from which my samples were drawn. Horn circumference is similar in both forms, averaging 23 cm (20–26 cm) in 15 trophies 5 years old and older. I am unable to distinguish an open-spiralled Sulaiman horn from a Kabul one. Subspecific distinctions here seem unwarranted on the basis of horn shape. The Kabul markhor in the Safed Koh and in Afghanistan are larger and have longer ruffs than most Sulaiman markhor, except for those on Takht-i-Sulaiman (Stockley, 1936). Stockley (1928) also felt that the southern animals are "much squarer in build." In spite of such differences I suggest that the straight-horned types be combined into one subspecies, *C. f. megaceros* (table 4).

Markhor with Kashmir-type horns are found in the Pir Panjal of India and then again in northwestern Pakistan and eastern Afghanistan, animals with Astor-type horns occurring in the intervening area. The extreme forms of each type are easy to recognize, but, as noted earlier, Kashmir-type horns are occasionally found within the range of Astor markhor. Of the 23 sets of horns seen by myself within Astor terrain none resembled the horns of Kashmir markhor closely but some were of intermediate shape, particularly in the amount of flare. Figure 11 illustrates not only the great variability in flaring, as measured by the distance between horn tips, but also the similarity in the amount of flaring between Kashmir and Astor markhors. Horn circumference averaged 25 cm (24–30 cm) in Kashmir markhor and 26 cm (24–30 cm) in Astor markhor in a total of 26 animals at least 5 years old. With the two forms being at times difficult to distinguish and other features in the two markhor being similar, it would seem that both types belong in the same subspecies, *C. f. falconeri*.

Although four subspecies have been reduced to two, a good argument can be advanced for reducing these to one. The mar-

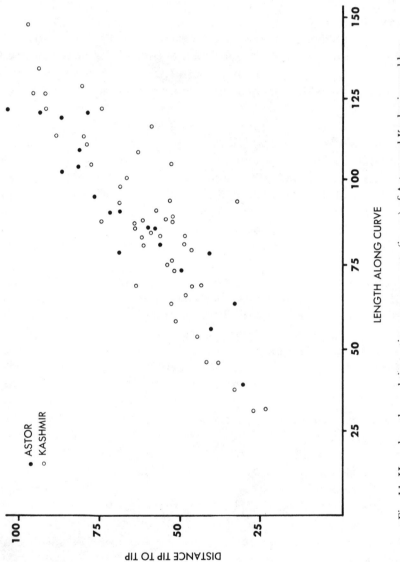

Fig. 11. Horn lengths and tip-to-tip measurements (in cm) of Astor and Kashmir markhor horns, showing the large amount of individual variation in spread.

khor in Pakistan represent a cline: the smallest animals with the shortest ruffs and straightest and most twisted horns are in the south, and the largest ones with the longest ruffs and most flaring and open-spiralling horns are in the north. Intermediate forms exist, which is not surprising since the ranges of some forms, such as the Kabul and Kashmir markhor, almost touch. Local populations are often isolated, and it is therefore surprising not that variation exists but that there is so little of it.

Ovis

The confusion in goat taxonomy is nothing compared to that in sheep: there are almost as many systems of classification as sheep varieties. Sheep have adapted to two different types of terrain, and this is reflected in their appearance. The North American sheep and snow sheep of Siberia are stocky creatures, designed for life on cliffs, whereas the others are rather lithe in form, built for speed on the open, rolling habitat they occupy. Recognizing this difference, most authors have divided *Ovis* into at least 2 species. Heptner et al. (1966) and some others designate the Siberian and North American sheep as *O. canadensis*, but American workers usually recognize 3 species: *O. nivicola*, the snow sheep, with 3 subspecies from Siberia; *O. dalli*, the thinhorn sheep, with 3 subspecies from Alaska and northern Canada; and *O. canadensis*, the bighorn sheep, with 7 subspecies from southern Canada to Mexico (Cowan, 1940).

The Eurasian sheep, excluding the snow sheep, consist of one species (*O. ammon*) with 7 subspecies according to Pfeffer (1967), one species (*O. musimon*) with 29 subspecies according to Haltenorth and Trense (1956), and 4 species (*O. ammon, orientalis, laristanica, musimon*) with 32 subspecies according to Ellerman and Morrison-Scott (1951). In addition there is the domestic sheep *Ovis aries*. To lump the small and distinctive urials with the large argalis negates the purpose of taxonomy by obscuring rather than clarifying relationships. These sheep seem to represent two evolutionary lines and deserve specific status, *O. orientalis* for the urials and *O. ammon* for the argalis.* The mouflon (*musimon*) of Corsica and Sardinia is similar to other urials, although differences in skull and other proportions exist, and it probably

---

*The Mongol word "arghali" means ram.

warrants only subspecific status. *Laristanica* has no significant characters distinguishing it from other urials.Thus I accept two species of Eurasian sheep.

Some 14 to 17 subspecies of urial (excluding *musimon*) are recognized in some recent publications (Clark, 1964; Nadler et al., 1973) even though some forms are known to be invalid. Pfeffer (1967) limits urial to two subspecies. The most variable and hence most useful physical characters from the point of view of identification are horn shape, ruff color and size, and saddle-patch color. The horns of rams have three basic shapes: with tips converging toward the back of the neck (heteronym winding), with the tips pointing at the neck, and with the tips pointing forward or sometimes diverging (homonym winding), the last-named being the usual sheep pattern. However, horns of intermediate shape may exist in any population.

Urials are so variable both within and between populations, and they show such genetic plasticity, especially in Iran, that no consensus of opinion on the number of subspecies can soon be expected. For the purpose of this report I recognize 9 subspecies including *musimon* (table 5). I am not certain of the status of *bochariensis*. Heptner et al. (1966) list it as a separate subspecies but their description of the animal is so similar to that of *vignei* that I have lumped the two. These authors also refer to the Kopet Dagh animals as *cycloceros,* but this may be an error. In addition to these subspecies, one hybrid population of *gmelini* x *arkal* stretches over about 200 km of the central Elburz Mountains in Iran (Nadler et al., 1971), another hybrid urial population of *laristanica* x *cycloceros* occurs in the Kerman region, and a third has been reported from the Kavir area (Bunch and Valdez, in press). Certain physical traits vary considerably from subspecies to subspecies. Animals with heteronym horns are found both in the western and eastern parts of the urial's range: horn types are not arranged in a tidy cline as pictured by Geist (1971a). The western subspecies have prominent ruffs only low on the ventral surface of the neck, whereas urial from central Iran eastward have long, luxuriant ruffs from the chin downward. Ruff color differs. Arkal have white ruffs and the other subspecies black ones but sometimes with a white bib. Nadler et al. (1973) stated that all eastern populations have white ruffs but this is not so. (On rare occasions, however, a Punjab urial may have a white ruff.) These authors further noted that the eastern forms have a poorly defined rump patch and a thin tail as compared to the western forms.

Recently Harrison (1968b) discovered an urial in Oman. A land connection to Iran may have existed as recently as 2500 years ago when crustal movements raised the Arabian Peninsula (Bibby, 1972). The subspecific status of this urial is unknown, but its ruff color and horn shape look like those of *cycloceros*.

Three subspecies of urial inhabit Pakistan (see fig. 18). The Afghan urial west of the Indus has homonym horns. Although two types of horns may occur within a population—one with a tight curl and one with a loose curl—the basic shape is consistent. The Punjab urial between the Indus and Jhelum rivers has sickle-shaped horns, some curving tightly and others loosely. Two urial caught shortly after birth at Kalabagh developed both horn types later, suggesting a genetic basis for this trait. Punjab urial in their winter coat have a more reddish pelage than the other subspecies in Pakistan, and they have a two-colored saddle patch. The color of the saddle may vary, with a few rams having no patch and others having only a white one. Ladak urial have heteronym horns, but according to Stockley (1928) a few individuals may have homonym horns or sickle-shaped ones. Heptner et al. (1966) made a similar comment about a Turkestan population of *vignei*. The horns of Ladak urial rise more steeply from the head and are more corrugated than those of Punjab urial. Stockley (1936) stated that Ladak urial lack a two-colored saddle, but I saw two rams with a dark line in front of a gray saddle patch. Horn circumference in a sample of 32 heads at least 5 years old was 21 cm (19–24 cm) in Punjab urial, 23 cm (21–27 cm) in Afghan urial, and 25 cm (23–26 cm) in Ladak urial, suggesting that the horns of Punjab urial are somewhat more slender than those of the other subspecies.

Haltenorth and Trense (1956) accept 14 subspecies of *Ovis ammon* and Nadler et al. (1973) 16. After measuring the skulls of various *ammon*, Pfeffer (1967) reduced the number of subspecies to 4 (*hodgsoni, poli, ammon, kozlovi*). The argalis have horns of two types. Those from the Tien Shan and Ala Tau westward to the Pamirs and Kara Tau—*severtzovi, nigrimontana, polii,* and *karelini*—have light and slender horns, whereas the others, in the Altai, Tibet, and Mongolia (*ammon, hodgsoni, darvini*), have heavy horns of large circumference. However, other characters do not separate the two groups with consistency, both *polii* and *hodgsoni*, for instance, having neck ruffs. These ruffs are different from those of urial in that they spread over much of the neck, not just in a narrow strip along the ventral side. Subspecies have been based on minor differences in horn shape, ruff length, and size of rump patch. Nasonov (see Heptner et al., 1966) relied on the degree of curvature of different parts of horns to distinguish subspecies, an unsatisfactory method. Other authors used shape. The horns of *karelini*, for example, are more rounded in cross-section, somewhat thicker at the base, and more wrinkled than those of *polii* (Miller, 1913; Wood, 1910). But the presence of angles on the front of horns seems to be correlated with bulk: as horns increase in circumference they become more rounded, producing in this instance a cline from *polii* to *karelini* to *ammon* (Geist, 1971a).

I have recognized the following argalis, the names in parentheses indicating other forms which have been here lumped:

1. Nura Tau	*severtzovi*
2. Kara Tau	*nigrimontana*
3. Tien Shan	*karelini* (*heinsii, collium, airensis*)
4. Marco Polo or Pamir	*polii* (*huemi, littledalei*)
5. Altai	*ammon* (*przevalskii*)
6. Tibet	*hodgsoni* (*dalai-lamae*)
7. Mongolia	*darvini* (*mongolica, kozlovi*)

So little is known about several argalis that this list is at best tentative. Pfeffer (1967) suggests that *kozlovi* warrants subspecific status mainly on the basis of its rather small skull, but Allen (1940) considered the population to be a part of *darvini*. *Karelini* may not differ enough from *polii* to justify a separation (Lydekker, 1898). Heptner et al. (1966) note that *severtzovi* is in various physical characters intermediate between urials and argalis and feels that the subspecies may be of hybrid origin. However, according to C. Nadler (pers. comm.), no evidence for hybridization exists in Turkestan.

Only morphological features have been considered so far because a field biologist needs to be able to identify animals on sight. According to the biological concept, a species is a genetically similar population capable of interbreeding. That all sheep readily interbreed and produce fertile offspring in captivity has been amply documented. Gray (1954) mentions hybrids of *canadensis* x *aries, musimon* x *aries,* and *canadensis* x *musimon,* and Nadler et al. (1971) report *karelini* x *aries* crosses. However, hybridization in captivity merely signifies that the animals are fairly closely related; contrary to the ideas of Uloth (1966), it does not help much in clarifying their specific status. Reproductive isolating mechanisms need not develop in the course of speciation if physical barriers prevent populations from meeting. I know of only two places where argalis and urials come into contact, one being the Turkestan area mentioned above. Ladak urial and Tibetan argali overlap in the Zaskar Range of Ladak. Blanford (1888–91) and Ward (1924) mention that hybrids exist, but proof of this is needed.

In recent years C. Nadler and his associates have made intensive chromosome studies of *Ovis* (Nadler et al., 1971, 1973; and others), and their stimulating work has renewed interest in the evolution of this complex genus. In brief, they found that all domestic sheep and the western urials (*musimon, gmelini, isphahanica, laristanica*) have a 2n chromosome number of 54. The karyotypes contain six large metacentric or V-shaped autosomes and 46 acrocentric autosomes, as well as a large acrocentric X and a small biarmed Y chromosome. The urials in the eastern half of their range (*arkal, cycloceros, vignei*) have a 2n chromosome number of 58, two autosomes being large metacentrics. Where the two types meet in the Elburz Mountains of northcentral Iran, a chromosomal hybrid zone is formed, with animals having 2n=54, 55, 56, 57, or 58. The hybrids around Kerman and Kavir have 2n=54 or 55 (Bunch and Valdez, in press). Two kinds of argali, *nigrimontana* and *ammon,* both have 2n=56, the karyotype consisting of four large metacentrics. For the snow sheep, Korobitsyna et al. (1974) found that *O. nivicola alleni* possesses 2n=52, including eight metacentrics. All North American sheep have a 2n=54, the same number as the western urial. In fact, the karyotypes of *musimon* and *gmelini* are so similar to those of *O. c. candensis* and *O. c. mexicanus* that the first three biarmed autosomes can be considered structurally homologous. Nadler et al. (1973) demonstrated this by meiotic pairing during spermatogenesis in *musimon-canadensis* hybrids and by comparing the G-band patterns on the chromosomes. Wurster and Benirschke (1968) suggest that the primitive karyotype is 2n=60 with 58 acrocentric autosomes and two sex chromosomes. When acrocentric chromosomes become reduced, there is an increase of metacentric ones, and the same number of chromosome arms is maintained. Centric or Robertsonian fusion is thought to be the main process of chromosome reduction in bovids. Thus chromosome evolution in *Ovis* proceeded from 2n=58 to 2n=52. Can a study of chromosome numbers help to clarify the taxonomy of sheep? On the basis of chromosomes, Vorontsov et al. (1972) felt that the Eurasian sheep should be divided into three species, *O. musimon* (2n=54), *O. vignei* (2n=58), and *O. ammon* (2n=56). They further concluded that domestic sheep (2n=54) were derived from *musimon* stock somewhere west of the Caspian Sea.

A difference in chromosome numbers does not necessarily indicate a difference in genetic material. Centric fusion may not affect the genes. Only minor morphological differences are shown by 2n=54 and 2n=58 urial, and at least four subspecies readily interbreed, there being no evidence of hybrid infertility (Nadler et al., 1971, 1974). As Mayr (1969) pointed out, many chromosomal mutations have nothing to do with speciation, and speciation can occur without chromosomal mutation as the goats well illustrate. This being so, the chromosomal evidence can at this stage of knowledge be used only to supplement other information when one is deciding taxonomic matters. However, the discovery of serum transferrins and hemoglobins unique to certain sheep forms may some day provide an important means of evaluating phylogenetic relationships (Lay et al., 1971).

Hemitragus

Tahr (often misspelled "thar") have short but massive horns, those of males and females being almost equal in size, a characteristic more typical of rupicaprids than caprids. In its glands and other features the genus is closely allied to *Capra*, though the chromosome number of Himalayan tahr (2n=48) differs from that of true goats (see fig. 43). Matings between male Himalayan tahr and female domestic goats resulted in miscarriages (Epstein, 1972). The tahr are divided into three forms: *H. jemlahicus*, the Himalayan tahr; *H. hylocrius*, the Nilgiri tahr of south India; and *H. jayakari*, the Arabian Tahr. Haltenorth and Trense (1956) and Charles (1957) felt that all tahr belong to the same species, but the evidence for this view is not convincing. Pohle (1944) designated a new subspecies of Himalayan tahr from Sikkim, *H. j. schaeferi*, based on minor differences in pelage color and horn shape. Since horns are variable and pelage changes with the seasons, and since Sikkim falls within the continuous range of the subspecies *jemlahicus,* a new taxonomic category is unjustified.

Ammotragus

In many respects the aoudad is like a goat. It has a stocky build with a rather long face, callused knees, and a tail that is flat and bare beneath. The enamel folds on its molars bifurcate and the surface of its horn core is quite hard, both characteristics of goats but not sheep (Epstein, 1972). The follicle arrangement of the pelage is more similar to that of goats than sheep (Ryder, 1958). Neither pedal, preorbital, nor inguinal glands are present, but there are subcaudal glands according to Dalimier (1954). Stripped of its distinctive ruff, the aoudad resembles the Dagestan tur (Petzch, 1957). Attempts to hybridize domestic sheep with aoudad have failed (Gray, 1954), but young have been produced by crossing an aoudad male with a domestic goat female (Petzch, 1957), and such a hybrid was successfully mated with an Alpine ibex (Haltenorth, 1961). However, unlike goats and like sheep, the aoudad lacks a beard. Its diploid chromosome number is 58 as in eastern urials. An electrophoretic study of blood serum proteins showed that *Capra* and *Ovis* differed and that *Ammotragus* and *Ovis* were almost identical, suggesting a close phylogenetic relationship (Schmitt, 1963). Biochemical tests place the aoudad closest to the sheep, morphological features closest to the goats, and from this no more can be concluded at present than that the animal is related to both. Four subspecies are generally recognized.

Pseudois

Bharal, like aoudad, show a bewildering combination of sheep- and goat-like traits. They lack beards and calluses on the knees, they have no strong body odor, and the females have small, almost nonfunctional horns, all characters typical of sheep. On the other hand they resemble goats in their flat broad tail with a bare ventral surface, the conspicuous markings on the forelegs, and the large dew claws. The horns are shaped like those of the Dagestan tur, and, as Lydekker (1898) pointed out, "the structure and colour of the horns are the same as in goats." There are no inguinal glands. Pedal glands are usually absent but some individuals seem to have rudimentary ones; the preorbital gland is also missing except for an occasional bare patch where the gland would normally be (Pocock, 1910). Haltenorth (1963) stated that pedal glands are present, whereas Pocock (1910) said they are not. One male I examined had no obvious pedal glands. It appears that these glands in bharal are either appearing or disappearing. Gray (1954) noted that bharal and domestic sheep will not interbreed, but a cross between a male bharal and domestic goats twice produced full-term, stillborn twins at the Omaha zoo (L. Simmons, pers. comm.). In general, bharal are considered to be aberrant goats with sheep-like affinities.

There are two subspecies of *P. nayaur* according to Ellerman and Morrison-Scott (1951), *P.n. nayaur,* mainly from Tibet, and *P.n. szechuanensis,* from Shensi and Kansu. Schäfer (1937) designated but did not name a new form from eastern Tibet. Since small size was the main character distinguishing this animal from others, and since size is often an environmental variable, the supposedly new form may not be acceptable.

3 DISTRIBUTION

The foundations of animal geography in the Himalaya were laid by early hunters and travellers who compiled faunal lists and noted the best places at which to shoot. I have drawn on these early reports and added my own data in an attempt to plot the current distribution of some Caprini in South Asia. Such an endeavor may seem anachronistic to someone familiar with the sophisticated level of wildlife work in Europe and North America, but parts of the Himalaya remain virtually unknown, zoogeographically speaking. The precise limits of species have seldom been defined, and the fascinating ecological problems raised when two related species occupy the same range have been largely ignored. I made intensive studies in several areas and conducted road and foot surveys in others (fig. 1). The road surveys consisted of stopping at intervals and interviewing villagers, local hunters, and officials about wildlife in the vicinity. Whenever possible, I examined trophy horns, measuring them in such diverse places as graveyards, mosques, private homes, and government offices. In addition to several short foot transects lasting a week or less, I made three long treks. One took me up the Yarkhun River in Chitral to its headwaters, and, after crossing the Hindu Raj Range, I traversed the Gilgit Agency and via the Dadairelli Pass moved south into Swat (see Schaller, 1975). A second trip, this one in the Karakorams, took me up the Braldo Valley and Baltoro Glacier to the vicinity of K2. In Nepal, I walked from Pokhara along the southern flanks of the Dhaulagiri massif before turning north across the Kanjiroba Range to Shey and other parts of Dolpo District and then continued west to Jumla.

A study of distribution should help to clarify centers of origin, routes of dispersal, and in general the evolution of a taxonomic group. *Ovis* and *Capra* would seem to be ideally suited to having their evolutionary history unraveled. Primitive members of the

subfamily such as the goral still exist, as do such seemingly intermediate forms as the tahr and aoudad. And it should be possible
to trace existing distribution patterns into the past. Unfortunately,
the Caprini arose and diversified during a turbulent period when
climates changed and the earth heaved and folded, when the
Lord "turneth it upside down, and scattereth abroad the inhabitants thereof" (Isaiah 24:1). Being essentially adapted to hills and
cliffs, the distributional patterns of the animals were seriously
disrupted, and since bones are seldom preserved in mountains the
Caprini have revealed little of their fossil past.

GEOGRAPHICAL DISTRIBUTION

Saigini, Ovibovini, and Rupicaprini

The saiga appeared in the Miocene. Its range today is relatively
small, stretching eastward from the Caspian Sea across Russia to
about the Altai, but in the Pleistocene it occupied a vast tract from
Europe to Alaska (Bannikov et al., 1967). The chiru is confined to
the Tibetan uplands.

The muskox, *Ovibos,* evolved in the late Pliocene or early Pleistocene. Confined now to the Canadian Arctic and Greenland, the
animal occupies only a vestige of its former range, which once
extended from England to Siberia and over North America as far
south as New York and Iowa. Several extinct muskox genera, such
as *Bootherium* and *Symbos,* also lived during the Pleistocene (Tener,
1965). The takin has a patchy distribution, most populations
being found in Yunnan and Szechwan but some reaching to the
borders of Kansu and southern Shensi and southeastern Tibet
(Allen, 1940). They are also found in northern Burma, in the
Tirap Hills along the India-Burma border, and in the Mishmi
Hills on the India-China border (Bailey, 1945). Farther west, the
species occurs in northern India just east of Bhutan (H. Dang,
pers. comm.), and in the Lunana and Lingshi districts of western
Bhutan (Ward, 1966).

It is generally agreed that the Rupicaprini were the ancestors of
the Caprini, that the sheep and goats represent the extreme development of animals which once resembled the goral. The separation occurred during the Miocene. *Oioceros* of the Miocene and
lower Pliocene of China, Kenya, and Yugoslavia (Gentry, 1968)
resembles the goral in size and in its small, pointed horns except
that the horns have an anticlockwise twist (Pilgrim, 1939). Pilgrim

(1947) considered *Oioceros* to be an ancestor of sheep, but Thenius and Hofer (1960) feel that the animal is too specialized for that. By the lower Pliocene, creatures intermediate between goral and goats had made their appearance in China, *Tossunnoria* having cranial features resembling *Nemorhaedus* but horn cores like those of *Hemitragus*. *Sivicapra,* a probable descendant of *Tossunnoria* in the upper Pliocene of the Siwaliks, was apparently the earliest goat. *S. sivalensis* had short, curving horn-cores, slightly twisted anticlockwise with anterior and posterior keels (Pilgrim, 1939).

The Rupicaprini once had an extensive distribution in Eurasia and possibly in Africa (Gentry, 1968). The tribe seems to have an Asian origin, *Pachygazella grangeri* of the Pliocene in China being a possible progenitor (Thenius and Hofer, 1960). The serow now extends only as far north as Japan and the goral as far as the Amur River, but some ancestor of these forms crossed into North America to give rise to the mountain goat. Both serow and goral are found widely in the forested areas of China, Indochina, and Burma, with the serow also having penetrated to Malaya and Sumatra as well as to Japan and Formosa (see Dolan, 1963). Both also moved westward in a tonguelike extension along the southern flanks of the Himalaya, the goral crossing the Indus to the Afghanistan border. This distribution represents a sharp range reduction in the Asian rupicaprids, for a large goral (*Gallogoral meneghinii*) existed in the Villafranchian of Europe. A rupicaprid gap now exists between Pakistan and the Caucasus, between the goral in the east and the chamois in the west. Chamois range westward from the Caucasus into parts of eastern Turkey and Europe, including various Balkan mountains, the Carpathians, Alps and Pyrenees. The earliest chamois fossils date only to the late Pleistocene in Europe and the evolutionary history of the genus remains a mystery (Kurtén, 1968). Other rupicaprids in Villafranchian Europe include a goatlike chamois (*Procamptoceras brivatense*) and the cave goats (*Myotragus*) of the Mallorcan and Menorcan islands. In their most advanced form these cave goats had extremely short cannon bones and protruding chisellike incisors, possibly used for rooting or stripping bark off trees.

Caprini

Figures 12 to 18 summarize the geographical distribution of several Caprini. For information outside of my area of study, I depended on published sources, especially on Heptner et al. (1966)

for Russia, Allen (1940) for China and Mongolia, Kumerloeve (1967) for Turkey, and Harrison (1968a) for Arabia. Maps such as these can only give an impression of an animal's distribution, for habitat requirements, human interference, and other factors force each species to use only a fraction of its total range.

Hemitragus. Tahr once ranged as far west as Europe, which, according to Kurtén (1968), saw two separate invasions of these animals during the Pleistocene. The first occurred during the second cold phase of the Günz glaciation and involved *H. stehlini;* the other, probably by *H. bonali,* took place early in the first Würm stadial. The latter animal was larger than the Himalayan tahr but otherwise resembled it, even to the extent of sporting a large ruff, if a cave painting dating to the second Würm stadial at Cougnac in France is indeed that of a tahr (Koby, 1956). The tahr, like the saiga, steppe bison, spotted hyena, and so many other species, disappeared from Europe toward the end of the Würm between 17,000 to 10,000 years BP (Philip and Fisher, 1970) as the cold, dry climate became warmer and more humid and the steppes were invaded by forests (Bonatti, 1966). In the Near East, in Turkey and Iran, where tahr presumably also lived at that time, the climate between 30,000 and 14,000 BP was colder and drier than now, with trees confined to refuge areas along the Mediterranean coast and the Black and Caspian seas. Although a sparse forest spread again after 14,000 BP, the climate remained drier than it is today (van Zeist, 1969). In any event, tahr vanished from most of their range for reasons not wholly explainable on the basis of climatic change. That tahr can tolerate extremely arid conditions is shown by the refuge population of Arabian tahr in the desert mountains of Oman (fig. 12).

Nilgiri tahr are scattered as isolated populations along the high tablelands and rolling hills that straddle the boundary for about 400 km between the states of Tamil Nadu and Kerala in south India, an isolated habitat over 2000 km from other tahr in the Himalaya and Oman. In historic times the total range of the species probably never exceeded 5500 sq km, and excessive hunting has reduced its present range to less than a tenth of that area, substantial populations now surviving mainly in the Nilgiri Hills, High Range, and Highwavy Mountains.

The Himalayan tahr inhabits a narrow strip along the southern

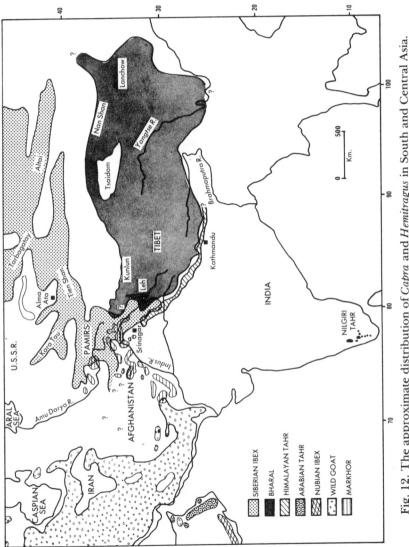

Fig. 12. The approximate distribution of *Capra* and *Hemitragus* in South and Central Asia.

flanks of the Himalaya, penetrating the main range only along some of the large gorges (figs. 13 and 14). The western limit of its distribution lies about 40 km west of Banihal Pass (Stockley, 1936) in the Pir Panjal, and the eastern limit somewhere in Bhutan (White, 1910). Within this area the animals once occupied all the main valleys and their tributaries, wherever precipices provided favored haunts. In the Western Himalaya, animals were once common in Kishtwar and Chamba and in the upper valleys of the Chenab, Ravi, Beas, and Sutlej (Kinloch, 1898); only a few were found in the Kulu valley (Stockley, 1928). In the Indian part of the Central Himalaya, tahr have been reported from Gahrwal and Kumaon (Stockley, 1928) and along the Rishiganga Gorge leading to Nanda Devi (Dang, 1964). The once continuous distribution of tahr in Nepal has become disrupted as settlers have penetrated the wilderness, clearing tillable land between cliff systems and killing off isolated populations. Figure 14 shows where I have seen tahr and where they have been reliably reported to me. The localities include the Bheri River and some of its tributaries upstream from Tibricot, the southern and western slopes of the Dhaulagiri massif, the upper Mohdi River north of Pokhara, Langtang National Park, several deep gorges north of Lukla in the Everest area, along some of the western tributaries of the Arun River, and in the Topke area east of the Arun River. I studied a population along the upper Bhota Kosi River northeast of Kathmandu, and tahr were also found along the tributaries to that river, in the valleys of the Kang Chu, Chyadu and Rongshar. Schäfer (1950) hunted tahr in the Tista Valley of Sikkim, northwest of Chungtang.

Capra. This genus first appeared in the mid-Pleistocene, having probably evolved from a tahr-like goat. Except for a fossil skull of questionable origin from Iowa (Reed and Palmer, 1964), no goat remains have been found in North America. Today the genus occupies most mountain systems between Europe and the Sayan Mountains near Lake Baikal (figs. 12 and 15). The most westerly form is the Spanish goat, which survives in the Spanish Pyrenees, the Sierra Blanca, Sierra de Gredos, Sierra de Cazorla, and Sierra de Ronda (Eichler, 1973). Almost exterminated during the past century, the Alpine ibex has been reintroduced into various parts of the Alps. The only other goat in Europe is the wild goat, now

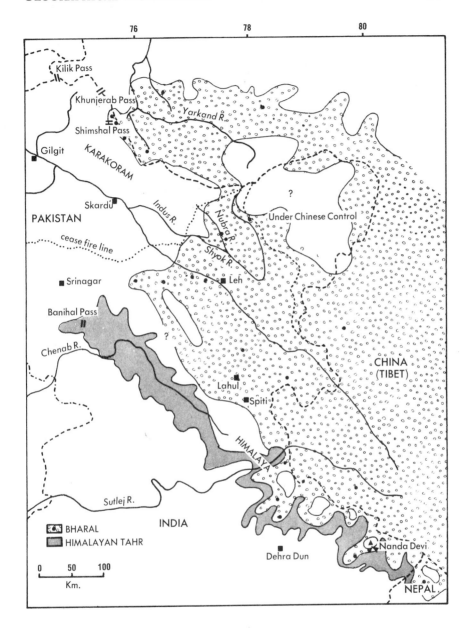

Fig. 13. The approximate distribution of Himalayan tahr and bharal in India and Pakistan. The black circles within the range of bharal indicate selected sight records from the literature.

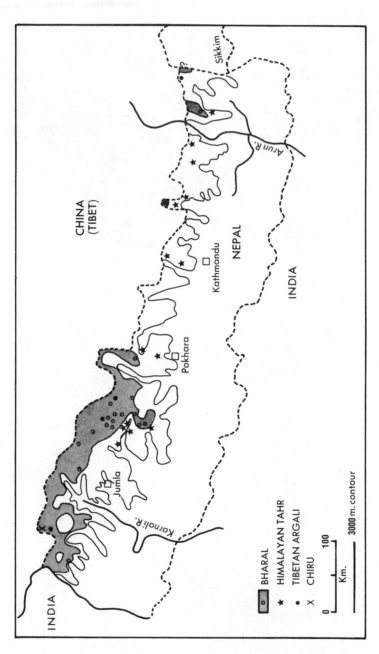

Fig. 14. The approximate distribution of several caprids and chiru in Nepal. The open circles within the continuous range of bharal indicate where I have seen the animals or from where they have been reliably reported to me. Since the total present range of tahr remains unknown, only selected localities are indicated.

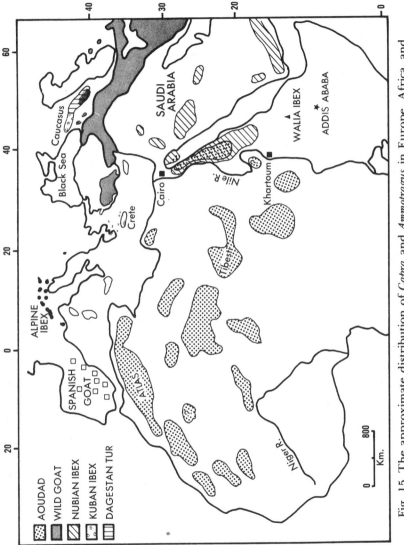

Fig. 15. The approximate distribution of *Capra* and *Ammotragus* in Europe, Africa, and the Near East.

confined to western Crete, to several small islands near Crete, and to a number of sites on or near the Greek coast (Papageorgiou, 1974; Schultze-Westrum, 1963). Wild goats also inhabited the mainland of Europe, surviving in Bulgaria until 1891 according to van den Brink (1967). Another species of goat, commonly called *Capra prisca,* is thought to have lived in Europe until historic times, ranging from Austria to the western Ukraine (Zeuner, 1963; Epstein, 1972). This goat had twisted horns, but the horn core cross-sections resemble those of wild goat. At least some of the specimens of *prisca* are nothing more than domestic goats from the Neolithic, as Thenius and Hofer (1960) have shown, but the status of others remains in doubt. It is, of course, possible that *Capra aegagrus* with a horn-shape variation of the type found in the Chiltan goat once lived in parts of Europe.

How the wild goat reached Crete and other islands, and how urial colonized Corsica, Sardinia, and Cyprus, remains a puzzle; it has been plausibly suggested that the animals were introduced. Two other alternatives, neither wholly satisfying, present themselves. During Pleistocene glacials, when much water was locked up in ice, sea levels were as much as 100 m lower (Kurtén, 1968), probably low enough to make some coastal islands accessible from the mainland. But Crete and Cyprus are separated from the coast by a water depth of at least 250 m. During the early Pliocene the Mediterranean was cut off from the other oceans when the earth's crust moved. The waters of the sea vanished due to evaporation, leaving the Mediterranean basin a lifeless desert until about 5.5 million years ago when the sea returned with a rush through the Straits of Gibraltar (Hsü, 1972). Some land mammals could have reached the islands during this waterless interregnum, but present-day *Capra* and *Ovis* had not evolved at that time.

The wild goat is found in southern and eastern Turkey and on the southern flanks of the Caucasus. During the late Pleistocene the species ranged south into Lebanon and Israel too (Zeuner, 1963). Today the animals occupy virtually all the hill ranges in Iran (Lay, 1967), extending northward into the Turkmenia province of Russia. An isolated population occurs in Oman (Harrison, 1968a). To my knowledge the only wild goats in Afghanistan occur in the Chagai Hills bordering Pakistan and in the western part near Iran (Couturier, 1962). In Pakistan the species has a wide distribution on the Baluchistan Plateau south and west of

Quetta, inhabiting most hill systems from sea level at Ormara to an altitude of over 3000 m in the Chiltan Range. The species may once have extended farther north and east than it does now. The northern boundary is usually given as the Bolan Pass along the Quetta-Sibi road (Stockley, 1928). However, I found evidence of a population, now possibly extinct, on the Gadabar Ghar 50 km west of Loralei and some 180 km northeast of Bolan Pass. Other populations may well occur in the intervening area, but at the time of my visit a survey was not politically feasible.

The Kuban ibex and Dagestan tur are confined to the Caucasus, and details of their distribution can be found in Heptner et al. (1966) and Vereshchagin (1967). The Nubian ibex occurs in the northern Sudan and in Egypt east of the Nile, on the Sinai Peninsula, in Syria, in Israel, and on the Arabian Peninsula (fig. 15). Confined to a small massif in the Simien National Park of Ethiopia, the Walia ibex has the most restricted distribution among the ibex subspecies. A gap of some 2000 km separates the ibex of the Caucasus and Near East from the nearest Asiatic ibex populations in Afghanistan. Most of the range of the Asiatic ibex is outlined in figure 12. Stretching from the Hindu Kush in Afghanistan to the Sayan Mountains in Russia, the range of the subspecies includes the Pamirs, Tien Shan, Kara Tau, Tarbagatay, and Altai mountains. As Heptner et al. (1966) noted, distribution is patchy with 100 km sometimes separating neighboring populations. Unsuitable habitat is the cause of such disjunction in some places, but in others the animals have been exterminated by man. Ibex occupy all the major ranges in northwestern India and Pakistan, making them the most widely spread and abundant large mammal in the area (fig. 16). The upper Shyok River and the vicinity of Leh mark the eastern boundary of the species in Ladak, but to the south, along the Himalayan chain, the animals extend eastward to the Sutlej.

Markhor have a limited geographical distribution, their range being squeezed between that of the ibex and wild goat mainly in Pakistan but also in India, Afghanistan, and Russia. The current distribution of the species was discussed in detail by Schaller and Khan (1975), and I summarize here mainly the information presented in figure 17. Three isolated markhor populations occur in Russia and one of these also extends into the Badakshan part of Afghanistan. Farther south, across a gap of 200 km which in the

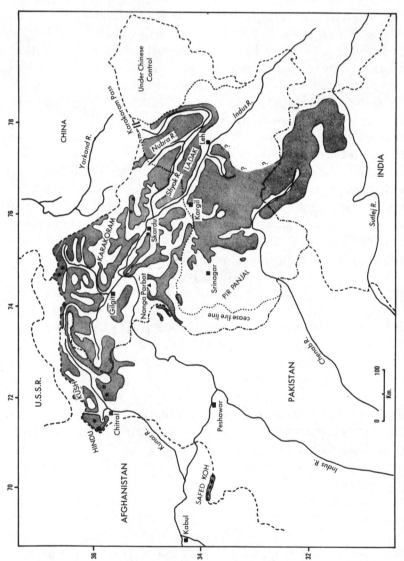

Fig. 16. The approximate distribution of Asiatic ibex in Pakistan and India. The black circles mark locations where I have seen ibex.

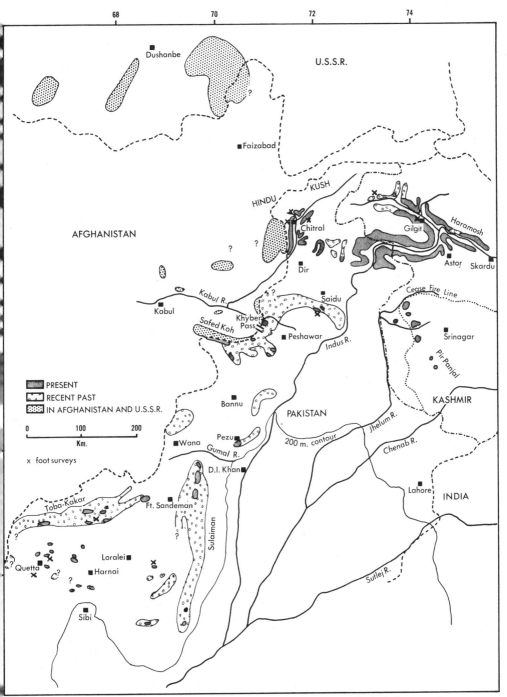

Fig. 17. The distribution of markhor (reproduced from *Biological Conservation* with permission from Applied Science Publishers).

past must have been occupied by markhor, lies the range of the flare-horned markhor of the Kashmir type extending from the Laghman province and the Bashgul River area of Afghanistan into neighboring Pakistan. There markhor still occur in many valleys bordering the Kunar River in the Chitral District, from Shogore to Arandu on the west bank and to Drosh on the east bank, and they also penetrate up the Mastuj River, a tributary of the Kunar, as far as the village of Barenis. In Dir District the markhor occurs along the upper Panjkora River, and in Swat District on the cliffs east of Mankial. Markhor with Kashmir-type horns also survive in a few small populations in the Shamsberi, Kaj-i-Nag and Pir Panjal ranges of India. The Astor markhor has colonized northern Pakistan by penetrating up the Indus and its tributaries, among them the Astor, Gilgit, and Hunza rivers. The animals inhabit both banks of the Indus from Jalkot upstream to about the village of Tungas near Skardu. Gakuch now represents the limit of their distribution along the Gilgit River, Chalt along the Hunza River, and the Parashing Valley along the Astor River.

Straight-horned markhor have a highly discontinuous distribution (fig. 17). This is partly due to the erratic location of cliffs and isolated massifs and partly to indiscriminate hunting which has brought the subspecies to the verge of extinction. Although some surviving populations were no doubt overlooked during the surveys, the map shows that the markhor have been eliminated over vast stretches. The Kabul markhor once ranged in a westerly arc from the Indus into Afghanistan to the vicinity of Kabul, and southward mainly in Pakistan as far as the Gumal River, which was generally considered its southern boundary (Burrard, 1925). Five small, isolated populations persist in Afghanistan, mainly in the Kohi Safi area northeast of Kabul (Petocz, 1973a), and scattered herds are also found along the Pakistan border in the vicinity of Khyber Pass and the northern flanks of the Safed Koh. In Pakistan, there are a few animals in the Sakra Range northeast of Mardan, west of the Peshawar-Kohat road, on the southern slopes of the Safed Koh, and near Pezu. The markhor in the Surghar Hills east of Bannu have probably been exterminated, and I have no information about the animals near Wana in Waziristan. South of the Gumal River, the Sulaiman markhor have fared just as badly. In the Sulaiman Range there are markhor on Takht-i-Sulaiman and on the cliffs north and south of that peak. Near the southern end of the range, a few animals survive about 25 km south of Fort Munro and in the Gurshani Hills; a few markhor may also persist in the Gadabar Ghar east of Loralei. At least two isolated populations occur in the Bugti Hills. In the Toba-Kakar Ranges along the Afghanistan border the animals are now confined to a few cliff systems. North and west of Quetta are isolated rugged hills broken by huge cliffs on which markhor were once abundant. These hills—Takatu, Murdar, Zarghun, Pil, Kalifat, and Shingar—now harbor only a few animals. At least three other populations are found in the hills south of Harnai in an area where access to outsiders is restricted.

Ovis. The distribution of sheep in North America has been accurately delineated in many sources (see Burt and Grossenheider, 1952). In brief, the thinhorn sheep range from Alaska to central British Columbia, and the bighorn sheep are found along the Rocky Mountains from British Columbia to Baja California in Mexico. Sheep similar to *O. dalli* appeared in North America in the Illinoisian glacial period, having moved across the Bering land bridge from Asia (Stock and Stokes, 1969). From 6800 to about 10,000 BP a sheep larger than the bighorn resided in Utah and Nevada, but to emphasize its close relation to *O. canadensis* Harris and Mundel (1974) classified it as *O.c. catclawensis. O. canadensis* of the modern type have been found in late Pleistocene deposits in Nevada, Washington, and British Columbia.

In recent years several authors have presented general maps of sheep distribution in Eurasia (Clark, 1964; Pfeffer, 1967; Nadler et al., 1973). Except for the snow sheep of northeastern Russia—which inhabits the Putorana, Verkhoyansk, Cherskiy, Stanovoy, Kolyma, and other mountains as well as various ranges on the Kamchatka Peninsula—the distribution of Eurasian sheep follows many of the mountain systems also occupied by goats. Taking urials first, the westernmost population, *musimon,* is found on the islands of Corsica and Sardinia, though the sheep has also been introduced into France, Germany, Poland, and other European countries where probably it was once native. Fossils of *Ovis* are rare, but there are Villafranchian specimens from France and China and several mid-Pleistocene remains from Europe, variously known as *O. antiqua* and *O. savini,* which could have been ancestral to *musimon.* "In the late Pleistocene, mouflon is not uncommon in Italy and North Africa and one specimen has been identified from a cave in the Franconian Jura" (Kurtén, 1968).

Outside of Europe, one subspecies of urial (*ophion*) survives on Cyprus, and another (*gmelini*) in parts of central, southern, and eastern Turkey and northern Iraq eastward into the Elburz and Zagros mountains of Iran with an extension northward toward the Caucasus. *Laristanica* and *isphahanica* are confined to the southern half of Iran, and *arkal* to northeastern Iran and the parts of Russia lying southeast of the Caspian Sea. The Afghan urial *cycloceros* has a wide range, covering most of the mountainous parts of Afghanistan, southeastern Iran, southern Turkmenia, and Pakistan. In the last-named country it occupies or occupied most of the highlands west of the Indus; that is, the whole Baluchistan Plateau and the hills bordering Afghanistan southward from the vicinity of Peshawar and the Khyber Pass. The Punjab urial is found only in the Salt and Kala Chitta ranges between the Indus and Jhelum rivers (fig. 18). The Ladak urial, like the markhor, advanced into the mountains of northern Pakistan and India mainly by penetrating the major river valleys such as the Kunar, Indus, Gilgit, and Shyok (fig. 18). The small populations in Chitral and in Swat, as well as those along the Indus and its tributaries downstream from Skardu, are almost extinct, but some animals still occur along the Shigar and Braldo rivers, along the upper Shyok, and possibly along the Indus to about 80 km upstream from Leh (Darrah, 1898; Cunningham, 1854). In Afghanistan and Russia this subspecies is said to occur in the southwestern corner of the Pamirs, the lower Wakhan Corridor, and along the northern bank of the Amu Darya (Flint et al., 1965).

The argalis are now confined to Central Asia, but Upper Pliocene fossils, referred to as *O. zdanskyi* and *shantungensis* by Heptner et al. (1966), suggest that argali-type sheep were more widely distributed in China in the past than at present. *Ammon* and *darvini* inhabit the northern part of the argali range, the former in the Altai Mountains and adjoining ranges to just south of Lake Baikal, and the latter in the southern Gobi of Mongolia and in China, where it occupies the Khingan and other ranges. *Karelini* inhabits the Tien Shan. Two subspecies are wholly Russian in distribution, *nigrimontana* in the Kara Tau along the eastern bank of the Syr Darya, and *severtzovi* in the desert ranges of the Kyzyl Kum between the Amu Darya and Syr Darya. Of special interest

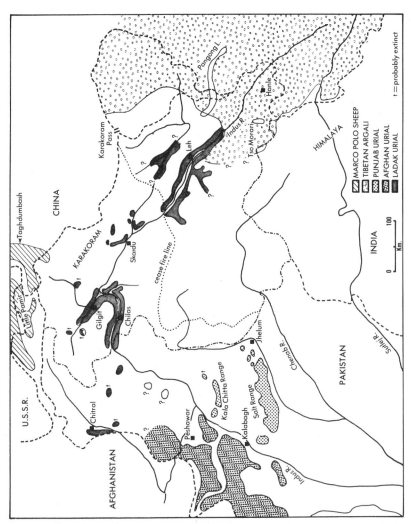

Fig. 18. The distribution of *Ovis* in northern Pakistan and India.

here are the subspecies *polii* and *hodgsoni*, which penetrate my area of study. The Marco Polo sheep has the center of its distribution in the Pamirs, the Bam-i-dunya or Roof of the World, as the Persians called this area of vast valleys flanked by high ranges whose outlines have been smoothed by glaciers. There are eight of these valleys or pamirs, most of them in Russia. Parts of the Great Pamir and the Little Pamir are in the Wakhan Corridor of Afghanistan. The Tagdumbash Pamir in China has two gentle valleys extending to the Pakistan border at the Kilik and Khunjerab passes. Marco Polo sheep cross into Pakistan at these two points, but suitable habitat is limited to fewer than 100 sq km (fig. 19).

The precise range of the Tibetan argali remains obscure. Bartz (1935) shows it as inhabiting almost the whole of Tibet. Travellers have often commented on its presence in the Outer Plateau from Bhutan westward along the Tsangpo, Sutlej, and Indus (Stockley, 1928; Rawling, 1905). The subspecies is also found in eastern and northeastern Tibet (Allen, 1940; Bailey, 1915) and in the Nan Shan, Altyn Tagh, and Humboldt ranges (Harper, 1945). These argalis cross the Tibetan border and penetrate Bhutan, Nepal, and India at a few places. In Bhutan they are said to occur in the Ha and Paro valleys (Gee, 1967a). Some lived around Tso Lhama in northern Sikkim until shot out by Indian troops (M. Ranjitsinh, pers. comm.). There are several reports from Nepal (fig. 15). Hooker (1854) saw these sheep in the Yangma Valley in northeastern Nepal and J. Blower was told that some animals may occur north of Charka in Dolpo District and in the northwestern corner of the country. I saw several skulls and was informed by villagers around Shey Gompa and upstream of Namdo in Dolpo District that though argalis were once fairly common they have declined so drastically in the past ten years that they have almost vanished. The reasons for their disappearance are unknown. Hunting was a factor but probably not the major one. Living at the edge of their range at low densities in an area where snowfall can be heavy, it is possible that the population was decimated by several severe winters coupled with heavy wolf predation. Farther west in India, argalis occasionally reach the border at such places as the Niti Pass north of Kamet. The sheep are rare visitors to Lahul and Spiti (Stockley, 1928). However, they are widely though sparsely distributed in eastern Ladak where hunters often shot them in the vicinity of Pangong and Morari lakes and north of the town of

Hanle (Darrah, 1898; Mcintyre, 1891). The western limit of the argalis' range is about 80 km upstream from Leh and just east of the Karakoram Pass (Stockley, 1928). I have not found records of Tibetan argali west of the Karakoram Pass, nor of Marco Polo sheep east of the Tagdumbash Pamir, indicating a gap of about 300 km between the two subspecies.

The most westerly form of argali today is *severtzovi*, south of the Aral Sea. Some evidence exists that these large sheep extended farther westward in the past, with, for example, Vereshchagin (1967) reporting middle and upper Pleistocene remains of argali-type sheep from the Caucasus. Zeuner (1963) noted that it is "conceivable that Asiatic steppe sheep penetrated as far west as central Europe." Of interest in this connection is the existence of a giant "sheep," in size and shape much like a muskox, whose Villafranchian fossils have been found in Europe and China (Kurtén, 1968). This animal, *Megalovis,* was powerfully built, more like a bighorn sheep than an argali.

Ammotragus. The aoudad confines itself to Africa north of 10° latitude (Fig. 16), being found in many of the ranges and isolated massifs that lie in and around the Sahara—Atlas, Ahaggar, Air, Tibesti, Dar Fur, Adrar des Iforas, and others (see Joleaud, 1928). It is not known whether the species immigrated from Eurasia or evolved in Africa from one of the primitive Caprinae that occupied the continent during the Miocene and later (Gentry, 1968). *Ammotragus* fossils date to the Pleistocene when *A. palaeotragus* was found in North Africa (Trouessart, 1904–5). Ancestors of the aoudad may also have occurred in Europe where several forms, variously known as *Ammotragus magna* or *Ovis magna* and *Ovis primoeva,* are possibly related to the African animal (see Ogren, 1965).

Pseudois. The bharal is essentially restricted to Tibet (fig. 12). The Kunlun and Nan Shan form the northern range boundary and the Himalaya the southern, although the species has penetrated the Tibetan marginal mountains and even moved through some of the gorges of the main Himalayan range. For example, the animals live on the southern slopes of several Himalayan peaks such as on Dhaulagiri in Nepal and Nanda Devi and Badrinath in India (figs. 13 and 14). The eastern boundary of the species lies in western Szechwan and Kansu in a rugged area broken by deep river gorges such as those of the Yangtse and Min (Schäfer, 1937; Allen, 1940). The western boundary is irregular, cutting northward from about the town of Kargil across Ladak up the Nubra Valley and then arcing around the eastern end of the Karakorams before continuing as far west as the Hunza District of Pakistan.

I attempted to collect information about the distribution of bharal along the southern edge of the animal's range, from Bhutan westward. White (1910) noted that bharal are found above 4850 m in Bhutan, and several people have told me that the species is common in the northwestern corner of the country. In Sikkim the bharal once existed "in considerable flocks at high altitudes" (Anon., 1894). The eastern half of Nepal has two known populations, though others may occur on the western slopes of Kangchenjunga. One inhabits the range on the eastern side of the Arun River and the other, consisting of no more than 75 animals, barely penetrates Nepal around the Lapche monastery in the upper Kang Chu Valley. In the western half of the country the animals are widespread along the Tibetan border. North of Annapurna they have been reported from Manang. In Mustang they are scarce, neither Peissel (1967) nor Caughley (1969) having seen them there. However, the species is common in neighboring Dolpo District where I have watched animals around Phoksumdo Lake and the Shey and Namgung monasteries. According to the reports of villagers and travellers they occur all along the upper Karnali River and its tributaries, as well as on the upper Bheri around Tarap. At this longitude bharal extend southward onto the western and northwestern slopes of the Dhaulagiri massif, in such places as the Gustung and Dogadi valleys, as well as in the Seng and Saure valleys where I have seen them. J. Blower told me of bharal near Dorphu north of Jumla, and further west populations exist on the southern slopes of Mount Api.

Bharal have penetrated the Himalaya in many places between the Nepal border and the Sutlej River—the slopes of Nanda Devi (Shipton, 1936), Bandarpunch and Trisul (Dang, 1968a), and the valley of the upper Buspa, a tributary to the Sutlej (Markham, 1854). From there the species extends westward across the Lingti and Tsarap valleys in Spiti and Lahul (Stockley, 1928; Mellon,

1972) as far as the Srinagar-Leh road, specific points being the vicinity of Kargil and near Lamayuru (Stockley, 1928). Farther north, the western limit lies along the Shyok River, about 40 miles below the junction with the Nubra (Stockley, 1936) and along the Nubra (Roosevelt and Roosevelt, 1926). Bharal have often been hunted in other parts of Ladak such as northwest of Tso Morari, near Leh, Karnak, and Rumbok, along the upper Shyok, and across the Tibetan border around Rudok (Darrah, 1898; Rawling, 1905; Dunmore, 1893; Stockley, 1928; Macintyre, 1891). The species ignores the southern slopes of the Karakoram but ranges along its northern flanks westward to the Hunza District of Pakistan (Etherton, 1911; Macintyre, 1891; Shipton, 1938). In Hunza the animals are found only in the upper Shimshal and possibly also in the Ghujerab Valley. While the distribution of bharal along the Himalaya and in Ladak is fairly well known, little precise information is available for the interior of Tibet, often because travellers did not clearly identify the animal. Harrer (1953), for example, merely notes "herds of wild sheep" near Lhasa, and Das (1902), travelling near Hang, saw "a broad steppe where wild goats and sheep and a few musk deer were grazing." But a few good records are available, among them those by Hayden and Cosson (1927), who found bharal on most Tibetan mountains during a trip from Darjeeling to Lhasa with detours around Lakes Nam and Tang Ra.

ECOLOGICAL DISTRIBUTION

Only the chiru and saiga and to a lesser extent the tundra-living muskox and argalis prefer level to undulating terrain; all others are partial to hills and mountains. But with respect to other features of the habitat, the Caprinae have adapted to a wide variety of altitudes, climatic conditions, and vegetation types. In assessing ecological distribution it should be remembered that the habitats of some species may have changed from what they were in the past. The ancestors of the cliff-dwelling Nilgiri tahr must have lived in relatively flat terrain during part of their history for they could not have reached southern India without crossing at least 300 km of the Indus or Gangetic plains. Hoofed animals may themselves modify their environment by foraging selectively on certain species. For example, Papageorgiou (1974) studied two islands near Crete, one heavily grazed by wild goats and the other untouched. The shrub cover on the grazed island was 88%, mostly spiny species not liked by the goats, in contrast to the other island where it was 22%. And man has had a tremendous impact on the environment, having in the past few thousand years removed forests from vast tracts. In Baluchistan there is no evidence for a drastic climatic change during the past 5000 years (Raibes and Dyson, 1961), and the current desiccation of the area can be attributed more to human improvidence than to natural causes. There flourished between 2500 and 1500 BC a well-developed urban civilization in the Indus valley whose ruins at Mohenjodaro and Harappa attest to its former glory. But,

> millions of well-baked bricks went to the building and rebuilding of Mohenjo-daro. Millions of tons of firewood went to the baking of them. With all allowance for the arrival of floating timber from the upper reaches, this implies a widespread deforestation of the surrounding region. This in turn, though partially compensated by growing crops, must have checked the transpiration of moisture and reduced rainfall . . . . Desert was encroaching on the sown. In rough terms, Mohenjo-daro was wearing out its landscape, whether by zeal or by indolence (Wheeler, 1968).

A similar tale with minor variations can be told for most arid lands in that part of the world.

Ovibovini and Rupicaprini

Tener (1965) describes the habitat of the muskox in the Canadian Arctic: "Part of the area consists of ridges and hills separated by depressions occupied by lakes or wet tundra, and part of drier, sandy or rocky plains . . . . In winter, on both the mainland and the islands, muskox ranges are gently rolling hills, slopes, or plateaus, wherever suitable forage exists and snow depth is kept low by prevailing winds." The importance of low snow depth to the survival of muskox was emphasized by Lent and Knutson (1971).

The takin in Szechwan occupies rugged terrain with jagged cliffs and granite precipices between the altitudes of 2500 or 3000 m upward to 4500 m, a range which includes thickets of bamboo and rhododendron as well as alpine meadows where the animals may feed in summer (Wallace, 1913; Schäfer, 1933; Sheldon, 1975). Farther west, in the Mishmi Hills along the Assam-China border, takin live in steep and dense evergreen forests at altitudes of 900 to 3000 m (Cooper, 1923).

The rupicaprids are essentially forest animals although some species spend much of their time in open terrain. The goral is found at sea level in the Ussuri area of Russia, and in the birch forests at timberline around 4000 m in Nepal. Within this altitudinal range, the goral's choice of habitat is liberal as long as the terrain is steep, rocky, and provides some cover. In Szechwan, Schäfer (1933) found goral common along canyons whose walls were covered with patches of grass and brushy clumps of juniper, rose, and *Cotoneaster*. In India Dang (1968b) reported them from grassy cliffs with *Berberis* shrubs and stands of oak, pine, and cedar. And in the Sikhote-Alin mountains of Russia Bromlei (quoted in Nasimovich, 1955) found goral on rocky slopes broken by small meadows and groves of deciduous trees. In summer, goral may wander up to a kilometer away from their cliffs, but in winter they restrict their movements to steep terrain where wind and sun soon remove the snow. All forest types above an altitude of 900 m harbor goral in the Indian Himalaya but the animals are not abundant above an altitude of 2500 m (Stebbins, 1912). In Nepal, I found the species fairly common in the upper Bhota Kosi and Kang Chu valleys at altitudes of 2500 to 3000 m. It is an area of huge cliffs with small meadows clinging to them (plate 1) and, on more gentle ground, with stands of oak, magnolia, and other trees; the understory consists of saplings, shrubs, and bamboo together with a chaotic litter of boulders. But in Nepal I have also seen goral on slopes denuded of pine forests, on arid grassy expanses broken by patches of brush and rocky nooks.

Although the serow often shares its habitat with the goral, it seems less tolerant of dry conditions, preferring damp and thickly wooded gorges. In Kumaon and Gharwal, Dang (1962) found them, "frequenting steep grassy hillsides from five to ten thousand feet high, with dense cover of Oak and Rhododendron close by. Less fond of cliffs than the Ghooral, far less versed in mountaineering technique than the Thar, the Serow is yet a mountaineer of much merit." Okada and Kakuta (1970) noted that Japanese serow live in oak, pine, and fir forests broken by grasslands around an altitude of 1000 m and that the requirements of the species include rocky overhangs beneath which animals can retreat during inclement weather and in times of danger.

In China the serow occurs between 2000 and 4000 m, often in dense spruce (Allen, 1940; Schäfer, 1933), in Burma between 200 and 2450 m, and in India between 1800 and 3000 m (Prater, 1965).

The chamois uses open terrain more than do the serow and goral. In the Caucasus the animals are found between the altitudes of 150 to 4000 m, in dry habitats as well as in damp ones with much snow. Some populations spend most of the year in forests at 700 to 2000 m, whereas others move for the summer to the subalpine and alpine zones at 1700 to 2500 m (Heptner et al., 1966). In Europe, too, chamois may live in oak forests as well as on the alpine meadows in the Alps and Tatras, though they often retreat into the forests in winter (Nievergelt, 1966a). Introduced into New Zealand, chamois have occupied treeless grasslands broken by cliffs and brush-covered ravines at altitudes between 800 and 2000 m, avoiding forests except to traverse them and to seek refuge in bad weather (Christie, 1964; C. Clarke, pers. comm.).

The Rocky Mountain goat, even more than its Eurasian relatives, is an inhabitant of cliffs. As Brandborg (1955) noted, "probably no other feature of mountain goat habitat is more apparent to the observer than the rugged and broken terrain with cliffs, ledges, projecting pinnacles and talus slopes that are characteristic of the goat country." Moving between summer and winter ranges, the goats in Idaho frequent several vegetation types, from the rather arid valleys at 600 m with their sparse cover of pine to the rocky ridges and expanses of scree at 3000 m where trees consist of isolated stands of fir and larch and grass is confined to tiny oases on ledges and in moist hollows.

Caprini

Hemitragus. From the standpoint of a preference for cliffs, the tahr seems to be the quintessential goat, for the animal "revels in the steepest precipices," as Burrard (1925) phrased it. However, tahr are also extremely adaptable animals, occupying diverse habitats. Thesiger (1959) described his search for Arabian tahr: "It was exhausting work hunting them, for the mountains rose four thousand feet above our camp, and the slopes were everywhere steep and unusually sheer, without water or vegetation." In contrast, the Nilgiri tahr, also inhabitants of treeless terrain, live in an intensely lush environment mainly at altitudes of

1200 to 2600 m. Grass-covered hills dotted with copses of stunted evergreen forest provide the tahr's foraging grounds adjacent to sheer granite cliffs that drop from the lofty tablelands to the tall evergreen forests below (plate 2). Rains are often torrential, up to 7500 mm per year, and clouds commonly hug the cliffs. The Himalayan tahr is a forest animal in much of its range, so much so that Kinloch (1892), Prater (1965), and others felt that the animals tend to remain in dense thickets, never ascending above timberline. On the other hand, Caughley (1970b), who observed tahr in Nepal rather than in India as did previous observers, found the animals only between 3800 and 4850 m, and he concluded that tahr occur "only above tree line, both in winter and summer. Previous reports therefore appear to be in error." Schäfer (1950) held still another opinion; namely, that tahr in Sikkim use many habitats seasonally, from mixed oak forests at 2500 m upward through the rhododendron and conifer zone to the alpine meadows at 5000 m. His observations agree with mine in Nepal. In the Kang Chu, I observed tahr and their sign between 2500 and 4400 m. In winter and early spring, tahr were mainly on the huge cliffs below an altitude of 3500 m, especially on those with a southerly exposure. The cliff faces are broken by many ledges and platforms that support small meadows and stands of bamboo, shrubs, and broad-leaved forest on which tahr forage and in which they retreat from cold and wind (plate 1). During summer many tahr move above timberline, judging by the numerous feces I saw on the alpine meadows. A herd of about 100 tahr live wholly within the alpine zone above 4000 m in the Langtang National Park of Nepal (J. Blower, pers. comm.), a situation similar to that in New Zealand where the introduced tahr have occupied the treeless uplands covered with snowgrass (*Chionochloa*), *Celmisia* daisies, and other low vegetation between an altitude of 1000 and 2100 m. The adaptability of Himalayan tahr is further shown by the survival of introduced populations in such diverse places as Ontario, Table Mountain in South Africa, and California. In the last-named area the animals restrict themselves to the coastal region with its summer fog belt and avoid the dry and hot inland hills (Barrett, 1966).

Capra. Goats are essentially cliff dwellers with a wide tolerance for altitude and to a lesser extent habitats, some living on treeless hills and others on wooded ones. The Spanish goat and Nubian ibex occupy rather barren crags from near sea level to about 3000 m, often in arid areas where the sparse forests have been largely destroyed by man. However, the Spanish goat may also inhabit oak and pine forests in some areas (Couturier, 1962). The wild goat is found mostly at low altitudes, though Stockley (1936) was wrong in claiming that they usually remain below 1500 m. In the Chiltan Range they may ascend to 3300 m and in the Elburz to 4000 m. Heptner et al. (1966) reported these goats from 550 to 3200 m. in Armenia, and at 1200 to 2200 m in the vicinity of permanent snow in Dagestan. In Iran I have seen wild goat in the temperate uplands of the Kopet Dagh where cliffs are sparsely covered with shrubs and trees and are surrounded by an *Artemisia*-bunchgrass steppe, and I have watched them in the arid Kavir Protected Region whose stark, serrated ranges are dissected by dry water courses and whose slopes are almost devoid of vegetation. The Karchat Hills of Pakistan, where I studied wild goat intensively, consist of a central convex plateau, 1030 m above sea level at the highest point, whose edges drop some 300 m to a stony desert either as steep slopes or abrupt scarps. A maze of deeply eroded ravines slash the massif into sections (plate 13). *Acacia senegal,* the dominant tree, grows thinly on the hills, as do various other shrubs and trees, among them *Euphorbia, Commiphora, Salvadora, Prosopis, Zizyphus,* and *Capparis.* Most of the ground is bare rock and sandy soil, with forbs few and grass cover limited to about 5%. Farther north, in the Chiltan Range and other massifs, there is more vegetation. *Artemesia, Nuheta, Daphne,* and other low shrubs provide a coverage of 30 to 40% on the more gentle slopes and an occasional *Juniperus* and *Pistacia cabulica* tree attest to the forests that grew there within memory of the oldest inhabitants (plate 11).

The markhor is essentially a goat of low altitudes. Its main requirements consist of cliffs in areas with little precipitation, deep snow being especially avoided. In the northern part of its range the mountains are high, often over 5000 m, yet the goat has failed to occupy the upper slopes permanently. Although markhor may ascend to 4000 m during the summer in Chitral and Gilgit, they require terrain below an altitude of about 2200 m in which to spend the winter; the animals do not occur where the valley floor is higher than 2200 m, even though suitable cliffs may be available. They are seldom exposed to temperatures below $-10°$ C, but at the other extreme they tolerate heat that may exceed 45° C. In general, markhor range from about 700 to 1000 m along the lower slopes of the Sulaiman Range upward to around 2700 m during winter in Chitral, to the timberline and above in summer. The habitat in the southern part of the markhor's range is arid and resembles that of the wild goat (plate 12). Although the hills look absolutely barren from a distance, closer inspection reveals some vegetation, mostly low-growing and subdued in hue. In the Toba-Kakar Ranges, for instance, scattered yellow-flowered *Sophora* shrubs grow in the valleys and an occasional *Pistacia* and *Fraxinus* tree break the contour of rock and earth. Three species of leafless *Ephedra* are quite common, and various grasses— *Cymbopogon, Stipa, Chrysopogon*—provide a ground cover of around 5%. The upper Indus, too, is arid and almost devoid of trees (plate 6). But in other areas markhor are associated with forest. Stands of pine surround some markhor cliffs in the northern part of the Sulaiman Range, and oak, spruce, and fir forests are found in the Safed Koh and in Chitral, Dir, Swat, and Indus Kohistan. In the Chitral Gol Reserve, my main study area, the slopes below an altitude of 2600 m are sparsely covered with an evergreen oak (*Quercus ilex*) which is usually no more than 8 m in height and whose estimated canopy coverage of 10–15% makes it the most prominent broad-leaved tree there (plate 4). Shrubs are scarce except for *Indigofera gerardiana*. Much of the ground cover consists of *Artemisia* with occasional forbs such as *Rumex hastatus* and a few grasses. Over 90% of the ground is bare, with rock outcrops and shale talus. On south-facing slopes some spruce and fir make their appearance above the oak and stretch from there to timberline at 3100 to 3300 m; on north-facing slopes tongues of coniferous forest descend almost to the valley floor at 2100 m. Markhor in the Pir Panjal favor "steep, stony land slips, precipitous grassy slopes, and rocky acclivities clad more or less with pine and birch" (Macintyre, 1891).

The remaining *Capra* are high-altitude animals confined to cliffs and alpine meadows above timberline, though animals may penetrate far down forested slopes along avalanche paths and precipices. In the Simien Mountains, forest once covered the massifs to 3700 m, and the Walia ibex lived from that altitude upward to 4500 m in a tussock grassland covered with shrubs and lobelias

(Nievergelt, 1970). Ibex in the Alps remain above timberline, generally between 2000 and 2800 m, visiting areas with sparse shrub and forest cover only occasionally (Nievergelt, 1966a). The Dagestan tur ascends to over 4000 m, whereas the Kuban ibex in the western Caucasus, where the permanent snow line is lower than in the eastern part, seldom ventures above 3300 m (Heptner et al., 1966). The Asiatic ibex may occur between the altitudes of 500 to 5000 m, the former figure being from the Altai where, as Heptner et al. (1966) point out, the goats live well within the forested zone. In the Tien Shan, ibex are often found as low as 1000 to 2000 m, in contrast to the Pamir, Karakoram, and other ranges where the animals seldom come below 3000 m even in winter. On their summer range in Chitral, ibex usually remain between 4000 to 5000 m, venturing even higher on occasion to cross from one valley to another (plate 7). However, above 5000 m the vegetation is scanty, confined to an occasional *Primula, Sedum, Draba,* or other restrained plant. In mid-winter, in the same area, herds were at around 3700 m, not on the alpine meadows but on precipices where winds and avalanches had exposed the scant herbage. In another part of Chitral, the Golen Gol, I saw ibex as low as 2400 m in February, as compared to the Kilik Valley in northern Hunza where in November most ibex were between 4100 and 4600 m, confined largely to south- and west-facing slopes from which the sun had partially melted the snow.

It is noteworthy that Asiatic ibex do not occur east of the Sutlej where suitable terrain seems to exist. The amount of precipitation, especially in the form of snow, may be one factor limiting distribution, in that ibex like most goats are generally found in areas that receive less than 100 cm per year. As shown in table 2, the amount of precipitation increases markedly between the Western and Central Himalaya.

Though their existence revolves around rocky rubble and cliffs, ibex do occasionally venture onto more gentle terrain. This willingness to leave a preferred habitat was essential to their success in colonizing many of the isolated ranges. Markhor also show a highly discontinuous distribution; to reach isolated cliffs the animals had to migrate across plains, through forests, and over rolling hills, sometimes for 30 km or more. S. M. Rizvi told me of meeting a small herd of wild goat on the plains heading from the

Khambu Hills to the Karchat Hills 10 km away. Geist (1971a) mentions similar behavior among American sheep.

Ovis. Eurasian sheep are all similar in their habitat requirements in that they prefer terrain to be gently to steeply rolling but not precipitous. Within this topographic limitation, sheep have adapted to a variety of conditions, their altitudinal range stretching from sea level to above 5000 m. Urial are generally found in arid country at relatively low altitudes, but the Ladak urial ascends to 4200 m. Most urial live in open habitats where trees are sparse or absent, but some notable exceptions exist. Mouflon in Corsica may be found in fairly dense forests (Pfeffer, 1967), and Cyprus urial occur in "forest growth much of which is a dense evergreen dwarf-oak" (Clark, 1964). The Ladak urial in Chitral was once found in oak forests, according to villagers who hunted them there, and the Punjab urial in the Kalabagh reserve may frequent dense thickets of *Dodenia* shrub. All this suggests that the urial as a species may once have been more of a woodland animal but that it readily adapted to wholly treeless hills as environmental conditions changed. Today urial even occupy deserts, such as the Kysyl Kum in Russia, where small eroded buttes rise a mere 100 to 200 m above the sands (Heptner et al., 1966). A similarly sterile environment occurs in the Kavir Protected Region where in the deeply eroded foothills of the Siah Koh a barren expanse of sand and crumbly rock is covered with only an occasional *Artemisia, Amigalus,* and *Ephedra,* almost all that was left when domestic stock was removed a few years ago. However, dense forests were no doubt major barriers to urial movements along some mountain ranges. For example, a forest on the Elburz probably enabled *gmelini* and *arkal* to evolve into subspecies before a change in climate altered the vegetation and permitted contact between populations. The Punjab urial is hemmed in by a forest belt and two large rivers. The animals must once have emigrated from the west, a herd perhaps crossing the Indus at a time when a landslide dammed the river in the mountains as described by Cunningham (1854).

I observed urial mainly in the Kalabagh Reserve. There the hills of the Salt Range rise gently from a plain before entering a rugged maze of small plateaus and tilted beds of rock, finally cul-

minating in a series of rounded ridges at an altitude of 1000 m
(plate 14). An acacia woodland with scattered *Salvadora oleioides*
and *Zizyphus nummularia* covers the slopes, and the shrub *Dodonea
viscosa* caps the ridges. Forbs are scarce. The grass cover is about
19%, but 12% of the 19% consists of *Cymbopogon* and *Aristida*
which neither urial nor livestock like.

Argalis inhabit terrain similar to that of urial although usually
at higher altitudes. However, they have not been reported in
woodland and they may use flatter terrain than do urial, animals
having been seen on the wide plains of Ladak and the Pamirs.
Ward (1923) noted that argalis never seek refuge among cliffs
whereas urials sometimes do so. Marco Polo sheep seldom descend below 3000 m.

> And when he is in this high place he finds a plain between two
> mountains, with a lake from which flows a very fine river. Here is
> the best pasturage in the world; for a lean beast grows fat here in ten
> days. Wild game of every sort abounds. There are great quantities
> of wild sheep of huge size. Their horns grow to as much as six palms
> in length.

Thus Marco Polo described the Pamirs after his visit there in 1273
(Polo, 1958). The preference of Marco Polo sheep for rolling
terrain is apparent in northern Hunza. Approaching the Kilik
and Khunjerab passes, one treks up deeply cut valleys flanked by
high jagged peaks until at 4200 m the slopes suddenly retreat to
reveal smoothly rounded hills with only their upper reaches consisting of rocky ramparts and glaciers (plate 15). Here live the
Marco Polo sheep. Below the Kilik Pass the vegetation consists of
Artemisia, Ephedra, and cushions of saxifrage, plants resistant to
aridity and intense cold, as well as of tiny sedge meadows where
moisture from melting snows has had a chance to collect. But
north of the gentle col and dark stone rubble that mark the top of
the pass conditions are more desolate: "There was no tree, no
bush, no lowly shrub to break the sweeping expanse of scree and
brown earth. Grass there was, but the blades were so sparsely
scattered that even at this time of high summer one's eyes received
but the vaguest impression of anything green" (Tilman, 1949).
The Khunjerab Pass area is more lush than the Kilik, with the
meadows larger and the carpets of sedges, *Potentilla, Pedicularis,*
and others deeper.

The habitat of the Tibetan argali in Ladak is more desolate than that of the Marco Polo sheep:

> On the wild bleak uplands of Thibet, where for hundreds of miles not a tree is to be met with; where in every direction, as far as the eye can reach, there is nothing but a vast expanse of barren soil, rock and snow; where there is no shelter from the glare of a cloudless noon, nor from the freezing winds that sweep the naked hills with relentless force toward the close of day; here, in the midst of solitude and desolation, where animal life has apparently to struggle for existence under every disadvantage, is the home of this great wild sheep (Kinloch, 1892).

Yet argali ascend the hills to 5750 m (Stockley, 1928) near the ultimate limit of vegetation. Farther east in the Dolpo District of Nepal, where I visited argali habitat, the land is not as barren as in Ladak (plate 16). *Caragana* and *Lonicera* grow profusely in the hills between 4500 and 5000 m and beneath these low shrubs there is much grass; moist swales are often thickly covered with sedges; and forbs and tufts of grass cling to rocky hollows and ledges. To the north, the habitat of *ammon* and other subspecies is similar to that already discussed although the animals are seldom found above 3600 m. A description by Demidoff (1900) from the Altai is representative: "... Opposite our camp stood fairly high rolling hills, which formed a marshy plateau .... The ground was stony and barren, with patches of grass here and there, and no trees were to be seen anywhere.... Its altitude, according to my aneroid, was 7800 feet above sea-level."

The snow sheep and American sheep occupy habitats totally different from those of urials and argalis. They frequent open rugged terrain in the vicinity of cliffs. There they forage on ledges and alpine meadows and in shrub-covered ravines, protected by the crags and slopes of scree that surround them. As Heptner et al. (1966) pointed out, this habitat preference is somewhat similar to that of Asiatic ibex; that, in fact, these sheep have filled a niche usually occupied by a goat. As is the case with Eurasian sheep, the American animals tolerate extremes in climate, ranging from that of the desert bighorn which lives in Death Valley below sea level where temperatures reach over 50° C to the Dall sheep in northern Alaska where the animals experience deep winter snows and temperatures of below −50° C. Welles and Welles (1961), Smith

(1954), Geist (1971a) and Heptner et al. (1966) describe and picture bighorn, thinhorn, and snow sheep habitats.

Ammotragus. Aoudad live in the desert mountains of the Sahara
from sea level to the edge of the snows at 3900 m in Morocco
(Joleaud, 1928). "The rocky slopes of the range are incredibly
rough. They are entirely covered with loose pebbles, stones and
boulders of all sizes" (Rodd, 1926). When introduced into new
habitats, such as in New Mexico and Texas, the animals select the
same type of rough terrain, shale bluffs and gorges (Ogren, 1965;
Evans, 1967). Vegetation in the aoudad's native habitat is sparse, a
few acacias, some shrubs, and a thin ground cover of grasses
(Brouin, 1950), a habitat similar to that used by wild goat and
Nubian ibex.

Pseudois. Bharal live above timberline, from 3500 m upward to the
limit of vegetation around 5500 m, except in the eastern part of
their range where in the gorge of the Yangtse some populations
descend to 2600 m (Schäfer, 1937). Hunting bharal in India, Kinloch (1892) found that the animals "delight in good grazing
ground in the immediate vicinity of rocky fastnesses." This was also
true in China (Sheldon, 1975) and in the areas of Nepal where I
saw bharal. In the upper Kang Chu and west of the Dhaulagiri
massif, bharal forage on grasses, *Berberis* and *Ephedra* shrubs, and
on *Polygonum* and other forbs, moving along the lower slopes of
the ranges as well as over cliffs, which, though steep, are broken
by many ledges, terraces, and small plateaus (plate 20). Around
Phoksumdo Lake in the Dolpo District the bharal concentrate on
the alpine meadows, with the crags and snow above and the
scraggly timberline of birch and conifers below (plate 17). At
Shey, the cliffs are surrounded by steeply rolling hills, some of
barren earth and scree, others covered with low shrubs and grasses
(plates 18 and 19). The bharal forage on the latter, using the
cliffs mainly as retreats in times of danger. Though living near
cliffs, bharal actually spend little time on them, a habitat preference similar to that of American sheep rather than of goats.

In comparing the ecological requirements of African antelopes,
Estes (1974) was successfully able to divide the many species into
those preferring open habitats and those using closed ones. Such

a clear segregation is not possible for many of the Caprinae. While it can be said that the goral and serow are usually found among shrubs and trees, and several species, among them chiru, saiga, muskox, bharal, and argali, are partial to open terrain, a number of forms either use both types of habitat or probably would do so if an ecological opportunity presented itself. Some takin populations move seasonally out of the forest, and Himalayan tahr, markhor, urial, and chamois are ecologically versatile with respect to their tolerance for cover. Born with the mountains and shaped by them, many Caprinae seem to have retained a certain plasticity that enables them to adapt to the constant environmental changes in their rocky realm.

ECOLOGICAL SEPARATION OF SPECIES

It has become axiomatic in biology that when two related species meet in the same region they tend to compete for the same resources and both can persist together only if they are separated ecologically either by habitat or food preference or both. When two species of different size occupy the same habitat, it may be inferred that they do select different foods, but if they are of about the same size and of similar morphology, then competition can be expected unless the habitat is partitioned.

The available information on Caprinae food habits is summarized in Chapter 6 and it does show that some species prefer certain types of forage, with goats, for example, being more partial to browse than are sheep. Here I limit myself to noting possible competitive situations on the basis of habitat selection.

Caprinae of somewhat different size often live in the same general habitat (see Chapter 4 for size of animals), but even so they usually specialize in different terrain or altitudes (fig. 19). Himalayan tahr, serow, and goral commonly use the same ground except that tahr are cliff dwellers who may venture above timberline and serow tend to remain in or near dense cover. In New Zealand, where Himalayan tahr and chamois occupy the same slopes, the former are usually at higher altitudes, and, according to K. Tustin (pers. comm), tahr are abundant where chamois are scarce and vice versa. A partial ecological separation between chamois and ibex is apparent in the Alps where the former usually remain at a lower altitude (Corbet, 1966). Where mouflon have been introduced in the Alps, the sheep live on the lowest slopes,

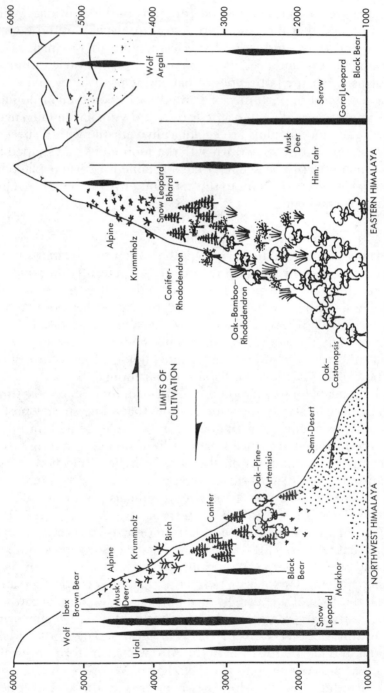

Fig. 19. The vertical zonation of the vegetation and the altitudinal distribution of large mammals in the northwestern and eastern Himalaya.

the chamois in the intermediate zone, and ibex in the highest portions, a separation which is particularly evident in winter (Pfeffer and Settimo, 1973). In the eastern Caucasus the chamois may be higher than the tur, and, when both are at the same altitudes, the tur are often found in shady valleys which chamois avoid (Heptner et al., 1966). Serow, goral, and takin inhabit the same terrain, as do takin and bharal at certain seasons. In the highlands of Tibet, kiang, chiru, and Tibetan argali may be seen feeding together (Macintyre, 1891), the sheep only partially separated from the chiru by a preference for more hilly topography; bharal, which roam the slopes above the plains, are to some extent isolated from the argali by their preference for the vicinity of cliffs, but the two species have nevertheless been seen near each other (Macintyre, 1891).

Habitat separation should be particularly evident if two species are of about the same size. There is a most striking ecological separation between *Ovis* and *Capra:* where the two species occur together, the sheep occupy mainly the undulating terrain and the goats the precipices. This is readily seen in Baluchistan. Mountains in arid climates weather into distinctive shapes according to the type of rock. Shale forms slopes of talus, but sandstone and limestone create cliffs which in Baluchistan often take the form of mesas. Wild goat or markhor use the cliffs and immediate surroundings, whereas urial occupy the tops of the plateaus and the eroded hills along the bases. In much of their range ibex and argalis exist together, the ibex usually on the cliffs above the sheep. Northwest of Kashgar, Shipton (1951) "saw hundred of both ibex and *Ovis poli* [*karelini*] .... The former were all on the northern flank of the valley and the latter on the southern .... The snow of the southern flank was far deeper than on the northern." In the Himalaya, where ibex or bharal occur above timberline, tahr confine themselves mainly to wooded gorges; where ibex and bharal are absent, as in parts of Nepal, tahr use the alpine zone seasonally, suggesting some form of avoidance. In the Altai, ibex occupy cliffs in the wooded zone, a habitat claimed by tahr in the Himalaya. However, the peak west of Banihal Pass in the Pir Panjal holds tahr and ibex, and in the next valley to the west are markhor and tahr (Stockley, 1936), but nothing has been published about the way the animals partition the habitat. According to Adams (1858) markhor sometimes feed with tahr, and, around Nanda Devi, tahr and bharal may be on the same pasture (Tilman, 1935). Aoudad and Nubian ibex overlap in parts of their ranges, and there is a fascinating situation in the mountains of Oman where wild goat, Arabian tahr, and Nubian ibex exist; but whether the species are geographically or ecologically separated is not known. Equally interesting is the overlap in range between bharal and ibex (fig. 12), two species with quite similar habitat requirements, but the literature provides no insights on a possible ecological separation except to note that bharal, ibex, and Ladak urial occur in the same valleys downstream from Leh (Stockley, 1928). I would expect ibex to feed mainly by cliffs, bharal on the nearby slopes, and urial in the valley bottoms. Snow sheep have no Caprinae competitors and American sheep have only the mountain goat, whose preference for cliffs exceeds that of sheep.

A separation of habitats should be especially apparent when two species of the same genus come into contact. Among sheep, the Ladak urial and Tibetan argali meet in the Zaskar Range of Ladak. There the "Shapo and *ammon* often feed in the same nullah near Gya (35 miles south of Leh . . .)" (Burrard, 1925). Blanford (1888–91), quoting H. Littledale, stated that hybrids exist, and Ward (1924) shows a photo of a skull of a supposed hybrid and states that "three or four of these cross-bred sheep have since been obtained." The existence of such hybrids needs to be verified, especially since Ward (1924) notes than in winter during the rut urials tend to be at lower altitudes than argalis. The slight overlap and the meager evidence for hybridization between sheep species is striking when compared to the large hybrid zone at the point of contact between two subspecies of urial in Iran. A perusal of figures 12 and 15 reveals that the goats (*Capra* and *Hemitragus*) and the goat-like *Ammotragus* and *Pseudois* show surprisingly little overlap in range considering the amount of contiguous distribution, and from this alone it may be hypothesized that these species tend to compete unless they partition the habitat geographically. The Kuban ibex and Dagestan tur have divided the Caucasus into eastern and western halves with occasional hybridization in the central contact zone. In the eastern Caucasus, where wild goat penetrate the habitat of Dagestan tur, the two species are separated at least in part by altitude, the tur living higher (Heptner et al., 1966). In Pakistan, markhor and wild goat overlap on at least two small massifs, Murdar and Gadabar Ghar. Villagers told me that on Gadabar Ghar the markhor occupy the western end of the range and the wild goat the eastern (plate 12), a situation similar to that of the ibex and tur in the Caucasus except that the Gadabar Ghar is only 20 km long and hybridization is unknown. In talking about wild goat and markhor, Blanford (1888–91) noted that "specimens of a wild hybrid between the two were obtained by the late Sir O. B. St. John on Takatu near Quetta." But in view of the confusion existing over the horns of Chiltan goat, at that time not known to science, this record is suspect. The markhor also touches on the range of the Asiatic ibex in Chitral, the Gilgit area, parts of Afghanistan, and in the Safed Koh. Ibex inhabit the slopes above timberline where markhor are only summer visitors (fig. 19). Haughton (1913) saw "a herd of markhor graze through a herd of ibex and pass on" in late May, a period when ibex often descend to lower altitudes to reach the first flush of green grass. Since the rut is fairly brief and confined to winter, when the two species are usually separated by altitude, the opportunities for hybridization are few.

The Caprinae also share their habitats with other potential competitors, and only a few Himalayan ones are mentioned here. The Tibetan gazelle, kiang, and yak associate with the chiru and argali. The Tibetan gazelle, the smallest member of the assemblage, is only 60 cm high at the shoulder; the kiang, whose habitat includes both "hill or plain, from 14,000 to 18,500 feet" (Rawling, 1905), is 142 cm high at the shoulder; and the yak is huge, one wild-shot bull reaching 203 cm at the shoulder and weighing 821 kg (Engelmann, 1938). The species may be partially separated by food habits and they may also divide their resources geographically when at certain seasons some migrate to more favorable grazing grounds. Several deer were the main associates of the Caprinae within my study areas. The most widespread of these was the muskdeer, a small antler-less animal some 60 cm high at the shoulder and with a weight of about 13 to 14 kg (Heptner et al., 1966), males and females being of about equal size. A skulking creature with hindlegs larger than the forelegs, muskdeer inhabit forest and thickets on rocky slopes from an altitude of about 2200 m to the upper forest limit, even penetrating the alpine zone on occasion (see fig. 19). Distributed along the slopes of the Himalaya westward to Chitral and Afghanistan, the species comes into contact with most Caprinae inhabiting the forests and cliffs below timberline. The muskdeer is the smallest of the Himalayan ungulates, whereas another cervid, the red deer, is among the largest. The hangul, a subspecies of red deer which is now threatened with extinction, once lived in most of the hills surrounding the Vale of Kashmir. Some 132 cm high at the shoulder with antlers over 120 cm long and a body weight that may exceed 200 kg, the hangul stag is the largest ungulate within its limited habitat (plate 28). In summer this habitat includes the alpine meadows and pine forests at about 3000 m and in winter the mixed deciduous and conifer forests in the valleys at 1800 m (Schaller, 1969a). Hangul avoid precipitous terrain and this reduces contact with their possible competitors, the ibex and serow. Other subspecies of red deer, among them the shou in southeastern Tibet and the maral in the Altai, occupy similar habitats.

This brief resumé shows that when two or more Caprinae occupy the same general area either they are of different size or they select different altitudes, types of terrain, or vegetation cover. And often both situations prevail. The separation is most evident among closely related species, where in fact it usually takes the form of a physical division of ranges with perhaps a narrow zone of overlap. Looking at the distributional patterns of ibex, bharal, and Himalayan tahr, it seems that one species has at times restricted another from expanding its range both horizontally and vertically. This is not surprising. When expanding their range, caprids often colonized unused habitats, terrain that had been recently vacated by ice, where no competitors existed. All being rather generalized feeders, competition for resources would be inevitable and coexistence impossible without a geographical partitioning of ranges. Sympatric survival is not possible in a simple habitat for caprids of equal size. Furthermore the fragmentation of habitat into widely scattered mountain oases undoubtedly has an effect on the number of species an area can support. I would find it difficult to visualize how bharal and ibex could share the few alpine meadows in the Karakorams, and it is not an accident of geographical distribution that ibex inhabit the southern flank of the range and bharal the northern. A complex environment such as a rain forest provides more ecological opportunities with a resulting greater diversification of species.

4 PHYSICAL ATTRIBUTES

From the lithe urial to the bulky takin, members of the Caprinae are physically diverse. The illustrations in this book (figs. 6 and 7; plates 21–43) give a visual impression of each genus; detailed descriptions can be found in various standard references (e.g., Lydekker, 1924; Walker, 1968). The size and appearance of an animal is influenced by several factors, such as the kind of terrain it inhabits and the social organization of its species. A summary of various physical attributes of the Caprinae is given here to provide a basis for exploring how ecological and social factors have interacted to shape the animals. Sheep and goats and some of their close relatives continue to grow larger throughout life, especially in horn size, an attribute which segregates members of a population into several age classes each with certain social roles. This chapter also defines the age classes of those species I studied.

BODY SIZE, SHAPE, AND COVER
Considering the hundreds or thousands of trophies that have been shot of various species and subspecies, there are surprisingly few exact body measurements except for horn size. Precise weights are lacking for several forms, and data on most others are limited to a few male specimens. Shoulder heights and weights for many Caprinae are listed in tables 6 to 12. Shoulder heights were drawn mainly from such sources as Stockley (1928), Harper (1945), Clark (1964), Couturier (1962), and Heptner et al. (1966), and the weights were gathered from many sources, as noted in the tables. With samples small and individual variation both within and between members of populations great, the figures are in most instances only rough approximations. For instance, male bharal usually weigh around 60 kg but in one area they averaged only 39 kg (Schäfer, 1937). Data from animals in their native habitat were used in most cases, for a species may increase in

83

average size after being introduced into a new area. For example, Himalayan tahr males are said to weigh about 90 kg (Burrard, 1925), but in New Zealand some reach 150 kg (Anderson and Henderson, 1961); I weighed a 5½-year-old male in New Zealand at 120 kg and a 6½-year-old female at 48.5 kg. Zoo animals frequently weigh more than free-ranging ones. Tener (1965) mentions captive bull muskox of 511 and 653 kg as compared to 369 kg for the heaviest of 3 wild bulls. One Burmese female takin at the Bronx Zoo weighed 70 kg at the age of 1½ years and 320 kg at 8 years, more than the published weights for males in the wild. Some weighed animals are not identified properly in the literature. Lydekker (1898), for instance, mentions 83, 92, and 108 kg as being weights of male markhor, but he does not designate the subspecies.

Caprinae males fall into several size classes of which the smallest consists of animals weighing fewer than 50 kg. This category includes chiru, saiga, goral, and chamois as well as several urial subspecies. Most other forms belong to an intermediate size range of from 50 to 150 kg with few straying far above 100 kg. Only certain argalis—Marco Polo, Tien Shan, Altai—reach the 150 to 200 kg size. Muskox and takin, both of which may scale over 300 kg, are in a class by themselves. Female Caprinae are generally small, all averaging fewer than 75 kg except muskox and takin. It may be conjectured that the size of each subspecies or species represents an optimal adaptation to a particular environment, an adaptation related to the logistics of traversing terrain, the availability of food, and the means of escaping from enemies and adverse climatic conditions. If so, then it must be assumed that in species with great sexual dimorphism it is the female which represents the ecological ideal in size, and that the male, whose great bulk is related mainly to competition for mates, is not as well adapted, even if his size opens up new food resources to him. From this point of view, the most favorable size range of Caprinae lies between 25 and 75 kg, animals of intermediate size.

Sexual dimorphism is pronounced in some species but less so in others (see tables 7, 9, and 11). Male and female goral and serow are of about equal size. Mountain goat, saiga, and muskox females are about three-fourths the weight of males, and the same may be true for takin, a species in which bulls are noticeably larger than cows but for which accurate weight samples are unavailable. Chamois females range from .62 to .90 the weight of males. Dimorphism among Caprini is great, females tending to be about half as large as males. However, the weights of animals in few populations have been adequately sampled, the data provided by Papageorgiou (1974) being a notable exception. He weighed 29 Cretan wild goat males and 24 females, aged 5 years old and older, and the mean for the former was 34 kg (26.1–41.7) and the latter 20 kg (17.5–22.5), females being .59 the weight of males. The fact that Heptner et al. (1966) published only approximate upper and lower weight limits rather than precise means makes it impossible to compute accurate ratios for the species mentioned by them. The ratio for aoudad is about .50, for bharal .65, for Himalayan tahr from Sikkim .58, and for Asiatic ibex from the Tien Shan .71, to mention a few goats and their close relatives. Armenian urial ewes are about .62 the weight of rams, and for two bighorn sheep subspecies the figures are .63 and .77. Geist (1971a) suggested that sexual dimorphism is greater in Eurasian than American sheep. So far there are too few data to provide confirmation on this point. Variation in dimorphism exists between subspecies, the ibex and bighorn providing good examples. Assuming that such differences are not due to sampling error, it is possible that they are based on habitat quality, on the nutritional level of the forage. Caprid males seem to gain proportionately more weight than females after reaching adulthood. Animals on poor range live on the average longer than those on good range, a point discussed in Chapter 5. From this it would follow that populations of poor quality, populations containing many old males, would tend to show more dimorphism than populations of good quality. This hypothesis needs verification, but some ibex data are suggestive. Couturier (1962) weighed 35 Alpine ibex males, 4 years old and older, whose average age at death was 12.8 (8–17) years, and Egorov (1955, in Couturier, 1962) weighed 15 Asiatic ibex males from the Tien Shan whose average age at death was 5.8 (4–8) years. Eleven Alpine ibex and 4 Asiatic ibex females were also weighed. Alpine ibex females are .38 the weight of males and Asiatic ibex females .71.

The body build of the various species falls into two broad categories based largely on the type of terrain the animals occupy: those partial to flat or undulating habitats are more lithe in build with longer and thinner legs than those inhabiting steep hills and precipices. Species whose existence depends on their ability to escape predators through speed need a slender frame. Animals of precipitous terrain require not speed but power; they need stocky legs and robust forequarters to climb and leap among the rocks. Saiga and chiru are built like typical antelopes except that they have inflated noses thought to be useful for warming and moistening air. During the rut a male's nose becomes extended and moves flaccidly like a proboscis. At that time chiru use their puffed noses "for bellowing challenges" (Rawling, 1905). Urials and argalis are rather stout-bodied but not heavily muscled, and they have light-boned legs. Most other forms are adapted to steep ground, although the rupicaprids show some variation in build. Goral and serow resemble each other in that both are stocky, rather goat-like in appearance, with large head and thick neck; in the words of Stebbins (1912) the serow looks like a cross between "a cow, donkey, pig, and goat." The chamois, being partial to a more open habitat than are goral and serow, has a less ungainly appearance; its head is smaller, its carriage is more erect, and its legs are not as heavy. In contrast, the mountain goat is a typical cliff dweller, stocky in build with powerful legs. All *Capra*, tahr, aoudad, bharal, snow sheep, and American sheep are similar in having massive shoulders, broad chests, and powerful legs. This stockiness is accentuated in some males by such secondary sexual characteristics as swollen necks during the rut, large humps, and Roman noses. Head shape differs between species, ranging from the elegantly narrow features of the Himalayan tahr to the blunt, rather triangular faces of Alpine and Asiatic ibexes. The basic similarity in build of all these forms is most evident among females, which look so much alike that species may be difficult to distinguish.

Huge and plump with stout legs and a bulging convex muscle, the takin looks ponderous, cow-like, a seeming exaggeration of traits first intimated by goral. Equally robust, with short legs and a broad face, is the muskox, which has a seemingly anomalous build for a tundra animal although it can climb rocky slopes with speed and agility (Tener, 1965). Habitat is, however, not the only factor shaping size and build. Large animals are more successful at repulsing predators than are small ones and they are also more efficient metabolically. Body surface becomes smaller as the volume of an animal increases, and thus larger animals conserve heat better. Living as it does in one of the world's most severe climates, the muskox needs to conserve energy. A bulky body requires strong legs for support. Its short legs also conform to Allen's rule, which states that mammals of cold climates have their heat-radiating surfaces reduced by a shortening of appendages.

To see if Allen's rule applies to one prominent appendage, the ear, I derived an index for several species by dividing shoulder height into ear length (tables 6, 8, and 10). Chamois, goral, serow, and some of the goats have proportionately the longest ears, with a ratio of from .15 to .19. The figure for Punjab urial is .13 (and for a chinkara from the same area .21), slightly higher than that for argali and North American sheep (.09–.11) Urials have more pointed, goat-like ears than other sheep and this contributed to their slightly greater length. Muskox and takin, inhabitants of strikingly different habitats, are similar in ear length. Thus there is no clear-cut application of Allen's rule to the ears of Caprinae.

Several minor physical structures are of use to animals living in rugged terrain. Most adult *Capra* and *Ammotragus* have a callus on the knee or carpal joint of the front legs, seemingly an adaptation for scrambling up steep inclines. The callus of one male wild goat was 3 cm in diameter. Himalayan tahr have a broad sternal pad covered with dense, short hair, a structure possibly of help in maintaining balance when resting and scrambling up cliffs. The dew claws of mountain dwellers are characteristically large—4 cm long in a male Himalayan tahr and 3.6 cm in a female bharal. The hoof pads are soft and bordered on the outer surface with a hard, horny rim, all adaptations for providing grip and traction on rock.

Most Caprinae live in a severe climate, whether it be hot, cold, or both. Large mammals can regulate their temperatures to some extent by behavioral means—for example, by seeking shade or avoiding the cold-air layer in the depth of valleys—but the most basic means of heat control is through insulation. The muskox is an extreme example of an animal adapted to cold. It has a dense and soft inner wool covered by outer guard hairs up to 50 cm long that hang like a curtain below the body, a pelage which in Tibet has its counterpart in the yak. The only bare skin is at the tip of the nose. The thick wool or pashm of Asiatic ibex was once an article of trade famous for its warmth and softness. Other species of cold climates, such as the chiru and Rocky Mountain goat, also have woolly coats in season, and the chamois has in addition a dense layer of guard hairs to provide insulation. The bharal has little wool but it has thick hairs which are brittle and look and feel as if made of plastic, a pelage similar to that of muskdeer. Sheep in cold areas have an inner layer of wool, and, according to Heptner et al. (1966), the guard hairs are hollow, filled with air, to provide warmth. The winter hair of the Tien Shan argali may be 5 cm long. Such warm winter coats pose problems of heat regulation. The poorly insulated legs of several species probably dissipate heat (Scholander et al., 1950), as do horns (Taylor, 1966). The Caprinae shed their pelage once a year in the spring or summer, leaving mainly the coarse and sparse outer hair which has poor insulating qualities. Animals inhabiting warm areas generally have little wool and a thin hair cover, as shown by Punjab urial and goral, though the length and density of the pelage varies with altitude. The takin has a peculiar coat, fairly short, dense, and so permeated with an oily substance that fingers become sticky after touching it. Possibly this is an adaptation to shed rain.

PELAGE COLOR AND HAIR LENGTH

The various Caprinae differ strikingly in color and length of their coats. Certain features are conspicuous only in males, indicating that these have little ecological relevance but represent social signals. Males usually display their most splendid adornments during the rut. Since coats may change their appearance with age and the seasons, the following descriptions apply mainly to adult males during the rut.

Chiru of both sexes are buff-colored with white undersides, and the male also has a dark face and black along the front of the forelegs. Saiga are nondescript, brownish above and whitish below with a sparse fringe of hair along the ventral surface of the neck. Takin are somewhat shaggy and brown to brownish-red in color (except for the Shensi subspecies which is a golden buff) with a dark dorsal stripe and a short ruff on chin and throat. Muskox are dark brown to black with whitish stockings, a grizzled saddle, and in some animals a white forehead. The hair on the shoulders forms an erect mop, giving the animal a prominent hump, a feature also found in yak. A scraggly beard drapes from the throat.

The gray to brownish to rust-colored pelage of goral is set off by a white throat patch, a dark spinal stripe, and a small crest of coarse hair on the neck. Some subspecies have white stockings.

Serow are variable in coloration. Those I saw in Nepal had a coarse black coat grading to rusty on the upper legs and sides. The edge of the lip and the throat patch are whitish as are the lower legs. There is a stiff, black mane which in Chinese animals may be white. The face of the Japanese serow may be framed with a grayish ruff. Chamois pelage is black with white on the chest, abdomen, throat, and rump; the insides of the ears are whitish too and so is the face except for a dark band between eyes and nostrils. A rusty spot marks the knees and brownish hair grows along the back of the lower legs. A dark dorsal stripe becomes visible in the brownish summer coat. The mountain goat is white all over, a hairy beast with a stiff ruff on neck and shoulders forming a hump, a prominent beard, and flowing hair on rump and upper legs.

In contrast to the species described so far, male and female Caprini differ greatly in their pelage adornments. The most prominent feature of the male Himalayan tahr is a huge shaggy ruff with hair up to 30 cm long that covers the forequarters and a long mantle of hair that drapes from the sides and rump. The pelage is coppery brown to black in color with the ruff often being straw-colored, especially in summer. A pale area stretches along each side of the dark mid-dorsal line. The abdomen is rusty. The back of the lower legs and the small rump patch are rusty too, the combination of these two colored areas making the rear of the animal conspicuous. The insides of the hindlegs and the chin are white, and the side of the muzzle shows a pale line. Females have only the rudiments of a ruff and their pelage is less rich in hue. Nilgiri tahr males are almost black, except for their grizzled white back and sides and sometimes rump, a feature responsible for their being called saddlebacks by hunters. The sides of the neck may be gray and the throat and abdomen are white; there is a pale facial streak and a white spot above the knee. These tahr have a short bristly mane, not a ruff. Females are grayish-brown with a white belly and a dark spot above the knee. Arabian tahr, according to Harrison (1968a), have a brownish coat, a dark dorsal stripe, black legs, a white belly, and elongated hairs around the jaw, nape, and withers.

All *Capra* have light-colored abdomens, dark dorsal stripes from nape to rump with hair either slightly elongated or extended into a short mane, and a beard on the chin. The beard may be short and broad as in Dagestan tur or long as in markhor. Female wild goat generally lack beards but about 2% of those in the Karchat Hills had wispy ones; some Asiatic ibex females also fail to grow a beard. Among the least colorful goats are the Kuban and Alpine ibexes and the Dagestan tur whose pelage is dark brown or gray-brown with a pale area on the belly. Contrasting with these is the Asiatic ibex, a colorful, dark brown animal with a silvery saddle, and in some males whitish areas also on shoulders and thighs. The abdomen is white and so are the lower legs and much of the face. A dark flank stripe is present in some animals. Down the front of each leg courses a black line which divides above the hoof to accentuate the white on the pastern. The white rump patch is surrounded by light-colored hair that extends down the back of the legs. Females resemble males except that their coats are gray-brown and the white on the bodies is less conspicuous. Spanish goat, Wali ibex, and Nubian ibex are similar in having blackish chests and upper forelegs, dark leg markings, whitish undersides, dark foreheads, and a body color ranging from brown to buff.

Adult wild goat males are almost wholly silvery-gray but with black faces and beards, a dark stripe down the front of each leg, black dorsal and flank lines, and a broad dark collar encircling the shoulders which merges with the black chest. The undersides are white as is the small rump patch. The females are plain, gray-brown with a grayish face, white undersides, a dark chest patch, and a stripe of gray on each leg. The most distinctive goat is the markhor. A Kashmir markhor has a flowing ruff of white to gray hair on chin, neck, shoulders, and chest, and often also whitish tufts on each foreleg and stifle. A dark flank stripe separates the white belly from the brown and gray sides. Some animals have a vertical slash of dark hair in front of each shoulder and a grizzled patch on the thigh. The rump patch is small and white and the lower legs are white too, except for a dark-haired wedge below the knee. The Sulaiman markhor has a short ruff, and its pelage is more grayish than that of the northern animals. Females have a gray face and fawn-colored coat except for their whitish chins, undersides, and lower legs; a dark line runs down the back and another from chin to chest; and the beard is short and wispy.

Among sheep, the urials are the most conspicuously pelaged members of the genus because of their distinctive throat ruffs, which in the Punjab animals may reach a length of 20 cm.

Unlike tahr, markhor, and aoudad, urial lose their ruffs during the spring molt. Ruff and saddle-patch colors are listed in table 5. Coat color varies between subspecies from a rich chestnut brown grading to black in *musimon* to a reddish buff in *arkal* to a pale fawn in some desert races. The face is often gray or whitish; the undersides are white too, as are the lower legs. There is a dark flank stripe and a gray wedge of hair extends upward and downward from the knee. Western races have a conspicuous white rump patch whereas the eastern ones have a white area only some 8 cm in width. Ewes lack both ruff and saddle patch and in general have a paler pelage.

Clark's (1964) account of Marco Polo sheep pelage is quoted below, a description which with minor emendations also suits most other argalis (see Miller, 1913, for *ammon;* Gordon, 1876, for *karelini;* Allen, 1940, for *darvini*):

> Poli carry a body-color of pale grayish brown or buff, which is sometimes sprinkled with whitish hairs. Along the sides of the body from elbow to flank runs a slightly darker, broad stripe, dividing the body-color from the lighter underparts. The face, throat, chest, lower legs, under parts, and rump-patches are a light cream-white. In winter the longer hair on the neck and front of the throat forms a suggestion of a white bib—not noticeable in the summer coat. The back of the neck bears a thick cape of long, woollike hair starting at the base of the horns and ears and tapering backward to the shoulders, where it fuses with a narrow strip of long hair running along the medial line of the back to the tip of the tail.

All argalis have whitish rump patches but these differ somewhat among the different subspecies in size and distinctiveness. For example, *ammon* seems to have proportionately the largest rump patch, and in *darvini* the "white of the buttocks is not sharply marked off" (Allen, 1940). Argali ewes lack ruffs and dorsal crests and the rump patches are less distinct than those of rams.

Male Kamchatkan snow sheep in winter pelage have "hair yellowish gray on back, lighter on belly, almost straw yellow on neck and head; legs rufous in front, yellowish gray behind; hind part of thigh and caudal disk yellowish white" (Harper, 1945). Other subspecies vary only slightly from this pattern. *O. n. lydekkeri* from the Verkhoyansk Range, for example, has a transverse dark band across the muzzle and the rump patch may be larger than that of the Kamchatkan race. The thinhorn sheep differ strikingly in pelage color with the white Dall sheep at one extreme and at the other the Stone sheep whose black coat is accentuated by white on the muzzle, white rump patch and belly, and a white trim down the back of each leg. Except for some desert races which are a light buff, bighorns are dark to grayish brown, and they have a white-tipped muzzle, a white rump patch which is proportionately larger than that of snow and thinhorn sheep, a dark dorsal stripe, and white along the backs of the legs. The color pattern of ewes is similar.

The aoudad with its pale tawny body grading to a whitish underside would be nondescript were it not for its distinct beard that begins near the throat and extends to the chest, and for its floppy, hairy leggings. Females also have these hairy appendages though in abbreviated form. Short-haired and brownish-gray to slate blue in color, male bharal are not conspicuous except for their black chest and black ventral surface of the neck. A black flank stripe marks the boundary between white underparts and the sides. There are prominent dark markings along the front of the legs, as in wild goat, and a dark dorsal stripe. The rump patch is white, as is the chin and tip of the muzzle. Grayish streaks mark the muzzle along the top and sides. The tail is black, and there is also a black spot at the tip of the penis sheath. Females resemble adult males, except that the black on neck and chest is less marked.

The tail has been seldom mentioned so far. None of the Caprinae have long tails, the aoudad and goral having proportionately the longest ones in relation to shoulder height (tables 6, 8, and 10). Most tails are .08 to .14 times the shoulder height of the animal. Argalis seem to have proportionately the shortest tails. Gordon (1876) measured a *karelini* tail of 14 cm, including the terminal hairs, and Macintyre (1891) noted that in Tibetan argali "there is hardly a vestige" of the tail, it being only about 10 cm long. Since the measurements in the tables do not include the terminal hairs, the figures do not convey actual tail dimensions. One chamois I measured had a tail 7 cm long without the hairs but 14 cm long with them; and one male takin had a tail 20 cm long without the hairs but 30 cm long with them. Furthermore, the width of a tail adds to its prominence. Most *Caprini* have tails of similar length but those of sheep are thin and rat-like compared to the rather broad flat tails in some other genera. Tails are generally dark, a color extension of the dorsal stripe. A dark tail is particularly conspicuous when surrounded by a large white rump patch as it is in Asiatic ibex, bharal, and bighorn.

Certain prominent features of the pelage appear repeatedly in various Caprinae tribes. A dark dorsal stripe, often elongated into a short crest, is common to many forms. Long neck ruffs have evolved independently in several species, among them muskox, Himalayan tahr, markhor, aoudad, and urials. Short ruffs are found in rupicaprids and argalis, and beards and bibs in takin, *Capra,* mountain goat, and others. Many species have markings on the legs, but only bharal and some *Capra* exhibit striking color patterns. Several rupicaprids, and to a lesser extent tahr, have white throat patches. Although a number of species have white rump patches, only the argalis, snow and American sheep have large ones covering much of the rump. However, some forms give the appearance of having white rumps because the small patches are surrounded by light-colored hair. This is the case in Asiatic ibex and in some urials, and the latter further increase the amount of white in the rump area by having large white scrota. One Punjab urial scrotum was 10 cm in width in late October, 2 cm broader than the rump patch. The aoudad has no rump patch but it does have a conspicuously large and pale scrotum.

HORNS

Horns are the most striking features for many members of the subfamily. Often long and massive with graceful twists and flowing lines, horns are objects of symmetrical beauty, ornaments which hunters constantly endeavor to remove from the owners. This quest for trophies has drawn hunters to the uttermost parts of the earth, the horns becoming Holy Grails for which incredible hardships are endured. As Rudyard Kipling wrote in "The Feet of the Young Men":

> Do you know the long day's patience, belly-down on frozen drift,
> While the head of heads is feeding out of range?
> It is there that I am going, where the boulders and the snow lie,
> With a trusty, nimble tracker that I know.
> I have sworn an oath, to keep it on the horns of Ovis Poli,
> And the Red Gods call me out and I must go!

With horns described in such sources as Lydekker (1898) and

Clark (1964) and pictured in figures 6 and 7, I limit myself to mentioning some aspects important to the animal's biology.

In comparing horns of species and subspecies it would be useful to know average length for particular age groups. Such data are not available, and to derive some comparable measure I took the mean length of the ten largest heads of each form as listed in Dollman and Burlace (1922) for Eurasia, and in Anon. (1971a) for North America. This method favors large horns, but all record heads are not listed in these volumes. Some exceptional horns were lost or not submitted and new records have supplanted old ones. For example, in 1922 the listed record for Altai argali was 159 cm, but Demidoff (1900) had measured and discarded a head of 160 cm and McElroy (1971) raised the record to 169 cm.* For many years the record for wild goat was 141 cm, but an animal measuring 144 cm was recently shot in Iran. During my travels in Pakistan I saw some magnificent heads hanging forgotten on dusty walls, a 102 cm Afghan urial, a 149 cm Kashmir markhor. Tables 6, 8 and 10 list the longest known horns insofar as I was able to locate them.

Chiru have long, slender horns rising almost vertically from the head, whereas those of saiga are short and slightly lyrate; the females of both species lack horns. Short, directed backwards, and sharply pointed, the dark conical horns of goral, serow, and mountain goat are similar in appearance; chamois also have horns of this type except that these rise vertically from the skull and have a hook at the end. Male and female rupicaprids have horns of about equal length, though those of the latter tend to be more slender. Muskox horns curve down and out, those of males being much more massive than those of females and joining on the forehead in a broad boss. Takin horns have a knob near the base, similar to that of wildebeest, and they then curve up, out, and finally back.

A tahr's horns are short, laterally flattened, and they curve backward, those of the Himalayan and Arabian species having a prominent keel in front in contrast to the Nilgiri tahr whose horns are almost flat there (see Schaller, 1971a). The aoudad with its heavy, corrugated horns sweeping out, back and in again is rather

---

*In these and other measurements I give the length of the longer of the two horns rounded to the nearest centimeter.

distinctive, especially since it and the tahr are the only Caprini in which females also have large horns. In all other forms the horns of females are spindly and rather insignificant, less than a third as long as the massive structures carried by males. Female horns generally resemble those of males, except that they tend to be more variable in shape. In some mouflon populations the females lack horns, and in a sample of 49 bharal females 41% had straight spikes, 49% had normal horns, growing first up and then laterally, and 10% projected forward at a sharp angle; 11% of the horns were broken.

The backward-sweeping horns of bharal and the horns of *Capra* were described in the chapter on taxonomy. So, too, were the horns of the urials. But a few more comments on argalis are needed. Severtzov's argali is small, its horns resembling those of Afghan urial in shape and size. At the other extreme is the almost mythical Marco Polo sheep, perhaps the most sought after trophy of all. The straw-colored horns form an open, outwardly extended spiral in which the tips arc up, out, and down again (fig. 7). Though slender in build, with an average basal circumference of 37 cm (34–41 cm) in 23 adult heads I measured, the horns are up to 191 cm long in the case of the record ram, an animal found dead and presented by Lord Roberts to the Amir of Afghanistan. *Karelini* horns are more rounded than those of *polii* and also larger in girth and with an intermediate amount of flaring (Carruthers, 1913). The horns of the other large argalis have the tips pointing forward and slightly outward. They are about 43 to 46 cm in circumference and may reach 53 cm in a large *ammon*. The horns of snow and American sheep are all homonym, and their lengths are listed in table 10. Taking the 10 largest heads as given by Anon. (1971a), the average circumferences of several subspecies of thinhorn and bighorn sheep are: Dall 36 cm, Stone 38 cm, Desert 41 cm and Rocky Mountain 41 cm.

Sheep and goat horns often have broken tips, as do the horns of old muskox bulls (P. Lent, pers. comm.). Such breakage is caused by fighting, as Shackleton and Hutton (1971) have shown for bighorn rams. Of interest is the possible inference that can be drawn from such breakage about the efficiency of horn design and the frequency of fighting, points which will be discussed in Chapter 10. The percentage of marred tips, either through heavy brooming or breakage, varied much between species. By checking animals at least 3 years old, using both trophies and horns picked up in the field, I found that of 182 Asiatic ibex tips examined, 6% were broken; of 66 wild goat (not including those from the Chiltan area), 3%; of 154 flare-horned markhor, 27% (including two horns snapped off at the base); and of 127 straight-horned markhor, 14%. Four percent of 30 bharal tips were damaged, and several also had large pieces of horn peeled from the tip. Urial often have damaged horns, though more than 5 cm of the tip is seldom broken off. Some 35% of 110 *cycloceros* tips, 57% of 56 *punjabiensis* tips and 50% of 20 *vignei* tips were damaged, indicating that horn shape in these urials has little effect on the amount of breakage. (Two living Punjab urial rams each had a horn snapped off at the base.) Marco Polo sheep horns often have a centimeter or two missing from the tips (49% of 68 tips), and horns may occasionally snap off near the end of the horn cores (Clark, 1964). Most hunters of Tibetan argali complain, as did Macintyre (1891), that "in almost all large specimens the tips are broken." Judging by museum specimens and photographs, heavy tip damage is prevalent in *darvini* and *ammon* too. Dall sheep seldom break their tips, Stone rams do so occasionally, and Rocky Mountain bighorns do so as a rule (Geist, 1971a). It would seem that large sheep with slender and light horns—the Marco Polo and thinhorns—damage their horns less often than do Tibetan argalis, bighorns, and others with massive, heavy horns.

Large horns may burden males with considerable weight, for not only must the animals carry the bony cores and keratinous sheaths but also the heavy bones of the skull that are needed to support these structures. For instance, the roof of the skull of a 5½-year-old Kashmir markhor male consisted of 4 cm of dense bone and sinuses, and Geist (1971a) gave figures of 5–7 cm for 3 old Rocky Mountain bighorns. I weighed several sets of dry horns with cranium: a 5-year wild goat, 1½ kg; a 10-year Asiatic ibex, 7 kg; 5½-year and 6½-year Kashmir markhor, each 3½ kg; an 8½-year bharal, 4½ kg and a 14½-year bharal, 6.8 kg; seven Punjab urial, 5½ to 8½ years old, 2.7 kg (2.4–3.3); and a 6½-year Marco Polo, 11.7 kg and an 8½-year Marco Polo, 10.8 kg. Tibetan argali heads may reach 18 kg (Macintyre, 1891), but one ram, 6½ years old, that I measured scaled only 7.2 kg. Altai argali heads weigh 20.3 kg (Demidoff, 1900), 22 kg (Clark, 1964), and up to 28 kg (McElroy, 1971). A large Kamchatka snow sheep skull weighed

6.8 kg (Heptner et al., 1966), and, as quoted by Geist (1971a), average figures for Stone sheep are 7.9 kg, for Rocky Mountain bighorn 11.3 kg, and for Nelson's bighorn 7.5 kg. Markhor and wild goat have proportionately light horns, 3–4% of their body weight, as compared to the 6 to 8% in large Asiatic ibex and bharal. Bharal horns seem particularly heavy in relation to their length and circumference (an average of 27 cm with a range of from 24 to 31 cm in males at least 5 years old). About 6.5% of the total weight of a Punjab urial ram consists of the skull and for Marco Polo sheep the figure is around 9.3%. In the Tibetan argali and others whose horns weigh nearly twice as much as those of Marco Polo sheep but whose body is not twice as heavy, horns contribute an even greater percentage to the total weight. About 10 to 11% of total body weight of American sheep consists of the skull, indicating that in these and other forms the social advantage of having large horns must be great indeed to compensate for the extra energy that is needed to carry them.

Horn length increases steadily with age (figs. 20 to 22), with the result that the longest and heaviest horns are generally carried by the oldest animals. However, horn growth is affected by several factors including nutrition, and the largest horned individual is not always the oldest. For example, six Kashmir markhor that I measured were older than the one with the longest horns (149 cm).

The average annual increment in horn growth decreases as the animal becomes older (figs. 20 to 22). Different species show different growth patterns. The short-lived urials and Marco Polo sheep reveal a steeply declining growth curve. After an initial spurt, markhor horn growth levels off between the ages of 2½ to 4½ years and then decreases steadily but gradually. On the average, the flare-horned subspecies grows more after the age of 4½ years than does the straight-horned. The wild goat shows a pattern similar to that of markhor. The size of horn segments of Asiatic ibex and bharal, the two species with the longest average lives, decreases slowly in curves reminiscent of those published for Alpine ibex by Nievergelt (1966a), who showed that different populations of ibex have different growth rates depending on their plane of nutrition. My samples of horns from specific populations are too small for such an analysis.

To elucidate trends in horn size among the Caprinae, I present figures of horn length relative to body size as measured by

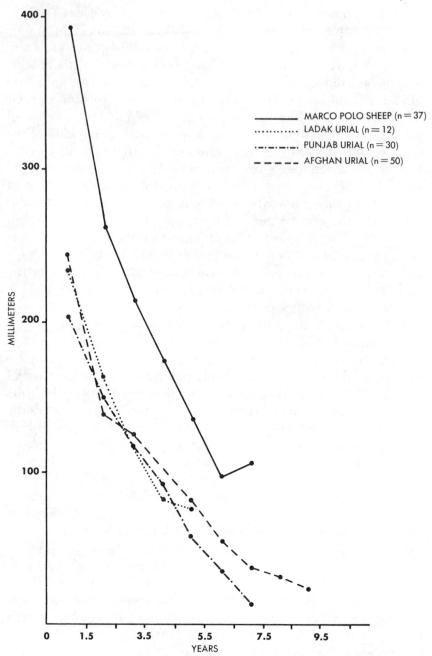

Fig. 20. Average lengths of annual increments in horn growth of several *Ovis* in Pakistan. In this figure, and in figures 21 and 22, the averages during the first 1½ years are somewhat low because horn tips are often broken.

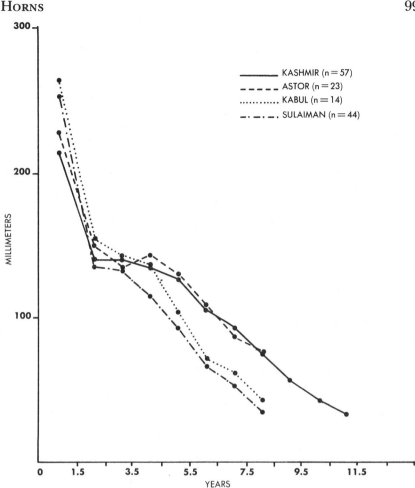

Fig. 21. Average lengths of annual increments in horn growth of markhor in Pakistan.

shoulder height in tables 6, 8, and 10. Figures for shoulder height are only approximations, for there are few actual data and some of these do not agree. For instance, Asiatic ibex in the Himalaya are said to have a height of 102 cm, but Heptner et al. (1966) give an estimated mean of only 92 (80–110) for ibex in the Tien Shan. Yet Church (1901) and others think that the Tien Shan animals are bigger in body than the Himalayan ones. All rupicaprids have relatively small horns, as do saiga, takin, and muskox. Tahr have horns proportionately slightly larger than the rupicaprids, length still being less than half the shoulder height. Bharal and aoudad have relatively small

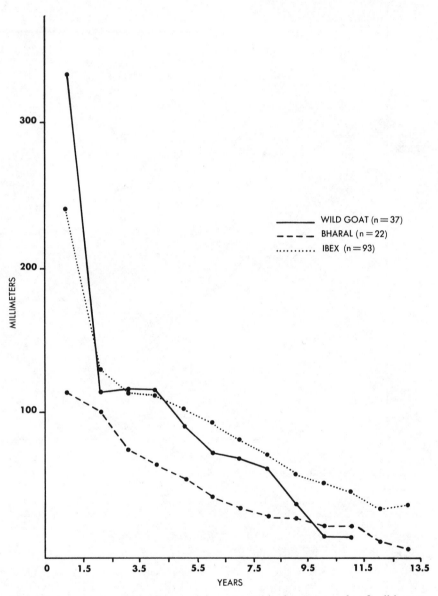

Fig. 22. Average lengths of annual increments in horn growth of wild goat,
    Asiatic ibex, and bharal.

horns too, but the others range from Alpine and Kuban ibexes, in
which horn length and shoulder height are equal, to wild goat,
Nubian ibex, and Asiatic ibex with horns around 1.3 times the

shoulder height. Sheep horns are less variable than those of goats. Most urial and argali have proportionately similar horn lengths (1.0–1.3), an exception being the Marco Polo sheep in which a few huge-horned individuals with unbroken tips have raised the index to 1.6. American sheep vary from 1.1 to 1.2. Geist (1971a) noted that "snow sheep appear to have relatively and absolutely smaller horns than thin-horn sheep while these in turn have smaller horns than [Nelson's] bighorns," but my data show no such clear trend (table 10). While snow sheep seem to have smaller horns than do American sheep, judging by average maximum size, it should be remembered that the total sample of trophies from Siberia is small indeed.

In this resumé of the physical attributes of the Caprinae one main generalization obtrudes: those species with little or no sexual dimorphism in body size and pelage characteristics, the rupicaprids, have proportionately the smallest and simplest horns, whereas among those species in which males are almost twice as large as females the males have striking secondary sexual adornments, the most prominent being the horns. The Ovibovini and Saigini are intermediate in dimorphism and the tahr are exceptional in that the sexes are quite different in size but the horns are small. There are ecological and social adaptations to the environment, and the two levels of selection work both independently and in unison to produce certain traits. Optimum body size and shape have no doubt been influenced largely by habitat, but most special features, such as the adornments of males were shaped by the social milieu of the species. However, the structure of societies is influenced by ecological conditions to produce an intricate connection between an animal's habitat and habits.

Age and Sex Classes
The males of sexually dimorphic species continue to grow in body size well after reaching sexual maturity and in horn size until death. An observer can readily place a male into one of several age classes which often correspond to the actual age of the animal. If one judges by behavior, these categories also have biological reality in that males of different ages behave differently toward each other and toward females. Actual age can often be determined by counting growth rings on the horns, although this method could

not be used on the rather obscure rings on the horns of Afghan urial in Sind and southern Baluchistan. The following categories of animals were recognized: young (0–1 year), yearling male and female (1–2 years), adult female (2+ years), and males of several classes. Since adult females were described earlier and yearling ones are similar except for their smaller size and more slender build, this discussion is confined to males in their rutting coat. Because horns and bodies may vary in their growth rates between populations, the descriptions refer chiefly to my main study areas.

Himalayan tahr

Yearling males resemble females except for being noticeably smaller and sporting a small neck ruff. Class I males, almost 3 years old, are as large as or slightly larger than females. Their pelage is not as dark as that of adult males, and they lack a mantle of hair along the back. The horns are smooth and yellowish rather than corrugated and dark as in adults. At the age of almost 4 years to perhaps 5 years (Class II), males have a rich brown to black coat with shaggy ruff and a short mantle of hair along the back. Though large, the animals are not as robust as full adults. Class III males, at least 5 years old, are handsome creatures with black faces, voluminous, often light-colored ruffs, and a flowing mantle hanging to the flanks and thighs.

A similar age classification can be applied to Nilgiri tahr. Yearling and class I males resemble females but class II males almost 4 to perhaps 5 years old, have a dark brown to black rather than the gray-brown pelage of young animals. Class III males are dark too, their most conspicuous feature being a silver saddle.

Bharal

Yearlings are two-thirds the size of females and have horns about 15 cm long, and class I males, 2½ years old, are as large as females with horns some 25 cm long. Class II males, 3½ years of age, are bulkier than class I males and have horns up to 35 cm long. At 4½ years, class III males have a powerful physique although they remain somewhat trimmer than full adults. Most class IV males are 5½ to 6½ years old with some 7½ years.* Bulky in looks, with

---

*In Schaller (1973b) classes III and IV were combined into a class III, and class IV comprised all animals over 7½ years of age.

black neck and chest and stout horns sweeping out and far back, these males are fully adult. Their necks are much swollen during the rut and the skin around the penis forms a prominent bulge. Class V males are at least 7½ years old, with horns 50 cm or more long. Their age is most readily ascertained by counting horn rings.

Wild goat

Yearling males look much like adult females except that they are smaller and their horns are more massive and longer, some 30 to 40 cm in length. Class I males, 2½ years old, are larger than yearlings and their horns tend to be 40 to 50 cm long. Males of this class have a wispy beard. Class II males, aged 3½ years, are about as large as adult females. They have a prominent beard and a black flank stripe separating the buff sides from the white abdomen. Horn length is roughly 50 to 60 cm, although much individual variation exists among animals of an age group. Class III males, aged 4½ years, are larger than females. Some have a faint vertical slash of black on the shoulder. The horns may be 75 cm long and have a knob or two along the anterior keel. Most class IV males were estimated to be 5½ and some 6½ years old. The animals are almost twice as large as females and the color of their pelage during the rut differs markedly from that of young males. The brown has been replaced by gray along neck and back. The face is dark, as is the ventral part of the neck and chest, and a broad, black collar encircles the shoulders. The horns are over 75 cm long. Class V males are fully adult, at least 6½ years old. The brown portions of the pelage have turned gray, a striking silvery gray, which is discarded with the spring molt for the more typical brown coat.

Kashmir markhor

Yearling males resemble adult females but they are still slightly smaller and their 30 cm long horns are darker and somewhat longer and broader. Class I males, 2½ years old, are the size of females with horns up to 45 cm long and a pelage which is dark brown with a grayish neck. There is no ruff. Males in class II, 3½ years old, resemble those of the previous age group but with the addition of a fringe of white hair on forelegs and across the chest, the first intimation of the ruff. The horns may be over 50 cm long.

Class III males at 4½ to 5½ years of age have a prominent black beard and a long ruff of white to gray hair flowing from neck, chest, and upper parts of the forelegs. There may be a vertical slash of almost black hair by the shoulder and before the haunch, and a conspicuous light-colored patch on the thigh. Horns are often over 75 cm long. Class IV males have accentuated these traits already conspicuous in class III ones. Their ruff is voluminous, their horns strikingly long, and their pelage has more gray than brown, except for the black face and upper parts of the legs.

Asiatic ibex
Yearlings resemble females except for a darker pelage. No saddle patch is evident, the beard is small, and the horns are about 18 to 30 cm long. At 2½ years (Class I) some males have a silvery saddle. The horns have lengthened by about 10 to 13 cm, an annual growth rate which is retained for about 3 years. At 3½ years (Class II) the animals are larger and the horns longer, and at 4½ years (Class III) they have the adult pelage. Class IV males, at least 5½ years old, have horns longer than 64 cm. A rough means of aging ibex at a distance, when annual rings cannot be seen well, is to count the number of prominent ridges along the front of the horn. Approximately 2 ridges a year are grown when the ibex is between the ages of 2 and about 9; older individuals often lay down only one ridge a year. I counted ridges on 87 trophies and the mean for each age was: 2–3 years, 5 ridges; 3–4 years, 6.6; 4–5 years, 8.3; 5–6 years, 10.4; 6–7 years, 12.2; 7–8 years, 13.4; 8–9 years, 16.4; 9–10 years, 17.7; 10–11 years, 17.8.

Punjab urial
Yearling rams are slightly smaller than ewes, the color of their coat is grayer than the pale buff of females, and the horns are bigger, some 15 cm long. A thin line of black hair runs down the ventral side of the neck, the first intimation of the ruff. At 2½ years, class I rams have a broad black line of hair on the neck and a dark flank stripe. Their horns are up to about 40 cm long. By 3½ years (Class II), rams have pelage redder in hue than the young ones have, and a short ruff. A few animals have the first sign of a saddle. Instead of being gray and quite small, the scrotum in this and subsequent classes is white, a structure as conspicuous as the small rump patch. The gray pattern on the

forelegs, typical of both females and young males, has turned white. The horns sweep outward and down for as much as 60 cm. Class III males are at least 4½ years old, and some are 5½ or more years, accurate age classification being sometimes difficult between this class and class IV especially if the horn tips are broken. A saddle is usually present, the ruff is large, hanging halfway to the knees, and the horns sweep in a semicircle. Class IV males are over 5½ years old with horns describing almost a ¾ circle. A similar classification was used for Afghan and Ladak urial.

These five age classes are partly based on horn size, and they correspond roughly to the ones established for American sheep by Geist (1971a). While this facilitates comparisons, the ages of animals older than class I may not be the same. American sheep reach their full size at the age of 8 to 9 years, in contrast to urials which do so some two years earlier.

5 POPULATION DYNAMICS

No ungulate population remains static. Annual changes in the weather have an influence on the food supply and physical condition of the animals which in turn affect birth and death rates. A change in these rates influences the age distribution of the population, as Caughley (1970c) has pointed out. For instance, a favorable season may cause many young to survive the first year of life. This may then cause a decrease in the birth rate the following year when conditions are more normal because there is a temporary increase in percentage of animals below breeding age. The rate of increase in a population depends on age distribution, sex ratio, and birth and survival rates, and one of the tasks of the project was to obtain information on these parameters. The need to know how large a population can be maintained in a protected area is obvious, but an even more urgent need is to analyze the processes that enable a population to reach and maintain a natural equilibrium. Most ungulates have a high reproductive potential—aoudad introduced into Texas increased from 42 to 600 in ten years (Evans, 1967)—but the scarcity of wildlife in the Himalaya attests to some possible problems there. As I was neither willing nor able to collect specimens for autopsy, almost all data are based on observations of living animals, a method which leaves some serious gaps in information.

POPULATION DENSITY
The numerical status of several Caprinae is critical; for example, only 100 to 200 Cyprian urial (Goodwin and Holloway, 1972) and "up to 300" Walia ibex (Anon., 1974) remain in existence. The status of straight-horned markhor, Punjab urial, and Nilgiri tahr is also delicate, as described below. However, most species and subspecies survive in fair numbers. An estimated 6500 Alpine

106

ibex remain in the Alps (Couturier, 1962), 17,000 muskox in North America (Lent, 1971), 32,000 mouflon in various parts of Europe (Pfeffer and Settimo, 1973), and 2 million saiga in Russia (Heptner et al., 1966), to mention four forms for which some estimates are available. Whatever the status of most animals, densities are now lower in most areas than they would be if man had not decimated populations through hunting and habitat destruction. Conservation measures have helped to rehabilitate animals like the Alpine ibex and saiga, but the trend for most species is downward, particularly in South Asia. This, of course, is not a recent development. Lydekker (1898) already considered the markhor in the Pir Panjal "in imminent danger of extermination" due to excessive hunting, and Carruthers (quoted in Harper, 1945) lamented in 1915 regarding Marco Polo sheep: "The colossal heads of 70 and 75 inches are no longer to be obtained, even on the Russian Pamirs. The Chinese Pamir has not produced many heads over 60 inches for several years, and one is lucky to get one over 50 inches now." Shooting ethics were then "in a crude state," as Stockley (1936) phrased it, and they have been just as good in recent years.

Far more detrimental to wildlife than trophy hunting has been meat-hunting by local people. "We were shown in the Chakmak fort the frozen carcasses of about fifty *Ovis poli* and black ibex stored as part of the winter food" wrote Gordon (1876). Ward (1923) describes how villagers in India drive serow into deep snow and kill them, and Dang (1964) relates how tahr and bharal are captured "through their habit of coming to lower rock-faces when snow lies deep on their summer haunts. Here they are cornered and chased by villagers and actually murdered with staves and spears."

Conservation matters aside, hunting has reduced most populations to a point where it is difficult to find localities with animals still living at natural densities and pursuing their existence in a normal social melieu. Constant persecution may affect the habits of animals (Geist, 1971c), and low densities and skewed population compositions may influence social interactions, making the choice of study area of more than casual importance. While searching for suitable study sites, I censused several species, and the results with some comments on general status are here presented.

Hemitragus and *Capra*.

Hunting had nearly exterminated the Nilgiri tahr in southern India by 1879, but the Nilgiri Wild Life Association, formed in 1877, interested itself in the species and afforded it protection. By the late 1930s 500 tahr were thought to be in the Nilgiris and a 1963 census revealed 292 animals. Most tahr in the Nilgiris inhabit the western escarpment from Nilgiri Peak south to the Bangitappal and Sispara Pass area, a stretch of cliffs about 37 km long. I spent a week in the northern third of the range in 1969 and counted 63 different tahr. The central third of the escarpment was not visited, but E. Davidar counted 66 animals there during a 1963 census. In this part of their range the animals seldom move as much as a kilometer inland from the cliffs. The hills along the southern third of the escarpment, an area of about 50 sq km of steeply rolling terrain, revealed 113 tahr in a week of searching. An estimated 300 tahr were in the Nilgiris. A gap of 100 km separates the tahr in the Nilgiris from those in the ranges to the south. The largest remaining concentration survives in the High Range, an uninhabited plateau about 80 sq km in size. These animals have been protected for many years by the High Range Game Preservation Association. I divided the plateau into seven blocks and counted all animals in one block on consecutive days. A total of 439 tahr were seen, and fresh spoor provided evidence for an additional herd, raising the total to about 500 (Schaller, 1971). In the Palani Hills perhaps 60 tahr exist and in the Highwavy Mountains about 100 (Davidar, 1975). Several other small populations may survive—in the Rajapalayam, Anamalai, and Nelliampathi Hills and around Top Slip, Siruvani, Srivillputtur, and Kodayar—but it is unlikely that the total world population exceeds 1500.

Excluding the 19,000 Himalayan tahr that had been shot commercially and for sport in 1973 and 1974, K. Tustin (pers. comm.) estimated in 1975 a population of 20,000 to 30,000 tahr in the 4,400 sq km of their breeding range in New Zealand, or 4.5 to 6.8 per sq km. Chamois in the same general area had a density of about 3.3 animals per sq km before commercial cropping began (Christie, 1964).

Wild goats were censused in two areas, one in Baluchistan, the other in Sind. The Chiltan Range near Quetta consists of a massif some 250 sq km in size. Having been told by villagers that the northern part of the range contains almost no goats, Z. Mirza and I divided the southern half into nine blocks and in November 1970 counted animals in six of these on consecutive days by scanning cliffs and canyons. A total of 107 goats were seen and we estimated a total population of 200. Mirza returned in August 1971 and with three men searched every canyon. No goats were found in the northern quarter of the range and most were in the southern quarter. A total of 169 animals were spotted, validating our earlier estimate of about 200, or .8 per sq km.

The Karchat Hills in the Kirthar National Park consist of a range about 20 km long and 6 km wide. Stockley (quoted in Harper, 1945) visited the Karchat Hills several times from before the First World War until the 1930s when the area was a shooting reserve. There were an estimated 400 to 500 wild goat and 80 to 100 urial there until 1929. But in 1931 all game watchers were discharged and the "local gentry swarmed in and slaughtered the animals," leaving fewer than 200 wild goat and 30 urial. In 1956, when the reserve was reestablished, only 168 wild goat remained, according to the Forest Department. No legal shooting has been allowed since 1967 and the hills were included in a national park in 1973. A census by the Wildlife and Forest Department in 1971 produced 1279 wild goats (Anon., 1971). Doubting this figure, I interviewed each game watcher as to the number of animals in his area and obtained a total estimate of 585 wild goat and 60 urial. My intensive work in the southeastern quarter of the range revealed about 150 animals. (A total of 157 were seen in one day during October 1974). Although I made no attempt to census goats in the other parts of the range, brief visits showed that my study area contained the highest density of animals. The total population was estimated at 400 to 500, or 3.3 to 4.1 per sq km. Heptner et al. (1966) give estimates of 3.7 goats per sq km in Armenia and .7 to 1.0 in the Kopet Dagh. On Theodorou Island, 97 wild goats lived on 68 ha or 143 animals per sq km (Papageorgiou, 1974). Feral domestic goats on the 3.2 sq km Macauley Island in New Zealand had a density of 1000 per sq km (Williams and Rudge, 1969). The goat biomass on Theodorou Island was 3,200 kg per sq km and on Macauley Island 3,400 kg per sq km. Both populations had seriously altered the composition of the vegetation.

Stockley's (1936) comments about the straight-horned markhor are as valid today as they were 40 years ago:

> Although the markhor of Kashmir has some sort of protection, his
> unfortunate relation of the Frontier hills is persecuted by all and
> sundry at all times of the year, while the local inhabitants are well-
> armed, and the peace which has lately invested that country has only
> given the tribesmen more leisure to hunt. Small wonder that the
> markhor have decreased almost to the vanishing point and are likely
> to decrease still further unless measures are adopted for their pro-
> tection. Such measures are difficult to enforce in country where my
> last four trips have had to be carried out with an escort of forty
> rifles, but at least the authorities might make some effort in places
> immediately under their control, instead of encouraging the local
> soldiery to shoot markhor and oorial for meat in lieu of meat ra-
> tions, using government ammunition to do it.

Schaller and Khan (1975) discussed the current status of
Sulaiman and Kabul—the straight-horned—markhor, but a few
points need repeating. In the hills around Quetta the Sulaiman
markhor are on the verge of extinction if one judges by figures
supplied by staff members of the Forest Department—who are
not given to underestimating numbers. S. A. Khan visited
Takatu in 1951 and estimated that there were 150 markhor on
that small massif then. Seven males were legally shot as late as
1968 and 9 in 1969. In the latter year hunting was banned and a
forest ranger made a census: in six days 30 animals were seen,
some possibly more than once. A seven-day survey on the nearby
Pil cliffs the same year revealed 5 markhor. A numerical estimate
for Kalifat in 1973 was 20 to 30, for Zarghun 20, and for Shingar
20. Roberts (1969) felt that the largest remaining populations
were to be found in the Toba-Kakar Ranges. Visiting that area, I
walked entirely around the Surghund massif, where, it was said,
herds of 50 markhor could be seen ten years ago. I saw the track
of a solitary male. Farther north, in the range of the Kabul markhor, a few animals survive in the Marwat Range near Pezu
around Sheikh Budin peak. After being nearly shot out in the
1930s, these markhor increased until in 1958 there were about
200, according to a resident forest officer. Shooting then again
decimated the goats to the vanishing point. Northeast of Mardan,
herds of 150 Kabul markhor descended from the Sakra Range to
feed in the wheat fields fifty years ago according to the old in-

habitants. Now fewer than 50 animals survive in the whole range. In Afghanistan perhaps no more than 50 to 80 Kabul markhor still exist (R. Petocz, in Goodwin and Holloway, 1972). I would guess that only around 2000 Sulaiman and Kabul markhor remain in existence.

The flare-horned markhor has fared somewhat better than the southern subspecies, although the status of animals with Kashmir-type horns has become precarious in such places as the Pir Panjal, where perhaps 250 to 300 persist (R. Wani, pers. comm.). The animals are also scarce in the Dir and Swat districts and in Afghanistan. In Chitral, my main study locality, at least fifteen main valleys once harbored large populations but few animals now occur in most of these. The Golen Gol, once famous for markhor, has perhaps 10 left, and Drosh Gol, which in the 1930s is said to have contained over 500 markhor, had a dozen survivors in 1972. The Chitral Gol, about 80 sq km in size, harbors the finest markhor population in Pakistan. In winter when animals are concentrated on the lower slopes of the main valley, herds are easy to tally. My estimate in December 1970 was 100 to 125, and in December 1972 it was 125 to 150 (Schaller, 1973c), or 1.3 to 1.9 per sq km for the reserve as a whole; in winter the animals concentrated at a density of 4 to 6 per sq km. Tushi, a private reserve, consists of a long, steep slope about 15 sq km in size north of Chitral town. The area contained about 125 markhor, or 8.3 per sq km, in January 1973, but some animals moved out of the area during the summer. Some 500 to 600 markhor exist in Chitral, and perhaps as many as 2000 markhor with Kashmir-type horns survive throughout their range. I made no censuses of markhor with Astor-type horns. In the Kargah Valley near Gilgit, Roberts (1969) reported "not less than 500 to 600" markhor, whereas informed local opinion now places the number at 50. In spite of an obvious decline in numbers, particularly as a result of shooting by the military while constructing roads along major valleys, Astor markhor may be still twice as abundant as Kashmir ones. Farther north, 1000 Turkmen and Tadzhik markhor exist in Russia (Heptner et al., 1966) with 125 to 150 also in the Badakshan part of Afghanistan (R. Petocz in Goodwin and Holloway, 1972).

Hunters have seriously decimated Asiatic ibex in the mountains of Pakistan. In February 1974, I looked for ibex along about 24 km of the Golen Gol in Chitral, a valley once known for ibex. Conditions for spotting animals were excellent, tracks being visible in the snow from far away, yet only 16 ibex were tallied. One week was spent in the Dorah Pass area in northwestern Chitral near the Afghanistan border in July 1972. The slopes of several valleys—Ustich, Artshu, Uni, Guzagjanir—were searched with the help of a member of the border patrol who had shot 20 ibex in the area during the past three years. Covering about 50 sq km up to an altitude of 4900 m, we spotted 10 animals. In the remote uplands of northeastern Chitral, at the headwaters of the Yarkhun, we devoted eight days to finding ibex in an area of some 200 sq km around the Chiantar Glacier and Karambar Lake. We saw four. The best ibex area in Chitral may now be near the village of Besti. There, in the Kulumbukht Valley, a total of 72 ibex were tallied in August 1972 within an area of 40 sq km, or 1.8 animals per sq km. Revisiting the region in February 1974 but covering a different part of the valley system, I spotted 40 ibex in 55 sq km of terrain or .7 animals per sq km. Taking the whole 140 sq km block of mountains, game watcher Gulbas Khan estimated 200 ibex or 1.4 per sq km. One census was made in Hunza, in a valley between Murkushi and Kilik Pass and westward from Haq halfway up the Harpuchang Valley, a distance of about 25 km. A total of 59 ibex were tallied, or roughly .4 per sq km, a not surprisingly low figure in view of the fact that the Gilgit Scouts near Misgar shot 60 animals in that area during the winter of 1972–73. I surveyed ibex along 30 km of north-facing slopes in the upper Braldo Valley between Bardumal and just beyond Urdukass at a time when ibex were concentrated along the lower edge of the snow. Only 49 ibex were seen. The dry, barren south-facing slopes offered little suitable ibex habitat, making it likely that no more than 100 ibex frequented the upper Braldo at that season. These animals actually represented the total population in at least 1400 sq km, most of it covered by the Baltoro and other glaciers and by sterile rock. Densities are considerably higher in Russia where, according to Heptner et al. (1966), 200 sq km of the Terskei Alatau contained 550 ibex in 1949 or 2.8 animals per sq km, and 100 sq km of the eastern Pamir harbored 600 ibex or 6 per sq km. Couturier (1962) presented several densities for Alpine ibex, among them 4 and 9 animals per sq km.

Pseudois

About 50 bharal frequented 35 sq km of terrain in the upper Kang Chu of Nepal or 1.4 per sq km, but animals moved back and forth across the Tibetan border. In November and December 1973, I attempted to estimate the number of bharal in about 550 sq km of the Kanjiroba Range in Nepal by studying the animals at Shey intensively and by transecting various valleys and slopes. There were probably about 500 bharal and possibly as many as 600 to 700 in that area or .9 to 1.3 per sq km. Of these, 175 to 200 were concentrated in winter around Shey at a density of 8.8 to 10.0 per sq km.

Ovis

For generations, the Kalabagh reserve has been the private property of the Nawab of Kalabagh. No special protection was afforded the Punjab urial there prior to the early 1930s and few sheep were left at that time according to Malik Asad Khan. With shooting prohibited, except by special permission, urial increased until in 1966 Mountfort (1969) was told that there were 500 animals. In October 1970, Schaller and Mirza (1974) visited the reserve. After dividing the area into five blocks, we counted urial in one block each day by walking along ridges and ravines. A total of 410 animals were seen. Some were no doubt overlooked and the most westerly part of the reserve was not visited, raising the estimated total to about 450–500. In April 1974 another census was made, this one covering the whole area in two days at a time when most animals were concentrated. I saw 426 urial, including 69 newborns. The two censuses produced similar results, and this, together with the 1966 estimate, suggests that the population has remained relatively stable. The census area included about 40 sq km of terrain, giving 11 to 13 urial per sq km. The Forest Department in the Punjab has made a rough census of urial in the Salt Range, excluding the Kalabagh reserve, and obtained a figure of 800 animals. No estimates are available for the Kala Chitta Range. The world population of this subspecies may not exceed 2000 animals.

One March day I clambered with a local guide up the slopes of Koh-i-Maran south of Quetta. Peering cautiously over the rim of a gorge, he suddenly readied his rifle. Below us were two Afghan urial females, both pregnant. To his annoyance I prevented him from shooting. These urial and one other were the only wildlife we saw in a long day of climbing over a massif once famous for the number of its urial and wild goat. The incident presages the fate of most Afghan urial in Pakistan. Though still widely distributed, most populations are now so small that censuses are seldom worth the effort. A few areas still contain or until recently contained a fair number of animals, judging by the fact that in the late 1960s a tribal chief and his party shot some 55 urial on Dhrun in south-eastern Baluchistan during one hunt, but political unrest prevented my visiting such areas. My contact with Afghan urial was limited to occasional sightings in such places as Khambu Hill in Kirthar National Park and the Gishk area south of Quetta.

The Ladak urial has fared even worse in Pakistan than the other subspecies. In the 1930s the ruler of Chitral had his subjects herd this sheep from the hills into the valley near Chitral town so that he could shoot them with ease. Today, the appearance of an urial near Chitral town is a noteworthy event. "Vast numbers of this species are driven down by the snow in winter to the branches of the Indus, near Astor" wrote Edward Blyth in 1841 (quoted in Harper, 1945), an area which now has few if any animals left. In fact, urial have virtually disappeared from the valley of the Indus and its tributaries downstream from Skardu. North of Skardu, a few animals persist around the mouth of the Shigar River and along the eastern side of the Shigar Valley upstream to the junction with the Braldo River. Far up the Braldo Valley, between the Biafo and Baltoro glaciers, I saw 38 Ladak urial. Fewer than 100 animals probably frequent this 25 km of valley. A few small and isolated populations are said to survive around the confluence of the Indus and Shyok rivers. A day-long search on a ridge northeast of Khapalu, where 100 sheep were said to live, revealed two sets of tracks. Probably fewer than 1000 animals remain in the country, and there the future of this sheep looks as bleak as its habitat.

The Marco Polo sheep is one of the rarest animals in Pakistan, found only around the Kilik and Khunjerab passes in Hunza. I visited both passes in November 1974 but saw neither sheep nor their tracks. Several hundred sheep once frequented the Khunjerab Pass according to villagers. An American hunter saw a herd of 65 rams there in November 1959, and shot at least 5 (Clark, 1964). G. Beg (pers. comm.), a member of the 1964 China-

Pakistan boundary commission, saw animals every month between May and October. However, the sheep were shot indiscriminately during the late 1960s and early 1970s, especially by the crews who surveyed and later constructed the Karakoram Highway. One military officer butchered some 50 animals for meat in 1968. The remnants spend most of the year in China where they are protected, venturing to Pakistan mainly in spring to forage on the new green grass. With the establishment in April 1975 of the Khunjerab National Park, this population of Marco Polo sheep will, one hopes, recover. The Khirgiz in the Tagdumbash Pamir once rounded up Marco Polo sheep with dogs and drove them into narrow valleys where hidden hunters shot them (Etherton, 1911). At that time, and until the 1930s, the ruler of Hunza offered the animals protection with the result that many frequented the Pakistan side of the Kilik Pass, using even the narrow valleys south and west of Haq. Gillan (1935) counted about 100 animals around Haq, and Cumberland (1895) saw them "lie about on the steppes" there. Then the situation was reversed: China prohibited hunting in 1949 and Hunza relaxed its restrictions with the result that the Kilik population was almost exterminated. According to Petocz (1973a) the Kilik animals belong to the Little Pamir population of Afghanistan whose movements up the Waghjir Valley are said to include visits to Sinkiang and Pakistan. According to my local informants, most Kilik animals are part of the Tagdumbash Pamir population in Sinkiang. At least 2900 Marco Polo sheep still occur in the Wakhan Corridor of Afghanistan (Petocz, 1973a).

Several sheep densities have been reported in the literature. In an area of 250 to 300 sq km of the Pamirs there were about 300 Marco Polo sheep in 1937, or 1.0 to 1.2 per sq km (Heptner et al., 1966). Geist (1971a) found 1.7 to 2.2 Rocky Mountain bighorn sheep per sq km and 0.2 Stone sheep per sq km in his study areas. Cherniavski (1967, quoted in Geist, 1971a) reported 0.9 to 1.8 snow sheep per sq km.

SEX RATIOS

Natural selection favors a 1:1 natal sex ratio in most circumstances, but when the female is in poor condition a lower ratio of males to females may be produced (Trivers and Willard, 1973). Unbiased ratios are difficult to collect because among young the sexes cannot be distinguished readily and among adults the males and females may segregate for part of the year. For example, the wild goat figures in table 13 indicate that within the study area males were always twice as abundant in autumn during the rut as in spring.

I have little information on sex ratios of young. Of 19 urial caught and tagged at Kalabagh, 8 were females and 11 were males. In a sample of 64 markhor young sexed in the Chitral Gol at an age of about 6 months, 29 were males and 35 were females. Heptner et al. (1966) reported an equal ratio at birth in saiga and Caughley (1966) reported the same in Himalayan tahr.

Table 13 lists yearling and adult sex ratios. In samples with fewer than 100 individuals, as well as in the samples for Nilgiri tahr and Punjab urial for April 1973, the figures are based on censuses with each animal recorded only once; other tallies may include repeated sightings. The ratios of yearling males to yearling females may vary even in samples from the same population. Some deviations from the expected 1:1 ratio may be attributed to random variation within a small population, some to sampling bias, as for instance the July 1973 Punjab urial data, and some to misidentification, especially when attempts are made to distinguish a yearling female from a stunted 2-year-old one. Given these limitations, no conclusions can be drawn about yearling ratios in most populations. Yearling wild goat males were consistently less abundant than yearling females yet adult ratios were close to parity, suggesting a disproportionately high death rate of young females. Among markhor, yearling females somewhat outnumbered yearling males, and among Punjab urial the sexes seemed to be present in about equal numbers.

The adult sex ratio in most bharal, ibex, wild goat, and urial populations was 1:1 or favored males slightly. Couturier (1962) noted a similar pattern in Alpine ibex. One Cretan wild goat population contained 6 female and 7 male yearlings and 33 female and 36 male adults (Papageorgiou, 1974). Spencer and Lensink (1970) reported a ratio of 127 males to 100 females, including yearlings, in a muskox population on Nunivak Island, Alaska. Male Nilgiri tahr and markhor were much less abundant than female ones. But male markhor in the Chitral Gol increased steadily in number (from 42 to 80 per 100 females) between 1970 and 1974 due to enforcement of shooting regulations. Himalayan tahr in New Zealand have an equal sex ratio, and in a California population there were 105 adult males and 137 adult females (Barrett, 1966). By contrast, the tahr in my small Nepal study showed a 1:2 disparity favoring females. However, males may wander widely, with for example Macintyre (1891) having shot a male over 50 km from the nearest tahr haunts. Heptner et al. (1966) reported a ratio of 1 male to 2 females among goral. Great variations in sex ratios may be found in American sheep (see Buechner, 1960). For instance, Dall sheep in Mount McKinley National Park showed parity, whereas a bighorn population on the Palliser Range in Canada contained 81 adult rams and 35 ewes (Geist, 1971a).

The factors causing such disparate sex ratios often remain unclear. Trophy hunting selects against adult males, and the effects of this when carried to an excess are evident among the Tushi markhor. Wolves and snow leopard kill a disproportionately large number of males, as noted later, yet most bharal and ibex populations retain a preponderance of males. Differential mortality is most likely caused by droughts and severe winters coupled with poor nutrition, those animals without sufficient energy reserves dying at such times. Feral Soay sheep showed a fluctuating ratio of 1 male to 3.4 to 10 females even though the sexes were equal at birth (Grubb, 1974c), and feral goats in New Zealand had a ratio of 1 male to 1.8 females (Williams and Rudge, 1969). Both of these populations are confined to islands which lack predators and from which animals cannot emigrate.

REPRODUCTION

Weather, nutrition, and population density may have a profound effect on reproduction. So many factors seem to conspire against a young animal between conception and weaning that one can only marvel that any manage to survive at all. In fact among muskox an entire calf crop may occasionally be missing (Lent, 1971). High or low air temperatures shortly after fertilization reduce pregnancy rates in domestic ewes, and high temperatures may also cause young to be born weak and small. Nutrition is critical to both mother and offspring. A low plane of nutrition not only decreases ovulation rates but also causes loss of ova, especially in those sheep with twins (see Sadleir, 1969). Thomson and Thomson (1949) placed some ewes on a good diet and some on a poor one during the second half of pregnancy. Of young whose mothers had had a high level of nutrition, 100% of the singles and 82% of the twins survived at least 4 days after birth, but the figures for young from mothers with poor diets were 64% and 14%. Ewes pregnant with one lamb need to increase their food intake by 50% and with twins by 75% (see Sadleir, 1969). Underfeeding of ewes, especially during the last two months of pregnancy, causes emaciation and retarded fetal growth rates. Consequently the birth weight of young is below normal. As Grubb (1974c) noted: "The singleton lambs of parous ewes with a low body weight had a poorer chance of survival than those born to ewes of higher body weight." Poor diet reduced the average birth weight of white-tailed deer fawns by nearly 1 kg (Verme, 1965). Poorly fed ewes not only lack interest in their lambs but also give less milk. About four days after lambing, ewes on a good diet produced 48 ml of milk per hour whereas undernourished ones gave only 14 ml (Alexander and Davis, 1959, quoted in Sadleir, 1969). Unable to control their body temperature, newborns may succumb to cold weather or to heat above 38° C. A heavy snowstorm during the birth season reduced survival of Himalayan tahr young one year by 24% (Caughley, 1970c). The social environment may also influence the reproductive capacity of a population in that a high density reduces ovulation and conception rates, fecundity, and survival of young, as noted by Teer et al. (1965) for white-tailed deer. These factors affecting reproduction need to be considered when evaluating the data below.

Seasonality of breeding

Ungulates in northern latitudes breed seasonally (table 14). The season of mating is relatively fixed, the shortening of daylight having brought animals into breeding condition. However, climatic and nutritional factors may affect the onset of the rut. In addition, some species, such as swamp deer in India (Schaller, 1967) and impala in Africa (Dasmann and Mossman, 1962) breed at different times in different areas in a pattern that is unrelated to obvious environmental conditions. Mating seasons in sheep and goats are not always precisely delineated in the literature. Bharal, for example, are said to rut from September (Prater, 1965) and October (Schäfer, 1937) to October-November (Stockley, 1928) and January (Wallace, 1913). Are such differences real or the result of careless observation? One problem lies in defining the rut. Males may show interest in and display to females weeks before the first one comes into estrus. This pre-rut is followed by the actual mating period, but only intensive observation can distinguish the two. Here I use the terms "mating season" and "rut" in the narrow sense as the period when females are in estrus.

Hemitragus. Judging by their size, most Himalayan tahr young in the Bhota Kosi were born over a period of at least two months. The rut seemed to have taken place between mid-October and mid-January, and one copulation was observed as late as March 25. In New Zealand, where the annual cycle is reversed, the "season of conception is centered in May and the season of births has a standard deviation of 18.5 days around a median of 30 November" (Caughley, 1970c). Caughley (1971) deduced a gestation period of 6½ months on the basis of testes weights. When I was in the Macauley Valley of New Zealand on May 20 the tahr were just entering the rut, and the following year K. Tustin (pers. comm.) observed courtship behavior there in early June. The peak of mating is probably during the first half of June, and with a median birth peak of November 30 this would indicate a gestation period of around 6 months. Nilgiri tahr are presumed to rut between June and August and give birth between December and February (Schaller, 1971).

Capra. "The rut of wild goats starts in late July and sometimes persists till mid September," wrote Roberts (1967). In 1972, the Karchat males paid little attention to the females in early September. Suddenly on September 18 and 19 the rut intensified and reached a peak in early October. The first young appeared in late February 1973 and most females had given birth by March 20 (table 15). In early September 1973, the rut was at a stage of intensity similar to the previous year. Returning to the area on March 20, 1974, a date on which in 1973 some females were still pregnant, I found the birth season well over. Assuming that the gestation periods were equally long in both seasons, the 1973 rut reached a peak about two weeks earlier than the 1972 one. When the population was checked on October 9, 1974, the rut was over. These data do not represent the complete range of variation in the Karchat Hills, for Roberts (1967) wrote that he has "seen very

young kids, only from mid-January to early February," and M.
Rizvi (pers. comm.) has also observed newborns in January. The
rut may thus reach a peak in mid-August during some years, at
least 1½ months earlier than the peak in 1972. In the Chiltan
Range, 500 km north of Karchat, wild goat were courting during
my visit in mid-November 1970. Wild goat also have a variable
rutting season in other parts of their range. Mating on some of the
Greek islands occurs between June and August and in the
Caucasus during November and December (Couturier, 1962).

Kashmir markhor in Chitral rut in December. In 1970 a female
was in estrus on December 17, and in 1972 the first one was in
estrus on December 12. The peak of the rut was around the third
week of December and thereafter courtship terminated rapidly;
when I returned to the Chitral Gol on January 9, 1973, the rut was
over. With a gestation period of about 155 days the young would
be born in late May and early June. According to Kennion (1910),
Astor markhor males join the females on December 22 and the
rut lasts 20 days. Straight-horned markhor rut in late October and
November (Stockley, 1928) and Russian ones in November and
December (Heptner et al., 1966).

Pakistan ibex mate in late December and January according to
villagers, and young would be expected from mid-June into July
after a gestation period of 170 to 180 days. This agrees with my
few observations. I observed no courtship in November and February, except for one young male which displayed to a female on
February 13. Females were neither conspicuously pregnant nor
had young at heel as late as May 15, but several young, some less
than a month old, were seen in late July. Heptner et al. (1966)
noted that in Russia the rut lasts 2 to 3 weeks and varies between
localities, some populations rutting in November, others in December and January. The Alpine ibex also ruts in December and
January, while the Walia ibex does so throughout the year, with a
peak from March to May (Nievergelt, 1974), and the Nubian ibex
in September and October and perhaps other months depending
on the region (Couturier, 1962).

Pseudois. Male bharal at Shey showed little interest in females until
about November 20. The main rut, with females in estrus, began
on November 29 and probably reached a peak around mid-
December. Most young would be born between mid-May and

mid-June after a gestation period of 160 days. P. Wegge (pers. comm.) observed a birth north of Dhorpatan on June 13, 1975.

Ovis. Punjab urial rut in September and October according to Roberts (1967) and Prater (1965). Rams largely ignored ewes in early October 1970, but between October 18 and 22 there was a marked increase in courtship activity, which however had not reached its peak by November 9. In 1972 the rut was subdued until early November and it then intensified, the rams still pursuing ewes on November 18. The birth season had begun a few days before I returned on April 14, 1973, and it was almost over by May 3 (table 15). An occasional young may also be born at other seasons: one heavily pregnant ewe was seen on July 13, the young having been conceived in February. However, the testes of rams had already decreased markedly in size by late December, indicating that little mating can be expected after the main rut. I did not visit the population during the autumn of 1973. Checking on lambs the following spring from April 21 to 28, I found that out of 130 females only 1.5% were still pregnant as compared to about 25% at that time the previous year. The peak of lambing was thus about two weeks earlier in 1974 than in 1973. The 1974 rut was at the same time as in 1970 and 1972. J. Eckert (pers. comm.) saw a number of pregnant females between April 15 and 20, 1975.

Ladak urial near the Baltoro Glacier mate later than do Punjab urial. No females were heavily pregnant in early May, suggesting that none would give birth before June. The rut had thus been in December and January. Afghan urial rut in mid-November around Quetta according to villagers.

Those species I studied, as well as most of those listed in table 14, have discrete mating and birth seasons, a good indication that environmental conditions are rigorous enough to favor their selection. Species that mate all year, such as the Walia ibex (Nievergelt, 1974) and aoudad in California (Barrett, 1966), still retain birth peaks. In general, breeding periods shorten as conditions become more extreme (fig. 23). Taking animals at northern latitudes or high altitudes first, it is obviously essential that newborns avoid severe winter weather. For example, hypothermy during inclement weather may kill many caribou calves (Kelsall, 1968). Births in Asiatic ibex, Kashmir markhor, and others occur a month after the snow retreats, after new green forage has been available for some time. This enables females to regain weight after the lean winter, it provides nourishment to the rapidly growing fetus, and it prepares the female for the nutritional drain of lactation. The species which lives at the highest altitude—the ibex—also gives birth latest in the spring. The short summer growing season provides ample forage. For most species the mating season is in winter but the main rut usually lasts less than a month. Courtship drains an animal's fat reserves at a time of year when these are least expendable and replaceable. A short rut is thus of advantage and it usually terminates before the heaviest winter snows. Annual variation in the reproductive cycle within a population is small, due probably to the predictability of the seasonal food supply.

Conditions are different in the arid, low-altitude environments of the Karchat Hills and Salt Range. Although animals have only extreme heat to contend with, the food supply is unpredictable and dependent on occasional showers. As figure 23 illustrates, rain falls most commonly during the monsoon from July to September. However, rainfall in deserts is best viewed in terms of number of showers than in number of centimeters. Precipitation is often local. Official records, usually taken far from the study area (75 km away in the case of Hyderabad), give only approximate levels. The relationship between rainfall, nutrition, and the onset of mating is not always clear, for in some species, such as the red kangaroo (Newsome, 1965), the rut begins almost immediately after a rainfall, whereas in others a response is detectable only after animals have been eating nutritious forage for a while.

The Karchat Hills had a drought in 1972, only three light sprinkles falling early in the monsoon; 1973 was more equitable, several heavy showers greening the land into September; and 1974 was dry again with rains once in August and once in September but the latter heavy enough to stimulate new growth. The wild goat rut in 1972 was later than in the following two years. This delay may have been due to a low level of nutrition prior to the rut. Work on domestic sheep has shown that good food may advance estrus by a mean of 20 days (see Jewell and Grubb, 1974). Nievergelt (1966b) noted that well-fed Alpine ibex rut earlier, and Pleticha (1972) observed that ibex in zoos give birth 20 to 30 days

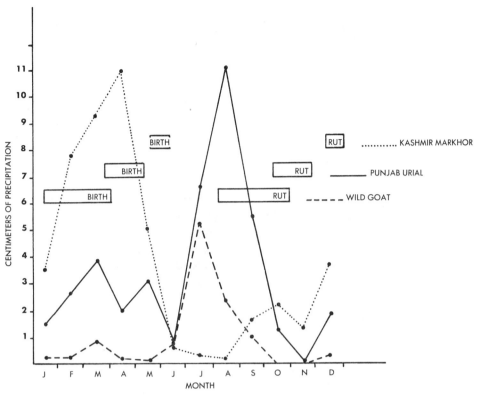

Fig. 23. The time and duration of the rut and birth season of Kashmir markhor, Punjab urial, and wild goat as related to average monthly precipitation. The rainfall data for the Kashmir markhor are from Chitral town (1967–74), for the Punjab urial from Daudkhel (1967–73), and for the wild goat from Hyderabad (1966–72).

before those in nearby mountains. The Punjab urial also mates shortly after the monsoon. The 1973 rut was earlier than usuai, but I was unable to correlate this with precipitation since about the same amount of rain fell each year. Possible delays in nutritional effects may also complicate interpretations. Working on domestic ewes, Smith (1965) found "that there remained a significant effect of the level of nutrition during spring 1962 upon the onset of oestrous activity during the summer 1963–64." Of course no single factor influences all species in an area equally. At Kalabagh the chinkara have not one but two periods of rut, a main one in autumn and a second in April (Schaller, in press).

Temperature may also provide a stimulus to the onset of mating. A drop in temperature seemed to precipitate the rut of Punjab urial between October 18 and 22, 1970. During that period the minimum night temperatures dropped from around 25° C to 17–20° C and daytime temperatures seldom strayed above 32° C instead of climbing to 35° C. The rut of bharal intensified suddenly when on November 19 the days became cool. The average daily minimum during the 6 days before November 19 was -10.5° C but during the following 6 days it was -13° C. Kashmir markhor suddenly courted much more on December 12, 1972, when the air became crisp.

The annual reproductive cycle in the Himalaya is importantly influenced by winter cold and snow. Summer rainfall is of little importance since melting snows provide moisture and hence green forage. But in a hot, arid environment life revolves around rains, and one can infer that natural selection has there timed the cycle to coincide with optimum nutritional conditions. In Africa, for example, the births of many species such as the buffalo can be correlated with the availability of good forage (Sinclair, 1974). Rainfall in wild goat and Punjab urial habitat is most probable in summer (fig. 23). This provides young with nutritious food not at birth but around the time of weaning, and, more importantly, it revitalizes the female after the stresses of late pregnancy and lactation during the hottest time of year and prepares her physically and physiologically for the rut.

Gestation periods
Most Caprinae have a gestation period of from 5 to 6 months as a sample of species in table 14 shows. Differences cannot be obviously correlated with habitat, with taxonomic affinities, or even with animal size, except that the large muskox and takin both have periods exceeding 200 days. The serow has the longest gestation time among the medium-sized Caprinae, the period lasting 180 days according to Okada and Kakuta (1970) and 210–220 days according to Komori (n.d.), with the latter providing the most reliable data. Length of gestation may be somewhat influenced by nutrition, poor fodder extending it for several days in domestic ewes (see Sadleir, 1969). Geist (1971a) suggested that the long gestation periods of American sheep and ibex are an adaptation to "cold, long alpine winters with their slim energy resources." In such an environment, he speculated, a long period of fetal growth

is necessary so that young are born large and able to withstand the climatic rigors. So far there is little evidence that high-altitude caprids produce proportionately larger young than low-altitude ones. Punjab urial ewes weighing about 25 kg have lambs weighing 1.5 to 2.0 kg at birth and Asiatic ibex females of 50 kg have young of 3.5 to 4.0 kg (Heptner et al., 1966). These two species, one from an alpine and the other from a desert habitat, have young of the same relative size. One wonders why low-altitude tahr have a 180-day gestation period and high-altitude bharal a 160-day one. The evolutionary significance of slight differences in gestation lengths remains obscure.

Fecundity

Under optimum conditions all Caprinae females can give birth for the first time at the age of two years. However, conditions are not always ideal. Caughley (1970c), for example, found that in a population of Himalayan tahr on good habitat 67% of the females produced their first young at 2 years of age, whereas in a population on a heavily grazed range the figure was 27%. All females can also have twins but the frequency with which this occurs varies from species to species, depending on both genetic and environmental factors. For instance, twinning is rare in American sheep but common in urials and argalis (table 14).

Yearling Karchat wild goat females took part in the 1972 rut, but the following spring they were neither pregnant nor had young at heel except for one presumed yearling. Nearly all females in this population gave birth for the first time at 3 years of age. Fecundity rates in adult caprids are normally high: 90% of Himalayan tahr had young each year (Caughley, 1971). However, as table 15 shows, only 72% of the wild goat females were pregnant at the beginning of the birth season, indicating that perhaps 20% failed to have young that year. Single young and twins are the rule but triplets have also been reported (Wahby, 1931). I observed no twins in 1973 and 1974 but T. Roberts (pers. comm.) saw some in 1976. Kashmir markhor yearlings also took part in the rut, but I do not know if any conceived and bore young. An average of 1.3 young accompanied each adult female in the Chitral Gol, a high twinning rate. Some Asiatic ibex may have their first young in the wild at 2 years of age but most have them at 3 years (Heptner et al., 1966). Alpine ibex may fail to bear offspring until 4 to 6 years old (Nievergelt, 1966a). Of 56 pregnant Asiatic ibex females examined from the Tien Shan and Pamir only two had twin fetuses (Heptner et al., 1966). In February 1974, I tallied 10 females with young at Besti; two of these, both separated from a herd, had two young with them. In their native habitat, both Himalayan and Nilgiri tahr appear to produce their first young at the age of 3 years, although the former species often gives birth at 2 years of age in New Zealand, as noted earlier. Single births are the rule. Caughley (1970c) found only one set of twins in the hundreds of tahr carcasses he examined. Zuckerman (1953) reported an 8% twinning rate at the London Zoo.

Visiting Lapche in March, I found that adult female bharal looked pregnant but yearlings did not. One female seemed to be accompanied by twins, my only such record for bharal, confirming Schäfer's (1937) observation that one young represents the usual number. In New Mexico yearling aoudad often conceive, and 58% of all females have twins (Ogren, 1965).

Punjab urial yearlings courted, and, judging by the fact that over 90% of all ewes were pregnant at the beginning of the birth season in 1973, a small percentage of yearlings probably conceived. Urial usually have twins, according to Jerdon (1874). Most females at Kalabagh were accompanied by single young. On 6 occasions in 1970 a ewe had two young at heel but some of these may have been lost ones who had attached themselves to a stranger. In April 1973, I found hidden newborns 12 times, among them one set of twins, and, in April 1974, 5 out of 64 ewes or 8% were accompanied by twins. Corsican mouflon sometimes conceive as yearlings. Twins are exceptional in these animals, but in an introduced population on the Crimea there were 1.5 young per female (Pfeffer, 1967). Of 152 ewes tallied by Decker and Kowalski (1972) in the Kopet Dagh, 55% were accompanied by single young, 42% had twins, and the rest had triplets. Valdez (1976) collected 57 pregnant Laristan urial near Shiraz in Iran. All 7 yearlings in his sample were pregnant and all had one fetus. The average number of fetuses in 2-year-olds was 1.00, in 3-year-olds 1.25, and in ewes older than 3 years 1.53. Among 66 Tien Shan argali ewes there were 65% with single young, 33% with twins, and possibly one set of triplets (Heptner et al., 1966). Macintyre (1891) reported that Tibetan argali "drop one lamb yearly." One captive Marco Polo sheep ewe gave birth to a litter of

5 and two years later to triplets (Lambden, 1966). American sheep rarely have twins and seldom conceive as yearlings (Geist, 1971a).

Domestic sheep and goats have a much higher reproductive rate than do their wild relatives. Among feral goats in New Zealand both males and females were fertile at the age of 6 months. Females often mated shortly after parturition and bore two sets of young a year, about half of the sets consisting of twins (Rudge, 1969). Soay sheep also mate as lambs, and twins are common (Jewell and Grubb, 1974). Only the saiga approaches the domestic forms in fecundity, females of this species conceiving at the age of 7 to 8 months and often bearing twins (Heptner et al., 1966).

Those sheep and goats living at relatively low altitudes and in arid environments have a high twinning potential whereas alpine forms such as Asiatic ibex and bharal have a low one. It may be conjectured that the species whose food supply is fairly predictable from year to year tend to have single young. In contrast, desert forms compensate for their high and irregular losses in droughts not only by adjusting the fecundity of yearling females, as do many ungulates, but also by attempting to produce twins when conditions prior to the rut are favorable. The tahr with their single births and fairly low-altitude habitats, do not fit this pattern, but the environment of the Himalayan and South Indian species is lush, not arid. Further exceptions are seen among American sheep in the southern part of their range where they inhabit deserts yet produce single young, and among most argalis, which may have multiple births in an alpine habitat. However, average lifespan may have an influence on the number of young born, in that long-lived species generally have fewer young. Ibex, bharal, and American sheep survive on the average longer than do argalis (see below). Furthermore, predation pressure may be more persistent on argalis, which inhabit open terrain, than on those species that can withdraw to the safety of cliffs. Thus, the incidence of multiple births can be attributed to several selection pressures.

MORTALITY

It is often difficult to discover why an animal has died. Usually one finds some bones or a set of horns and can then only conjecture which of several possible factors—disease, accident, predation, malnutrition, or a combination of these—caused the animal's death. For example, most travelers to Marco Polo sheep country comment on the large number of horns lying about. "At one spot before reaching the Alichur Pamir, I counted seventy *Ovis poli* horns within a quarter of a mile," wrote Younghusband (1896); Gordon (1876), Dunmore (1893) and others noted similar concentrations. Miller (1913) attributes such concentrations to wolf kills in the Altai: "We came across a great quantity of derelect horns that day; in one small valley below a cliff I counted fifty in about half a mile.... At that season [winter] packs of wolves are continually harrying the sheep. A herd, in its mad rush for safety gets caught in the drift; the females and young rams, unencumbered with 40 lb. weight of horns, make good their escape, while the old rams get stuck fast and are killed." Cobbold (1900) implicated disease as cause of death. Traveling past Karakul Lake in the Pamirs, he found that "there had formerly been vast numbers of *Poli*, but that a disease had set in among them during the past winter and thousands had died." I obtained some information on death rates in my study populations and occasionally discovered clues to the fate of animals, but my comments on causes of mortality remain somewhat conjectural.

Death rates

Subadults. So many factors have an influence on the survival of young that only one season of population data could lead to wrong conclusions. For example, Pfeffer (1967) noted that one of his mouflon age classes contained exceptionally few animals, and he attributed this to a forest fire that swept through his study area. Table 13 shows subadult mortality in several species. Kashmir markhor reveal consistent death rates in that there were about 50 to 60% fewer yearlings than young, indicating considerable mortality between the ages of 7 and 20 months. By contrast, wild goat have an erratic pattern. An estimated 80% of the adult females were pregnant in March 1973. At least half of the young died at birth or within a few days and by October half of the survivors had also vanished, leaving 19 young to 100 females. By the spring of 1974 there were still 21 young per 100 females, the slight difference from the October figure being a sampling error. The death rate of these animals remained low and the October 1974 count still revealed 21 yearlings to 100 females. The survival pattern of young was somewhat different in 1974. About half of the young again died shortly after birth, but the rest thrived so that twice as many reached the age of 7 months as had the previous year. In a dense population of feral goats Williams and Rudge (1969) calculated a prenatal mortality of 33% and a postnatal one of 34 to 57% during the first 6 months of life.

Punjab urial have a fairly consistent lamb survival rate. About one lamb per adult female is born but a third of these die within a month or so. Table 13 does not give female-young ratios for April because many young still remained hidden at that time. The figure of 57 young per 100 females for July 1973 is lower than it would have been if the Nawab of Kalabagh had not captured 25 young for a zoo in April that year. The ratio of yearlings to 100 females was between 30 and 62 depending on the season, 1973 representing a year with particularly good survival. Between July 1972, and July 1973, when 3 to 15 months old, only about 5% of the subadults died, but between July 1973 and April 1974 about 25% of them died and a further 30% had vanished by October.

Assuming at least a 90% pregnancy rate in Nilgiri tahr, fewer than 25% of the young died between birth and the age of nearly 1 year, but then around a third of the survivors disappeared before the age of 2 years. The Himalayan tahr showed a similar pattern. Death rates in Asiatic ibex were variable, depending on the population, with young-to-female ratios of from 64 to 80 in winter and early spring, and yearling-to-female ratios of from 35 to 64. Two bharal populations revealed quite different statistics: most Lapche young survived through the yearling stage (82–88%), whereas the young-female ratio at Shey was low (40%) and the yearling-female ratio even lower (29%).

Many wild goat, Punjab urial, and probably Shey bharal young died soon after birth. A number of these fatalities can probably be traced to undernourished females giving birth to small and weak young and then lacking enough milk to support them. The only disastrous wild goat reproductive season followed a severe drought. The young I saw were tiny and lacked vigor; one new-born was found dead. Conditions were not as extreme at Kalabagh. Several lambs seemed lethargic and one that weighed only 0.9 kg was near death. The high death rate of Kashmir markhor young during their first winter, rather than shortly after birth, suggests that in this population the regulating mechanisms work directly on subadults, in this instance through the stresses imposed by snow, cold, and lack of good forage. A similar situation was noted among chamois in Bavaria by Knaus and Schröder (1975). Of 360 chamois found dead after a hard winter, nearly half were young and over a quarter were yearlings. By contrast, death rates in young urial and wild goat were to a large extent predetermined by the nutritional state of the females. One would thus expect the young of high-altitude species to be more susceptible to the direct means of control than those living in low-lying arid environments where the food supply is unpredictable.

Adults. The number of yearlings entering the adult class of a stable population roughly equals the number of adults disappearing through death and emigration. Several study populations were thought to be relatively stable at the time of my visit. The yearling percentages in these were: Baltoro ibex, 11%; Karchat wild goat, 8–20% depending on the year; Shey bharal, 10%; tahr in the Nilgiris, 19%; and Kalabagh urial, 11–18% depending on the year. Thus, for example, about 20 adult bharal out of a population of 200 at Shey would be expected to vanish per year if reproduction and survival rates of young remain fairly constant. Several study populations were probably increasing, and in these the yearling percentages exceeded the death rate of adults: Besti ibex, 14%; Chitral Gol markhor, 13–16% depending on the year; and Lapche bharal, 21%.

Since females are difficult to age and their horns are not often recovered after death, my comments on mortality rates are based on males. Horns can be aged accurately by the number of rings or segments, one of which is deposited each winter after the first one (Caughley, 1965; Geist, 1971a). Table 10 lists the age at death of males in two categories: one consists of trophies from homes, mosques, and so forth, and the other of horns picked up in the field, the animals having died of natural causes. Although the former reflects the choice of hunters for large-horned individuals, the data are useful in that they reveal the maximum ages reached by individuals.

Eurasian sheep have short life spans: only one urial and one Marco Polo sheep in my samples reached an age greater than 9 years. Similarly, of 87 urial skulls picked up in the Kopet Dagh none were older than 10 years (Decker and Kowalski, 1972), and of 187 urial males shot in Turkmenia, 50% were younger than 4 years, 28% were 4 years old, 15% were 5 years, 6% were 6 years, and 1% was 7 years (Heptner et al., 1966). Few mouflon live longer than 9 years in the wild (Pfeffer, 1967), although they may survive up to 19 years in captivity (Crandall, 1964). Soay sheep seldom exceed 10 years of age (Grubb, 1974c). Of 200 argalis in the Pamirs, 14% were 2 to 3 years old, 40% were 4 to 5 years, 26% were 6 to 7 years, 17% were 8 to 9 years, and only 3% were greater than 9 years (Heptner et al., 1966). Urials and argalis, irrespective of habitat, tend to live fewer years than American sheep, which may reach an age of 15 and even 20 years in the wild (Geist, 1971a). *Capra, Hemitragus,* and *Pseudois* may attain an age of 15 years or more too. For example, Nievergelt (1966a) noted an occasional Alpine ibex living to an age of 17 to 18 years and Caughley (1966) reported the same for Himalayan tahr.

The age structure of males varies considerably between populations (table 17). About half of the living Baltoro ibex and Shey bharal were 5 years old and older, whereas only 5% of the Besti ibex and 9% of the Lapche bharal reached that age. The horns of males found dead at Baltoro and Shey reflect the high proportion of elderly animals in these populations (table 16). A comparison of the percentages of living males in each class with those found dead in the field reveals that disproportionately few animals (except Marco Polo sheep) die between the ages of 1 to 4 years, or as yearlings and young adults. Most deaths are among prime animals, 4 to 10 years old, a trend especially noticeable in ibex and bharal.

Causes of death

My observations on causes of death consist mainly of dramatic incidents such as predation, whereas most animals probably die subtly, the victims of poor nutrition and disease.

Disease. Only an occasional animal in my study populations seemed to be in poor condition. One Karchat wild goat was emaciated. Both a young and an adult female markhor died of unknown causes in Chitral, and a bharal female did so at Shey. Zugmayer (1908) met a blind bharal female accompanied by a blind young in Tibet. Urial at Kalabagh are said to die occasionally when after a drought they gorge on new green grass. Asiatic ibex in Russia die periodically of scabies, Marco Polo sheep succumb to anthrax (Heptner et al., 1966), and rinderpest and hoof-and-mouth disease probably affect wild ruminants just as they do domestic ones.

I collected fecal samples and had these checked for parasite ova. Markhor, urial, and wild and domestic goats are often infected with intestinal round worms such as *Trichuris* and *Trichocephalus* (table 18). The last-named feeds on blood and may therefore be debilitating to an animal. The coccidian *Eimeria* was found in the mucosa of the intestinal tract of most species, and it also occurs in Asiatic ibex (Heptner et al., 1966). Heavy infestations of this parasite have been known to kill young domestic animals in large numbers (Smith, 1954). Many Punjab urial are infected with lungworm (*Cystocaulus*), a type of parasite which is known to affect bighorn sheep seriously (Buechner, 1960). Wild goat had few parasites possibly because these are not readily transmitted in rocky and dry terrain. Domestic and wild caprids tend to share the same parasites in an area. Lungworm at Kalabagh was an exception to this, probably because veterinarians there treat the domestic stock.

Hoogstraal et al. (1970) collected the tick *Dermacenter everestianus* from a bharal in Nepal. I picked several *Hyalomma marginatum isaaci* ticks off Kalabagh urial. Chiru commonly carry insect larvae beneath the skin (Adams, 1858; Rawling, 1905). Punjab urial were often bothered by a warble-type fly, probably *Oestrus ovis*, which lays eggs in the nostrils of animals. Sometimes urial dashed along singly or in herds only to halt suddenly with their heads lowered in patches of tall grass. They also sought cover and cowered motionless as if trying to hide from flies. This interrupted their feeding, and during the headlong flight an occasional lamb became briefly separated from its mother. Karchat wild goat behaved similarly at times in that they ran and shook their heads vigorously. Ogren (1965) also observed such behavior in aoudad.

Accidents. Goats are probably much prone to accidents, living as they do on cliffs where a misstep or rockfall may plunge them to their death. In New Zealand, C. Challies (pers. comm.) once watched three tahr slip on a steep icy slope and severely injure themselves, and I observed a male with a broken hindleg in the same country. Two Karchat wild goat females limped, as did an urial ram. Rutting bharal males took such long leaps down cliffs that it seemed only luck prevented them from being injured. Avalanches are a major cause of death among ibex. Of 19 ibex remains found near the Baltoro glacier, 5 animals had definitely died in avalanches and several others had probably done so. Up to 10% of the ibex in the western Pamir may die in avalanches during a snowy winter, and wild sheep may also succumb in this manner (Heptner et al., 1966). Chamois, Kuban ibex, and Dagestan tur are often swept away by avalanches in the Caucasus (Nasimovich, 1955). A markhor female died when her hoof became wedged in the fork of a tree 2.5 m above ground (Kennion, 1910). Muskox show a propensity for falling off cliffs (Tener, 1965; P. Lent, pers. comm.). And saiga males fight so hard during the rut that they occasionally kill each other (Bannikov et al., 1967).

Predation. Predators vary in their impact from area to area but in general they are now so scarce that they have little effect on most prey populations. The last leopards at Kalabagh were shot during the 1960s and wolves and caracal have also vanished from there, leaving only a few foxes and jackals. Cheetah are known to hunt urial in Iran (F. Harrington, pers. comm.) but this cat may have followed the lion and tiger into extinction in Pakistan. Striped hyenas are widespread though uncommon and they might kill an occasional caprid young. A markhor in Chitral was seen to hold a lynx at bay with lowered horns (S. A. Khan, pers. comm.), but lynx are rare in the mountains. Temminck's cat—of which I saw one in the Kang Chu—could prey on goral and young tahr. Nilgiri tahr in the High Range were mainly exposed to leopard and jackal after tiger and wild dog had been reduced to infrequent transients. Birds of prey may kill newborns. A lämmergeier is said to have carried off a young goral (Baldwin, 1876), two lämmergeier attacked a young ibex until it fell off a cliff (Stockley, 1928), and golden eagles have been implicated as predators of Tibetan argali lambs (Kinloch, 1892) and young ibex (Heptner et al., 1966). However, predation by these and other species is incidental. Only foxes, of which *V. vulpes* is most common, Asiatic black bear, brown bear, leopard, snow leopard, and wolf could have significant local impacts in my study areas. Fox and jackal are potential predators only on young. Bears, though capable of killing scavenging (table 21). This leaves leopard, snow leopard, and wolf as the main predators. Predation by these species will be discussed in detail in Chapter 6 but a few comments are relevant here. Snow leopard and wolves killed bharal at Shey (tables 20 and 22). However, an analysis of bharal population structure indicated that nutrition rather than predation had the greatest impact on numbers. Both of these predators also frequented the Chitral Gol, yet the markhor population increased in spite of this. Livestock and marmots acted as important buffers in both areas by providing alternative prey. A disproportionately large number of males also appear to be killed by these predators, but polygamous species have an excess of males whose loss has little effect on population levels.

Conditions were different in the Karchat Hills where at least 2 resident adult leopards obtained about three-fourths of their food from wild goat. Assuming that a leopard has to kill an average of about 1000 kg per year (Schaller, 1972b), that 750 kg of this represents wild goat, and that the average wild goat weighs 25 kg, then 30 wild goats, subadult and adult, are taken annually by each leopard out of a total goat population of 400 to 500. This estimated kill level of 13% may be significant enough in a drought year to prevent the goat population from increasing and it may even cause a decrease. In a year with good reproduction such predation would only dampen the rate of increase.

Nutrition. Earlier I discussed the critical role of nutrition on the development and survival of young and on fecundity of females. The effects of nutrition actually pervade an animal's whole society. Populations of low density on good habitats are usually increasing, and they contain many young animals and few old ones, whereas stagnant populations on overgrazed range reproduce poorly but animals live a long time. Figure 24 illustrates the age structure of an increasing Himalayan tahr population in New Zealand, based on 125 females shot by Caughley (1971), and a stable Cretan wild goat population on an overgrazed island, based on a total count by Papageorgiou (1974). The age structure of several of my study populations reveals something about population quality and by inference about habitat conditions.

The bharal well illustrate the difference between an increasing and a static population. At Lapche, 9 out of 10 adult females had

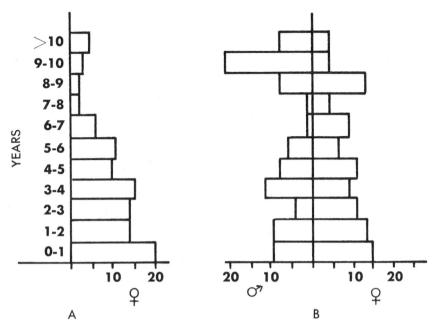

Fig. 24. Age structure (A) of an increasing female Himalayan tahr population in
New Zealand (Caughley, 1970c), and (B) a stable wild goat population
(Papageorgiou, 1974).

large prey, usually confine themselves to vegetal matter and to
offspring at heel and there were almost as many yearlings as
young, indicating excellent reproduction and survival. Over two-
thirds of the males were 1 to 4 years old, and only a few (9%) were
in the oldest age classes. By contrast, 4 out of 10 females at Shey
were with young, the yearling percentage was less than half that at
Lapche, and only a quarter of the males were aged 1 to 4 years,
most (59%) being in the oldest classes. The Lapche population was
small, hunted occasionally by man, and it lived on a mountain with
many luxuriant grassy ledges which even in winter provided
much forage. The bharal at Shey crowded into a small, overgrazed winter range where they were protected by religious sen-
timent. The ibex figures show a similar though less clear pattern.
As indicated by the number of young, fecundity was about the
same in the Besti and Baltoro populations, but survival to the
yearling stage was poorer in the latter area (table 13). Thereafter
the Baltoro animals lived a long time, nearly half of the males

being in the oldest classes. Relatively few males at Besti reached an age greater than 5 years, but trophy hunting contributed to their deaths. These ibex inhabited large alpine meadows whose condition was good except near villages, whereas the Baltoro animals occupied frugal habitat at the altitudinal limit of their range. The markhor in the Chitral Gol seemed to maintain a high plane of nutrition, at least in summer when most foraged at high altitudes in areas often inaccessible to livestock. The twinning rate was high, higher in the Chitral Gol than at Tushi, where the goats remained on the bleak lower slopes all year.

The Karchat wild goat population was probably stable during my study, although recent prohibitions on hunting and livestock grazing and at least one good rainy season may provide a stimulus for increase. Many newborns were weak and died, twins were rare, and yearlings seldom bred, all indications that the range was nutritionally defective. With the drought conditions during the late 1960s and 1970s the population had grown aged, nearly half of the males being over 5 years old. Similarly, Papageorgiou (1974) found that on an overgrazed island only 16 out of 29 female wild goat, aged 2 to 9 years, gave birth and that there were many old individuals (fig. 24). The population was stable, with a mortality rate of 15.7% and a replacement rate of 14.4%. Censuses at Kalabagh indicated a relatively stable population. The twinning rate was poor (less than 10%), and a moderate number of males were in the oldest classes, all symptoms of rather poor habitat conditions.

The fact that animals in a population on a good habitat usually die younger than do those living on a poor one was reported for Alpine ibex (Nievergelt, 1966a) and mountain sheep (Geist, 1971a). Some of my data suggest a similar survival pattern. The seemingly paradoxical nature of this fact needs an explanation. Before entering a physiologically demanding period, animals often store fat reserves. The winter is the most stressful time of year in cold climates, the dry season in arid ones. Caughley (1970d) found that among Himalayan tahr "mean fat indices for all age groups reach a peak in mid-winter followed by a decline that arrests and reverses over the summer months." Adult males reached a peak a month earlier than females, apparently a hormonally controlled means of preparing them for the rigors of the rut. In the Serengeti National Park, pregnant wildebeest deposit much fat before giving birth and then lose it during lactation (Sinclair, 1974). I have no information on fat levels of species in my study areas, except that two urial rams were fat in October and one was lean in May, but I assume that the animals followed a pattern similar to that of Himalayan tahr. Animals which expend energy on matters other than maintenance may find their fat reserves exhausted before the plants are again nutritious enough to support them, and they may then die. Not all animals enter a stressful period with the same amount of fat. Adult males, for example, feed less and are more active during the rut than the females, with the result that they may face the winter with depleted reserves. Feral Soay sheep rams lost 25 to 35% of their weight in winter and ewes only 20 to 25% (Grubb, 1974c). Poor condition predisposes animals to disease and predation. Young animals who need to invest much energy in body growth also enter demanding periods with low fat reserves. Geist (1971a) noted that in American sheep populations on good range the rams begin to rut at an earlier age and interact more frequently and more intensely than do rams on poor range. A higher level of activity raises the cost of living, and thus the probability that the animal will expose itself to malnutrition. Early maturity also means that the ram reaches an advanced physiological age earlier than a slowly growing individual. Thus males in an expanding population on good habitat would be expected to have a short life. This is true also for females but for a different reason. Females, in contrast to males, dissipate little energy in social interactions, but their body reserves are drained first by the fetus and then by lactation. A female who has failed to conceive or whose young died at birth has a better chance of surviving stressful seasons to live a long life than a reproductively successful one.

6 NOTES ON THE HABITS OF SEVERAL PREDATORS

This chapter outlines the status of the major predators in my study areas, gives some information on their habits, especially on their food habits, and relates some of my encounters with wolves and snow leopard.

CANIDS

Fox and Jackal

The red fox is found over much of Pakistan and northern Nepal even in the high mountains, where it is quite common at extreme altitudes. For example, I saw fox spoor in winter at 5100 m above Khunjerab Pass and at 5300 m in the Kanjiroba Range. Driving one night along the upper Indus River between Besham and Sazin, I spotted 4 foxes and a leopard cat. A pelt count in the Chitral bazaar in January, 1974, gave some idea of the relative abundance of furbearers: 28 red fox, 3 wolf, 2 stone marten, 1 snow leopard, 1 leopard cat, and 1 jackal. Fox tracks are seen in most valleys, around villages, along glaciers, almost everywhere within the limits of vegetation, except that the animals seem to avoid rolling uplands. For example, they seldom came around Shey even though such rodents as *Alticola stoliczkanus* and *Pitymys irene* were abundant there. Perhaps open terrain exposes foxes to wolf and human predation.

The fox has a varied diet as table 19 indicates. Chitral foxes ate much vegetal matter and fruit (in 88% of the droppings) during winter. Rodents were the staple food at all seasons, at least 42% of the feces in every sample including rats and mice. I made no attempt to identify the species of rodents in the droppings but *Apodemus flavicollis, Mus musculus,* and *Rattus turkestanicus* were present in the Chitral Gol. Ungulate remains were found in up to 24% of the droppings, and these probably represented for the

138

most part meat scavenged from such sources as predator kills and avalanche victims. Feathers and eggshells, mostly of snowcock, were common only in the spring sample from the Baltoro area. I found a site where a Chitral fox had just killed a blue rock pigeon with 25 acorns in its crop.

In the Sang Khola, west of Mount Dhaulagiri, I observed a fox hunt for 29 minutes on an alpine meadow. During this period it tried to catch something 8 times either by lunging or pouncing. Three times it missed, twice it grabbed a vole, which after a few shakes it swallowed, once it snatched up a lizard, and twice it devoured small unidentified items, possibly grasshoppers.

Jackal were uncommon in my study areas except in the High Range of India. There an analysis of droppings showed a diet consisting mainly of rodents (94%), lizards and snakes (29%), and miscellaneous items such as land crabs and snails (table 19). Visiting the same area at a different season, in April, Daniel (1971) noted many beetle remains in the droppings.

Wolf

Wolves occur over most of the Himalayan uplands, avoiding only dense forests and steep gorges. Animals smaller and lighter in build than the northern ones are found in the arid lowlands, especially on the Baluchistan Plateau, but I saw no such desert wolves in Pakistan. According to villagers, they usually travel alone or in twos or threes and prey on domestic stock and whatever else they can catch, including fox and jackal. Wolf tracks are quite commonly found in the mountains of northern Pakistan, the animals roaming alone and in small groups up and down valleys and other routes used by man. However, being shy, sparse, and often nocturnal, wolves are seldom seen. I observed two in the Chitral Gol and another two as they followed an ibex trail at 4800 m near Kilik Pass. With their propensity to wander, wolves pass through areas only at long intervals. I spent 66 winter days in the Chitral Gol, and wolves visited on three.

One pack passed through Shey several times. Five wolves came near the village on November 2, large beasts with luxuriant coats, one almost white, another a silvery gray, and 3 grayish with tones of muted gold. The following day I met 4 of the 5 again, and spoor suggested that one pack member had become solitary. The same pack of 4 was seen for the last time on December 5. Wolves, singly and in groups, visited Shey on 6 of my 35 days there, as determined by sightings and tracks. (A track of average size was 4.8 cm in pad width and 8.2 cm in paw length.) This pack also traveled eastward across a 5150 m pass, moving along human foot-trails and using the stone Buddhist prayer walls as signposts around which to urinate and defecate.

Ibex and livestock are the main foods of wolves in northern Pakistan, and bharal and livestock are the principal prey around Shey (table 20). Several wolves killed a markhor at Tushi in February 1974, a prey species which failed to appear in my sample of droppings. Marmots constitute an important food supplement in summer. Stockley (1928) twice watched as a wolf stalked marmots and made futile rushes. At Shey the marmots began their hibernation around mid-October after the first heavy snow, and at the same altitude in Baltistan they had just ventured above ground in mid-May, having been unavailable to predators for some 7 months. Heptner et al. (1966) reported that in the western Pamir 11 wolf-killed ibex were found along 14 km of valley. All were males about 6 to 11 years old. Argali rams in the western Pamir also fall prey to wolves more readily than do ewes. In a sample of 61 rams, 7% were 2 to 3 years old, 21% were 4 to 5 years, 39% were 6 to 7 years, 28% were 8 to 9 years, and 5% were over 9 years old (Heptner et al., 1966). Argalis are much handicapped by deep snow when trying to escape from wolves. Stockley (1928) noted that during the winter of 1910–11, when snow depths exceeded 46 cm in Ladak, the wolves found it easy to capture argalis and Tibetan gazelle.

Wolves are cordially disliked by herdsmen for their habit of killing goats, sheep, cattle, horses, and even yak. Around Shey an average family has 30 to 40 sheep and goats, of which it expects to lose 3 or 4 per year to wolves; the village of Zhuil in northeastern Chitral maintains about 250 sheep and goats, of which an average of 15 are killed each year. Attacking a herd unexpectedly, wolves slay an animal and bolt their food so rapidly that the shepherd often has no time to prevent the loss. Stockley (1928) witnessed such an attack: "He then spurted suddenly in the most amazing manner into the middle of the flock and pulled down sheep after sheep with such wonderful speed and dexterity that there were five lying on the ground within a distance of thirty yards.... He came up on the right side of each sheep ... and, seizing the galloping sheep behind the right ear, jerked its head downward and inwards so that it pitched on its nose."

Wolves killed several tethered goats of mine, all but one of them at night. One had each side of its throat ripped open, exposing the neck muscles. Although the two wolves were driven off before they could eat, the goat nevertheless died in spite of its trivial wounds. Another goat had its throat deeply lacerated and it had been disemboweled. The two wolves abandoned the carcass after eating less than a kilogram. A third goat, also attacked around the neck, had its larynx slashed open and a jugular vein cut.

Two instances of wolves hunting wild prey were observed. On January 30, at 1440 hours, I watched 9 markhor in the Chitral Gol foraging in an open stand of oak. Suddenly all animals except a young bolted toward a nearby cliff, as two wolves, seemingly a male and female, swept side by side along the slope. Each wolf concentrated on one markhor, but the female gave up the chase after 130 m when her intended victim reached the cliff. The male continued his pursuit onto the cliff and at one point was within 3 m of a markhor when one of his feet slipped on a snow-covered ledge and he halted. Neither wolf saw the young markhor bounding belatedly toward the cliff through deep snow. The two predators joined, angled up a trail leading above the cliff, and passed out of sight over a ridge. When I left at 1520, the markhor still stood on the cliff, except for one 2½-year-old male who walked alone uphill across the route taken by the wolves.

Four female and two young bharal foraged at 0805 hours on a slope covered with low shrubs. Suddenly they startled and bunched up. Running downhill were two wolves, the speed of their attack hampered by stunted junipers over which they had to bound. The bharal fled, at first slowly and then when the wolves were within 40 m at full speed. Three bharal veered to one side and the predators continued their pursuit of the others. Running parallel and about 50 m apart, the wolves seemingly tried to overtake their prey before it could reach the cliffs below. They almost succeeded with one lagging female. However, she angled sharply downhill and escaped, with the nearest wolf 20 m from her. After a chase of about 300 m the wolves halted, then trotted uphill where they joined two other wolves who had not participated in the hunt. All four moved to a nearby grassy spot where they rolled on their backs and milled around until at 0845 they moved from sight. Six hours later the bharal still remained on and near the cliffs.

BEARS

Asiatic black bear

The Asiatic black bear has a wide distribution, from Russia and Korea to Indochina, and in the forests of the Himalaya below an altitude of 3750 m westward as far as Afghanistan and Iran. The bears in the Himalaya (*Selenarctos thibetanus laniger*) are thought to be subspecifically distinct from the so-called Baluchistan black bear (*S. t. gedrosianus*). I saw one of the latter bears in a Tehran zoo and was unable to distinguish it from the Himalayan form. The Baluchistan bear is said to occur south of the Kabul River, but it now survives only in isolated relic populations. Recent sightings have come from the Safed Koh, the Takht-i-Sulaiman massif, the vicinity of Ziarat, the Tor Khan Range in Sibi district, some hills southeast of Khuzdar, and possibly the Shodar Hills in Waziristan. Black bear also persist in southeastern Iran. The forests in these areas consist of pine, juniper, or pistacia, and also of sparse acacia woodlands. My observations on black bear were confined to the Dachigam Sanctuary in Kashmir where the species was abundant, and to the Chitral Gol where at most two animals resided intermittently.

Black bear males generally weigh 100 to 150 kg, with some close to 200 kg, and females weigh about 50 to 100 kg (Bromlei, 1973). Carnivores of such size are potential predators on even the largest ungulates, and indeed Prater (1965) and others note the occasional predilection of bears for livestock. J. McNeely told me of a bear in eastern Nepal who preyed persistently on domestic cattle and buffalo. In October 1968, I met bears on 17 occasions in the Dachigam Sanctuary—including 7 different bears in one morning. All were at altitudes of from 1750 to 2000 m. Sixteen of these bears were feeding, 13 on the fruits of *Celtis*, 2 on walnuts, and 1 on acorns. Bears at that season ate mostly *Celtis* berries (in 40% of the droppings) and walnuts (in 33%), and they supplemented their diet with acorns, grape, apple, maize, and apricot fruits (table 21). In November 1969, the bears ate rosehips almost exclusively. Such dietary changes continue throughout the year. Maize and mulberry are eaten in July and August, according to the sanctuary staff, and in May and June the bulk of the diet consists of grass (Schaller, 1969b). The bears eat little animal matter in Dachigam. Droppings around a winter den in the Chitral Gol showed that grass was a major food, probably in early spring. Markhor was also well represented (table 21). Black bear in Russia have a diet similar to those in the Himalaya in that 85% of their food consists of plants, principally acorns, cedar nuts, cow parsnip stems, sedge, and so forth (Bromlei, 1973), and that they also eat whatever meat they can obtain including such items as voles, mollusks, and fish (Stroganov, 1969).

Schaller (1969b) describes feeding behavior of black bear in Dachigam:

> Black bears obtained much of their fruit by climbing into trees, sometimes into the upper branches 10 or more metres above ground. After choosing a horizontal branch or a fork on which to lie, sit, squat, or stand, they reached out with their forepaws and pulled fruit-bearing twigs toward them. Small twigs simply were hooked with the long, curved claws of one paw and broken inward, but large branches required more effort. One bear bit repeatedly into the base of a branch, then bent it inward with a paw until it snapped and the fruit could be reached. Another bear broke a branch by pulling with both paws and then with its mouth as well. Afterward it bent twig after twig toward its face and plucked the *Celtis* berries with the lips, the usual method of detaching fruit. In one instance, a bear broke several branches in a walnut tree, but most nuts fell to the ground. The animal descended, ate the fallen fruit, then climbed back up and bent in several more branches. Virtually all walnut, oak, and *Celtis* trees had several broken branches, attesting to the heavy use of these species by bears in the sanctuary.

> Although Prater (1965) stated that black bears are primarily nocturnal, those in the Dachigam Sanctuary were frequently active during the day. Bears were seen feeding 4 times between 0600 and 0900 hours, twice between 0900 and 1200 hours, 4 times between 1200 and 1500 hours, and 6 times between 1500 and 1800 hours.

> On one occasion a feeding bear broke several branches inward while standing in the fork of a *Celtis* tree. Each discarded branch was pushed and trampled into the fork, forming a crude platform which resembled the sleeping nests of the great apes and the Malayan bear (*Helarctos malayanus*) as reported by Schaller (1964). After the bear finished eating, it rested on this platform in the morning sun. Novikov (1962) noted that a feeding bear "drags down and breaks

numerous branches, which give rise to the assumption that the black
bear builds special 'arbors' for resting."

Bromlei (1973) found that "in the course of eating, a lot of
branches gradually accumulate under the bear in the form of a
clumsily constructed 'nest'." Both statements suggest that nest-
building is an inadvertent concomitant of feeding rather than an
activity in itself.

Black bear in Russia become dormant for 4 to 5 months beginning in November (Stroganov, 1969). Stebbing (1912) stated that
bears remain active all winter in the Himalaya, but my informants
insisted that the animals enter dens for at least some time between
December and April depending on the severity of the winter. I
have followed bear tracks as late as January 15 through .3 m or
more of snow in the Chitral Gol, the animal meandering from oak
to oak apparently in search of acorns. One den was found in a
limestone cliff at 2600 m. A cavern, over 20 m long and 10 m
high, gave way to a tunnel some 10 m long and just high enough
for a person to squirm along it. This tunnel opened into another
chamber about 5 m long and 10 m high where shed hair and other
sign indicated that bears had spent some time. There was no nest
material. Almost all droppings were in the outer chamber.

Brown bear
The brown bear is found widely over Eurasia and northern North
America. Its southern limit in Asia is in the Himalaya where it is
largely confined to rolling uplands and alpine meadows above
timberline, ecologically separated from the forest-dwelling black
bear. The subspecies *Ursus arctos isabellinus* has a sporadic distribution in the Pakistan and Indian Himalaya eastward to near the
western border of Nepal. The subspecies *U. a. pruinosus* is found
in Tibet, and this animal may be responsible for the rumored
records of brown bear in Nepal and Bhutan (Gee, 1967b). The
brown bear was once abundant in the Himalaya, with for instance
Kinloch (1892) seeing 28 in one day and shooting 7 of these. The
only bears I saw in Pakistan were dancing in the streets of large
cities (plate 52). Herdsmen in the mountains shoot the adults and
sell the cubs to itinerant entertainers. This, together with the fact
that much of the terrain is too rocky and barren to support bears,
has made the species rare in that country. The bear is virtually

extinct in Chitral: one animal was reported in the northwestern part in 1973, and I found old sign near the headwaters of the Yarkhun. Fresh spoor of one bear was discovered near Karambar Lake in Gilgit Agency. Villagers in Hunza told me that a bear or two may visit the Kilik area, and at Khunjerab Pass I saw several holes dug by bear in pursuit of marmot. One bear, judging by fresh tracks, roamed the Braldo Valley between the Biafo and Baltoro glaciers in May 1975. These were my only sightings of bear spoor in eleven months in northern Pakistan. The last stronghold of the species in the country may be the Deosai Plains southeast of Skardu.

Brown bear in the Baltoro area subsisted mostly on grass and various roots including those of *Artemisia* (table 21). They ate ibex too, no doubt scavenging on avalanche victims. Of 6 droppings found around Karambar Lake, 5 contained grass, 2 roots, 1 a fly, and 1 several bees. At one spot 3 vole nests had been excavated. The omnivorous habits of these bears have received comment from many authors. Darrah (1898) observed bears "cropping grass like cow or goat," and Hedin (1903) found a marmot and "several herbs" in the stomach of a Tibetan bear. Bromlei (1973) noted that two-thirds of the diet of Russian brown bear consists of plant food, particularly berries, acorns, and *Pinus koraiensis* seeds. Ants and other insects are commonly found in the droppings, as are fish and the remains of muskdeer and pig. Brown bear remain dormant during winter from about December to April or May.

CATS

Leopard

Stretching from Russian Manchuria and China to central and western Asia and Africa, the range of the leopard includes Pakistan where until recent years this animal was widespread, even extending into the mountains north of the Indus Plain. There it confined itself to forests and the major valleys. Leopard and snow leopard were sympatric in southern Chitral, at least in winter (Asad-ur-Rehman, pers. comm.), but the former has been exterminated there. Leopard still occur on many of the ranges on the Baluchistan Plateau. I saw droppings on the Zarghun, Surghund, and Koh-i-Maran massifs, but my only frequent contact with spoor was in the Karchat Hills. Probably two leopards roamed through the southern half of the range, often using certain routes along the crests of ridges and human foot-paths. They left sign of their presence in the form of droppings and scrapes, the latter made by raking the hindpaws on a spot of sand or gravel. Several scrapes may be found together at conspicuous places such as a bend in the trail or on a promontory. Of 42 scrapes examined, 17 also had some feces on or beside the scrape, a dual means of marking also found in tiger. Interestingly, I never saw a scrape marked with feces in the Serengeti National Park (Schaller, 1972b). The leopard's main food was wild goat (70%) and livestock (14%). The Karchat hills have been closed to livestock grazing since 1970, making this food source almost unavailable now. In general, leopards eat whatever they can catch, and they may even include berries in their diet (table 22).

Leopard occur in most forests of Nepal, penetrating the Himalaya to within a few kilometers of snow leopard country along such gorges as the Kang Chu. There, at an altitude of 2600 m, I watched a leopard stalk a goral unsuccessfully. The goral stood on a low cliff, snorting and stamping a foreleg as it stared at some shrubs below. After 5 minutes it bolted, and 10 m away a leopard rose and swiftly walked away. One leopard-dropping in that area contained langur hair.

Snow leopard

In 1572 the College of Arms in England granted Thomas Fludd of Barsted a patent for a heraldic shield with what appears to be the head of a snow leopard as a crest (P. Till, pers. comm.). Yet the snow leopard remained unknown to the scientific world until Georges Buffon pictured the "L'once" in 1761 in his *Histoire Naturelle*—and mistook it for a cheetah when he stated that it occurs in Persia and is used for hunting. Although in 1779 the naturalist Peter Pallas accurately defined the snow leopard's range in the Altai and carefully distinguished it from the leopard (see Guggisberg, 1975), the cat remained a creature of mystery. Rare, shy, and withdrawn into one of the most remote regions on earth, it has been encountered by few outsiders. I ventured into the mountains with the hope of studying snow leopard but my attempts failed, as almost perversely the animals eluded my efforts to observe them. In my chosen study area around the Chitral Gol most snow leopard were shot before I could begin intensive work.

The following pages present my own notes and some published observations to indicate how little has so far been revealed about the habits of this lovely cat.

Status and distribution. The snow leopard inhabits the mountains of Tibet and the Himalayan chain westward through Pakistan into Afghanistan and northeastward over the Pamir, Tien Shan, and Altai ranges to the Sayan Mountains near Lake Baikal. The animal may be found in conifer and oak forests as well as above timberline. In the Dzhungarian Ala-Tau it has been reported at 900 m, and in Chitral and Kashmir it may descend to 1500 m, although it usually remains above 3000 m with some sightings having been made as high as 5600 m.

Snow leopard density in the Himalaya was fairly low even fifty or more years ago, judging by how seldom hunters encountered them. Of course man's attitude toward predators has not been one to promote contact. Ward (1923) felt that snow leopard "should be classed as vermin" and indeed the cat was so treated legally in Kashmir until 1968. India, Pakistan, and Nepal now give snow leopard official protection, but residents continue to kill them, sometimes in defense of livestock and at other times merely because an animal presents a suitable target. The skins often find their way to the fur shops. For example, I counted 14 skins in one Srinagar shop in 1968, and 18 in various shops in Lahore during 1972.

The snow leopard has a wide distribution in northern Pakistan. Definite reports were obtained from northern Dir, from most of Chitral, and from such places in the Gilgit Agency as the Yasin Valley, the vicinity of Gupis, the Kargah Valley, the slopes above Bunji, the Nanga Parbat massif, and the Haramosh Range. The cats also occur throughout Hunza, especially in the Khunjerab Valley and north of Misgar where near the Kilik Pass I saw the tracks of two animals. Further east, animals persist in the Baltistan area and there I found the spoor of one animal near the Baltoro Glacier. All villagers agreed that the snow leopard is rare, with usually no more than 2 or 3 animals frequenting a particular valley at intervals. My attempt to census animals was confined to the vicinity of Chitral town, an area which probably had the densest population in Pakistan until the late 1960s. In December 1970, I saw a female with small cub in the Chitral Gol, and tracks

indicated the presence of another animal, most likely an independent subadult. That winter at least one snow leopard was shot
near the reserve. During the winter of 1971–72, Prince Muta-ul-
Mulk shot 2 females at Shogore and a villager killed 3 in Kesu Gol.
Another death brought the known tally that winter to 6. When I
visited the Chitral Gol again during the winter of 1972–73, one
snow leopard passed through the area and a second animal is also
said to have spent a few days there. In mid-January 1973, Prince
Burhan-ud-Din notified me that a snow leopard was at Tushi.
Checking at his home, I found its fresh skin hanging on the ver-
anda: his gamekeeper had just shot the animal, a male. Later that
month I saw a female at Tushi. During a visit to the Chitral Gol in
January 1974, I found no snow leopard spoor and the sanctuary
staff told me that no animals had been there all winter. None
came later that season either, although one animal hunted around
Chitral town for awhile. In February 1974, when snow lay low on
the slopes and wildlife had concentrated in the valleys, I attempted to census snow leopard by searching for spoor and interviewing persons in many villages. After covering an area of about
3000 sq km, I had evidence for 4 or possibly 5 animals (Schaller,
1976b). In four years a viable snow leopard population had been
here reduced to the vanishing point through lack of protection. It
is not coincidental that the decline occurred during a political and
legal interregnum, a time when the semiautonomous state of
Chitral was absorbed into the North-West Frontier Province.
Snow leopard seem to exist in similarly low densities over much of
northern Pakistan. An extrapolation from the Chitral census to the
cat's total range in the country gives a figure of 104–130 animals.
Anon. (1972) placed the number at about 100, but until more precise data are available a rough estimate of fewer than 250 seems
realistic.

   Snow leopard are widely but sparsely distributed along the
higher reaches of the Himalayan chain in India (see Dang, 1967).
In Nepal they occur along the northern fringe of the country
above the evergreen forest belt. I attempted to estimate their
numbers in a block of about 500 sq km centered around Shey and
Phoksumdo Lake in the Kanjiroba Range. No cats were wholly
resident at Shey. During my 35 days there, one animal hunted the
area from November 12 to 16 (without catching anything), and
the tracks of two transient animals were found on the morning of

November 25. I approached Phoksumdo Lake from the east in early December, having walked up the Namgung Valley across the 5350 m Namdo Pass and down the Deokomukh Valley. The whole route was covered with fresh snow. At the northern end of the lake I spotted one snow leopard, and tracks indicated that a second one had vanished unseen. On the slope above the lake was a third set of tracks, that of a large male. Probably at least 6 snow leopard used the 500 sq km area. In March 1972 one large and one or two medium-sized animals patrolled the slopes around Lapche, but they spent most of their time across the border in Tibet. I am unable to give an estimate of the number of snow leopard in Nepal. Dang (1967) felt that "in the region of 400, give or take two hundred" survive in "the Himalayan complex of mountain ranges" but this figure is much too low.

With snow leopards becoming increasingly rare in South Asia, it is fortunate that zoos have improved their techniques of raising animals. The *International Zoo Yearbook* for 1966 reported 34 snow leopards in zoos but of these only two (6%) had been born in captivity, whereas in 1974 there were 127 animals in zoos of which 46 (35%) had been bred in captivity (Kitchener et al., 1975). Snow leopards in zoos also tend to live longer now, with Crandall (1964) reporting a maximum of almost 9 years, Marma and Yunchis (1968) of 11 years, and Kitchener et al. (1975) of 17 to 19 years.

Social relations. I encountered a female with one small cub, a solitary female, and an unsexed medium-sized individual, which, judging by tracks, was accompanied by another animal of similar size. In addition, I came across 29 sets of recent tracks of which 25 were of solitary individuals and the rest of pairs. Among the tracks were 7 large sets, probably those of adult males with forepaws measuring at least 9.5 cm in length and pads at least 7 cm in width. Dang (1967) met 12 solitary snow leopards and 4 pairs, and Ward (1923) saw 5 single individuals and a group of three—and he shot 5 of the 8 animals. Haughton (1913) observed a pair, of which one was a male, and three solitary individuals. Other lone animals were reported by Baldwin (1876), Dunmore (1893), Darrah (1898), Stockley (1928), Kennion (1910), and Ward (1966). Meinertzhagen (1927) watched two animals, and Kusnetzov and Matjushkin (1962) observed a single animal and the tracks of a pair. Elliott (1973) gave second-hand accounts of a pair on an ibex kill and another pair on a bharal kill. Snow leopards obviously travel alone much of the time. When two or three large animals associate they may represent an adult male and female, a female with almost-grown offspring, or independent cubs in a litter which has not yet split. Schaposchnikoff (1956, quoted in Hemmer, 1966) reported on a group of 5 snow leopards, the largest on record. The sexes of sighted or slain animals are seldom reported. In 1972 a villager in Chitral shot a group of 3 snow leopard which consisted of a female with two large cubs, a male and female. The sexes of some other snow leopard, all solitary animals, killed in the Chitral area included 8 males and 4 females. The *International Zoo Yearbook* for 1975 lists 57 males and 70 females in zoos, and the yearbooks published between 1966 and 1975 report 48 male cubs and 37 female cubs among sexed litters born in captivity.

An animal may be solitary but not necessarily asocial. Residents in an area may know each other and associate occasionally, as in the case of tigers (Schaller, 1967). "One fact has been repeatedly noticed, and that is the incidence of pairs working valleys in co-ordination, the prey, generally bharal, being chased from one part of the valley into the area where the other animal of the pair hides in waiting . . . . Often the pair will eat together at a kill," wrote Dang (1967). Neither my few observations, nor those in the literature, can support or refute this viewpoint. A male and at least one medium-sized animal used the same stretch of terrain for several days at Lapche, their tracks crossing on several occasions, but to my knowledge they did not meet even when the smaller animal spent about four days on a bharal kill. At Kilik Pass a male followed the two-day-old route of a medium-sized animal for 3 km before the tracks veered apart. It thus remains unresolved whether the snow leopard has a social life like that of the tiger or whether it remains solitary except when courting or when a female has cubs, as is the case among puma (Hornocker, 1969).

Snow leopards use several direct and indirect means of communication. Their vocal repertoire is varied. Hemmer (1966) noted that they purr and produce a soft puffing sound through the nose, a vocalization which is also found in tiger and clouded leopard and serves as a greeting. Snow leopards growl, snarl, hiss, and spit, as do all cats, and they may also emit a "low rumble" and a coughing roar (Freeman, 1975). Two animals shot by Dang

(1967) "let out startled growls" and one he met "snorted in surprise." Small cubs at the New York Zoological Park gave long-drawn, high-pitched miaows when hungry, grunted when they anticipated food, emitted bleats when held against their will, and hissed and spit when annoyed. Males miaow harshly while copulating. Snow leopards do not roar like other large cats, their only long-distance call being a "piercing yowl," as Ulmer (1966) phrased it. One March evening at Lapche, as I huddled against a boulder during a snowstorm while waiting for a snow leopard to pass along the trail below, I heard a loud miaow at 1830 hours and another at 1855. Villagers there and in other parts of Nepal told me that the cats call like that when searching for a mate.

Both males and females mark their range by depositing scent and feces and by making scrapes. These three methods are often combined by males, as shown by the behavior of one at the New York Zoological Park: he first rubbed his cheek sinuously against a log, then scraped the cement floor of his cage several times with his hindpaws, squatted to deposit a small amount of feces, and finally swiveled around and with the base of his tail quivering sideways squirted several jets of fluid up against the log. Later he sniffed the site and grimaced, with his tongue hanging far out in the manner of tigers. Scent marks are difficult to discern in the wild, except for occasional pungent odors on boulders, but the 20-cm-long scrapes remain conspicuous for a long time. Eight out of 59 scrapes had been marked with feces and at least two with urine. Most scrapes had been scratched into gravel or dust but two were in snow and two in pine needles. The presence of one mark may stimulate the same or another animal to use the site again, and it was not unusual to find several scrapes or fecal deposits of different ages together. The effectiveness of the marking system is enhanced by the fact that the animals often use the same routes and leave their signs on prominent locations. In traveling along, the cats tend to trace the base of cliffs where the precipices give way to scree, in contrast to wolves which usually follow valley bottoms. Livestock trails along slopes are also favored. Snow leopard often mark sharp bends in the trails, crests of ridges, promontories, and other such sites. For instance, near Phoksumdo Lake I followed a much-used trail along the contour of a slope for about 5 km. Six of the 11 scrapes along that trail were on a crest or rocky spur and 3 others were within 50 m of such a site, as were 3 out of 5 sets of droppings. The snow leopard's marking system closely resembles that of leopard in the Karchat Hills and of tiger.

Virtually nothing is known about interactions between adults in the wild. Ognev (1962) reported how two unsexed individuals reared on their hindlegs in play and exchanged blows, and then "arching their backs at one another" they parted. With estrus lasting 2 to 8 days (Frueh, 1968; Kitchener et al., 1975) contact during courtship remains brief. One estrous female at the New York Zoological Park permitted the male to mount her only on February 9, 10, and 11. I observed the pair for 2⅓ hours on the second day, and during this time the male mounted 22 times. The two courted in typical cat fashion. The female initiated 18 of the copulations by circling the male and rubbing herself against him before crouching in front of him. On 13 of the copulations he grasped the skin on her nape with his teeth and sometimes bit it repeatedly while he was mounted. He also licked her neck on three occasions. Toward the climax he often emitted long-drawn, harsh miaows, and the female sometimes growled deeply. After the male dismounted the female rolled on her back on 5 occasions or immediately began to pace the cage restlessly until she presented herself again. The same pair mated again from March 4 to 6. The female gave birth to three cubs between 1430 and 1700 hours on June 12.

In Russia births occur in April according to Novikov (1962). With a gestation period averaging 96 to 105 days (Kitchener et al., 1975; Marma and Yunchis, 1968), the young would have been conceived in late December and January. Snow leopard in the Himalaya usually court in March and April and give birth in June and July according to my local informants. Dang (1967) found 3 small cubs in a den in July, and the cub I saw was probably born in August. Captive snow leopards are born between April and August with a peak in May (Freeman and Hutchins, in press). One to four but usually two to three cubs are born, each weighing from 500 to 650 g on the average (Kitchener et al., 1975). Ward (1923) caught two litters with two cubs each, Baldwin (1876) mentions that a female was killed near Neti Pass in Tibet and two small cubs were found nearby, and Pocock (1939) reports a litter of 3 cubs from Gyangtse.

The young are weak and have their eyes closed at birth. Of two cubs born at the New York Zoological Park, one opened its eyes at the age of 7 days and the other cub did so at 8 days. At the age of 3 days they could creep and at the age of 11 days they could sit and stand shakily. The first teeth erupt between the ages of 17 to 23 days (Frueh, 1968), and at the age of 5 to 6 weeks they may venture from their den (Calvin, 1969; Freeman and Hutchins, in press). Since cubs are relatively immobile for two months the female needs secluded haunts and a readily available food supply if she is to raise a litter successfully.

I observed a female with a cub, about 4 months old, for several days in the Chitral Gol as the two fed on domestic goats we provided. A second cub had apparently vanished earlier that month, for the locals insisted that they had seen two young with that female. On December 14, at 0835 hours, the two cats reclined side by side on a spur. Behind and slightly downhill of them was a rocky cleft. Into this shelter the cub soon retreated and it still had not reappeared by 1530. The following morning, at 0700, the cub clambered about on the rocks 5 m from its mother. Suddenly it ran to her and touched its forehead against her cheek. It then fed on the goat for 40 minutes while its mother rested on a nearby boulder. Approaching her after its meal, the cub rubbed its cheek against hers, licked the top of her head, returned to the kill for one minute, and finally vanished from sight into its rocky retreat. Although the female remained in view all day, the cub did not venture forth until dusk at 1650. For three days after that the cub did not show itself in daytime, but at 0700 on December 19, I saw both lying 3 m apart on the crest of the ridge. They left the site during the night of December 20 to 21.

Food habits. The snow leopard is a cat of moderate size, about 200 cm long of which 90 cm consists of tail, males being somewhat larger than females (Ward, 1923). Dang (1967) weighed one male which "turned the spring at 110 pounds" (49.5 kg). One 4-year-old female at the New York Zoological Park weighed 31.7 kg and an adult male 39.4 kg. Big cats can generally subdue prey weighing at least three times their own weight (Schaller, 1972b). Thus most wild high-altitude ungulates except adult kiang, takin, and yak represent potential prey. In the Aktau Mountains the cats feed on Persian gazelle, and in the Trans-Ili Alatau on wild pig (Novikov, 1962). Kennion (1910) met a snow leopard on a Marco

Polo sheep kill. Dang (1967) mentions snowcock, takin, tahr, serow, goral, and musk deer as food items, and Ward (1923) includes monal pheasant and chukor as well. Ibex and bharal are the two main food species in the Himalaya, and in fact the snow leopard is known in parts of India as "Bhurel hé" or bharal killer. My Chitral fecal samples were collected mainly in the Chitral Gol and at Tushi with the result that markhor appear prominently (40%) in the diet (table 22). The Nepal samples contained at least 50% bharal. Livestock, mostly domestic sheep and goats, was a principal food in all areas but especially in Chitral, where 45% of the droppings contained this item. Villagers told me that a snow leopard may kill bullocks weighing as much as 135 kg. Meat is also scavenged when available. A yak died at Lapche and during the night a snow leopard ate from the carcass. Marmots contributed importantly to the diet at Shey. Vegetation was found in several droppings, those from Chitral containing the forb *Rheum emodi* and those from Shey some *Polygonum* and grass.

Novikov (1962) stated that snow leopards are active mainly at night, but judging by the literature they often hunt in daytime too. Dang (1967) watched them stalk bharal three times, tahr once, and snowcock once; Haughton (1913) saw them hunt ibex three times; and Ward (1923) and Stockley (1928) also observed them pursue ibex. With their smoky-gray fur tinged with yellow and white and broken by dark spots and rosettes, snow leopard blend so well into the background, whether it be scree or snow, that they are difficult to spot even during the day, at least by a human observer (plate 50).

Attempts to capture prey have been described several times. "We were lying behind a boulder watching the thar climbing leisurely up the scree and rock overhangs," wrote Dang (1967), "when a flash of white and grey fur dived into the spread out herd, and rolled down some hundred feet, all the time hanging on to a young thar ewe. The thar was in a very bad shape when we reached it, but still breathing. The snow leopard, of course, had vanished as soon as we rose to view." Stockley (1928), while hunting, was surprised when "a snow leopard suddenly raced across the hollow in which they [ibex] were feeding and made an attempt on a buck, which started away just in time. The leopard's outstretched claws raked a great lump of hair from the ibex's coat as it wheeled away." The herd then halted 100 m away and watched the cat. Haughton (1913) observed another unsuccessful attempt during which a snow leopard "followed two ibex, bounding along behind them for a good three hundred yards until he lost them in bad ground where he could not follow." Another time this author watched a snow leopard chase several male ibex around and around a small meadow for about 20 minutes, the prey circling without escaping until the cat gave up its pursuit. In the Tien Shan a snow leopard was seen to walk casually toward a herd of ibex which watched it approach to within 60 m before bolting (Kusnetsov and Matjushkin, 1962).

With prey scarce, a snow leopard must travel far in search of food. The size of its home range is unknown, but judging by the long intervals between an animal's visits to certain valleys it must be quite large. In spite of the many bharal at Shey, snow leopard came there only twice in over a month. M. Sunquist waited in the Golen Gol of Chitral from February 5 to March 6 with the hope of live-trapping one of the cats. Only one animal walked down the valley during that period. I tracked various snow leopards for a total of about 43 km. One attempt at hunting was made in that distance and it culminated in a bharal kill. Several African reserves contain a variety of predators and prey at a ratio of about 1 kg of predator to 100 kg of prey (Schaller, 1972b). At such a ratio the Shey bharal could support only one snow leopard. Since several snow leopards and wolves inhabit the region, they must by necessity roam widely even though they eat livestock in addition to bharal.

Traveling snow leopards usually plod on at a steady pace, stopping only to mark the trail. One male scraped once in 5 km of travel, and another made two scrapes and one fecal pile also in 5 km. The routes may traverse snow-covered boulder fields, the cat leaping a meter or two from rock to rock, zigzag through a maze of cliffs, and angle across slopes, valleys, and streams. One individual on reaching a stream walked along the edge until it came to some ice-covered boulders. It leaped 2 m onto one boulder, 2.5 m to another, its claws digging in to get a grip on the smooth surface, and a final 2.5 m to the opposite bank. Meinertzhagen (1927) observed how two snow leopards simply waded through a stream, one pawing the water before entering, and then shook themselves afterwards. Ognev (1962) claimed that snow leopard can leap 15 m uphill, no doubt an exaggeration. Occasionally an individual tarries along its route and investigates various sites. I followed one such trail for 5 hours along a fir-covered ridge at 3000 m and then down a slope. The snow was about 20 cm deep. When I found the track it meandered within a small area as if the animal had been hunting black-naped hares whose spoor was common there. Then the track went to a large, solitary fir at the base of which the snow leopard had made a scrape. Further on it had halted by a fox dropping. A few minutes later it left the crest of the ridge to detour to a gnarled fir which it circled, and after backtracking to a second fir it returned to the main path. On the top of a spur stood two old firs, and in the needles beneath them the snow leopard made a scrape over one meter long. Traveling down the slope, it checked a hollow tree and soon after that the snow vanished and I lost the tracks.

When stalking prey, the snow leopard uses broken terrain for cover until it is close enough for a rush (plate 51). Tracks in the snow revealed the course of events leading to the death of one bharal male almost 4 years old. The bharal had wandered alone around the village of Lapche, which in March was still devoid of summer residents. Angling into a shallow valley, he had gone to a rivulet, his descent no doubt observed by the snow leopard. Its advance hidden by a boulder, the snow leopard stalked closer and, as the bharal stood by the water, it attacked, pulling its victim down at the point of impact. After eating some viscera and part of the ribcage, the snow leopard dragged the carcass 150 m up a slope and there deposited it on a rock. Much of the meat had been eaten when I found the site, and the cat, having observed my approach, fled unseen into a chaotic mass of boulders. At Tushi I examined four places at which markhor had been killed in January, 1973: one was at the bottom of an 8-m-deep ravine, another at the base of a small cliff, a third in a shallow gully bordered by scrubby oak, and the last in an eroded depression at the base of a rocky spur. These kills consisted of a yearling male, a 2½-year-old male, a 3½-year-old male, and a young—a rather disproportionately large number of males from a population in which this sex is poorly represented (table 13). The sanctuary staff at the Chitral Gol collected for me the horns of 7 markhor which they claimed had been snow leopard kills; of these 6 were males and 1 a female. Dang (1967) reported 6 male bharal kills, and I found two such kills at Lapche. Thirty ibex kills were counted in one valley in the western Pamirs and of these 22 were males, two-thirds of them older than 4 to 5 years (Heptner et al., 1966). These figures suggest that males are more vulnerable to snow leopard predation than are females.

Snow leopard subsist partially on livestock in most areas. Domestic animals are usually guarded during the day and locked into corrals at night. With a cat's habit of eating slowly, the snow leopard is often driven from its kill before it has had a chance to eat much. The villagers in Buddhist Nepal avidly scavenge all kills, but those in Muslim Pakistan leave the carcasses for the vultures. In either case the snow leopard seldom has an undisturbed meal and even more seldom does it have time to lead cubs to the kill. Sometimes a snow leopard invades a livestock shed and slaughters several sheep and goats, as many as 15 in one instance related to me. In Chitral snow leopard sometimes lurk around villages after dark and snatch unwary dogs. I followed one track at dawn as it traced the narrow, stone-walled alleys of a village before ascending a slope beyond. However, snow leopard are now so rare that few villagers report consistent losses or complain about them.

I tied out several goats as snow leopard baits and one of these was killed at 1628 hours by a female: "She advanced slowly down the slope, body pressed to the ground, carefully placing each paw until she reached a boulder above the goat. There she hesitated briefly, then leaped to the ground. Whirling around, the startled goat faced her with lowered horns. Surprised, she reared back and swiped once ineffectually with a paw. When the goat turned to flee, she lunged in and with a snap clamped her teeth on its throat. At the same time she grabbed the goat's shoulders with her massive paws. Slowly it sank to its knees, and, when she tapped it lightly with a paw, it toppled on its side. Crouching or sitting, she held its throat until, after eight minutes, all movement ceased" (Schaller, 1972a). This attack proceeded rapidly, but during another stalk the cat moved only 50 m in ten minutes, alternately advancing in a crouching walk or sitting. The whole stalk lasted 45 minutes, but in the end it was too dark to observe details. These two goats, as well as three others, died of strangulation, judging by the superficial puncture wounds on the throats of the animals.

Dang (1967) asserted that of "34 natural and domestic kills ... most were neatly killed, either with the neck or spine broken," a statement whose accuracy I question.

The snow leopard begins to eat either around the chest and forelegs or around the lower abdomen and thighs, usually leaving the digestive tract intact. It guards the carcass closely for the usual gathering of Himalayan griffon, lämmergeier, jungle crows, and magpies would soon strip the bones. One female, obviously uneasy about being in the open near her kill, left the carcass 3 times within 15 minutes and walked 50 m away only to return at a fast walk when as many as 8 crows and magpies landed near the meat. Four times she lunged at the scavengers but these hopped nimbly aside. She then reclined by the kill for 25 minutes before abandoning it.

Snow leopards have never been known to prey on man, and even attacks, such as the seemingly unprovoked clawing of a woman described by Burton (1926), are exceedingly rare. Villagers may show their disdain for the cat by beating it to death with sticks and axes after cornering it in a livestock shed. A snow leopard may at times be surprisingly tolerant of man. Darrah (1898) shot 3 times at a snow leopard on a bullock kill and missed, as was his wont, yet the cat continued to return to its meal. Kusnetsov and Matjushkin (1962) described how a snow leopard watched two men at 25 m without being apprehensive. I observed a female for 28 hours over a period of 7 days in the Chitral Gol. This animal behaved with remarkable nonchalance toward me. When I first approached her slowly, casually, she remained lying on a boulder, watching as I sat down 75 m away. After that she dozed, but when two villagers climbed up the slope toward her, she slid backwards off the boulder and sneaked uphill, slinking from rock to rock in such a way that she was usually screened from view. Once she halted behind a shrub and looked down at the men before vanishing. Yet on the following day she permitted me to approach again. When I advanced to within 50 m of her she retreated 20 m and watched me from behind a boulder, only the top of her head visible. Then she reclined in full view on a ledge for 3½ hours. Toward dusk she snaked toward her kill below, a step at a time, keeping her gaze on me as she placed each paw carefully without looking, moving almost imperceptibly until after 10 minutes she finally reached the carcass. I stayed there that night, lying in my sleeping bag on the crest of a spur some 60 m from the snow leopard as she fed and wandered around near her kill.

7 MAINTENANCE ACTIVITIES

An animal's most important daily activity revolves around obtaining enough to eat. In quantity and quality this food must not only maintain the individual but also provide enough surplus energy to allow it to reproduce. For example, Papageorgiou (1974) found that a wild goat needs about 2.5% of its body weight in forage (dry weight) each day. While a habitat may contain much vegetation, only a certain percentage of it is both available and preferred by a species. Many studies have shown that food preferences are related to the nutritive quality of the forage, animals choosing those parts of a plant which contain the highest protein and lowest fiber levels. For instance, as forage quality decreases black-tailed deer abandon feeding sites and move to better ones (Klein, 1970). Forage with less than 4% crude protein is generally of too low a quality to maintain the body weight of hoofed animals.

Several factors influence the quality of forage, the most important of these being the growth stage; young growing parts have the highest protein content and dormant or dead ones the lowest. Range condition also affects quality. Overgrazing reduces the protein levels of plants as soil becomes depleted of its nutrients through erosion and leaching. *Pistacia* leaves on ungrazed land contained 8.4% crude protein whereas the level of those on grazed land was 4.7% (Papageorgiou, 1974). To maintain an adequate nutritional intake animals must forage on those species which at the moment offer the highest protein levels, and this in turn means either sampling many species or moving to different feeding areas. From this it follows that the greater the plant variety the better the diet. Many kinds of plants are unavailable to a herbivore in an area: some are too high above ground, others protect themselves with thorns, and still others have chemical defenses in the form of toxic substances (called secondary compounds) which inhibit digestion. If eaten, these chemicals must be detoxified and excreted to prevent physiological impairment or death. The gut flora of an animal can degrade only so much toxic material in a given time, limiting the amount an individual can eat. Although an animal can function on a diet containing up to 50% of plants with essential oils and phenols—found in such genera as *Juniperus* and *Artemisia*—other foods must also be eaten. Herbivores "should prefer to feed on foods that contain small amounts of secondary compounds, and their body size and searching strategies should be adapted to optimize the number of types of foods available with respect to the total amount of food that can be eaten" (Freeland and Janzen, 1974).

Exposure, altitude, and degree of slope affect the growing season of plants and hence their nutritional quality. A great variation in topography will increase the length of the growing season and thus the period when good forage is available. Sheep and goats in the Himalaya have adapted to this movable feast by migrating seasonally up and down the slopes, but species living on low desert ranges have little opportunity to prolong their high-quality food intake. However, mountain species must contend with extreme cold as well as snow so deep that the vegetation may be wholly covered. The means by which various sheep and goats maintain their energy level is the subject of this chapter.

One further point needs comment. Man usually harasses wild ruminants either directly by hunting or indirectly by livestock grazing and other such disturbances. This may affect the nutritional balance of the animals. By being forced to flee, the animals not only waste valuable energy but also may be obliged to forage in poor habitats rather than in preferred ones. The resulting weight loss may weaken individuals, causing them to lose fetuses and in other ways affecting them adversely (Geist, 1971c). Disturbance also influences patterns of behavior with the result that observations may not in fact reveal a species' preferred routine. Red deer and chamois in New Zealand began to forage on grasslands in daytime rather than along the forest edge at night after hunting ceased (Douglas, 1971).

FOOD HABITS
Food habits can most readily be ascertained by observing feeding animals, by checking sites at which animals have foraged, and by analyzing rumen samples. I used all three methods and table 23

lists some of the main food plants of several species. To obtain an impression of the type of forage available, several vegetation transects were made in selected habitats. A 20 x 40 cm rectangle was placed on the ground at 5 m intervals and the amount of ground cover was estimated within it.

Hemitragus and *Capra*

Himalayan tahr had few green plants available along the Bhota Kosi toward the end of the winter:

> Oak leaves were eaten whenever a group was in a forest patch. To reach low-hanging leaves, an animal may rear up on its hindlegs and bend and hold down a branch with one or both forelegs while browsing rapidly. Once a subadult male leaped 2 m into the fork of a tree, behaviour common in foraging Kashmir markhor but not tahr. Bamboo was also an important food, but the abundant leaves of rhododendron were seldom sampled. The tahr's principal food was dry grass. In late February animals spent hours foraging on *Danthonia schneideri, Cymbopogon thwaitesii, Arundinella nepalensis,* and other species. After obtaining a mouthful in one or more bites, an animal characteristically raised its head and chewed. I recorded the type of vegetation selected by two male tahr on two days in February. Of 155 mouthfuls, 75% consisted of grass, mainly wads of dead leaves and stems bitten off at the base, 7% of twigs and leaves from several shrubs and saplings, 6% of bamboo, 4% of dry forbs, and the rest of unidentified material which was at times obtained by first pawing the ground. By early March, green grass shoots became conspicuous, and tahr nibbled these. Yet dead grass continued to be eaten, even late in March when much green forage was available. Newly sprouted leaves of *Polygonum molle, Leucoceptrum canum* and other forbs and shrubs were at that time also a part of the diet. One tahr ate the blossoms of *Daphne gracilis* and several others appeared to lick crustose lichens off rocks (Schaller, 1973a).

Snow grass (*Chionochloa*) is a principal food of tahr in New Zealand (Caughley, 1970c), and I have also seen animals there browse on *Podocarpus* and *Dracophyllum*. In California the tahr eat grass supplemented with a little oak and other browse (Barrett, 1966).

Nilgiri tahr ate mainly grass during October and November, but a few forbs (*Heracleum*) and some browse (*Strobilanthes, Acacia*) were also taken.

Karchat wild goat have catholic tastes, eating the leaves of most shrubs and trees as well as various such grasses as *Pennisetum* and *Dichanthium* (table 23). Forbs are scarce in the area, as they are in most arid environments. In September 1972, A. Laurie and I timed individual goats as they fed for a total of 245 minutes, and of this time they devoted 79% to grazing and 21% to browsing, principally on acacia, which was in leaf then. In March 1973, at the height of the dry season when grass is yellow and most trees are leafless, the goats devoted 54% of their time to grazing and 44% to browsing mostly on green *Leptadenia* twigs and the fleshy leaves of *Salvadora* and *Capparis*. On 7 occasions an animal climbed 2 to 3.5 m into a tree to obtain leaves. Of two rumen contents examined in September, one contained an estimated 80% grass by volume and 20% browse, the other 90% grass and 10% browse. Ninety vegetation plots near the cliffs and on the plateau revealed 94% bare ground and a 6% grass cover, isolated tufts growing an average of 25 cm apart in depressions and cracks where soil had accumulated. Trees and shrubs provided a canopy of about 5% except in ravines where stands may be dense. Much of the sparse groundcover is palatable. *Cymbopogon*, a grass not much liked by hoofed animals, makes up less than 1% of the cover near cliffs but becomes more prominent toward the center of the plateau. The preferred food of Cretan wild goat consists of various shrubs (*Olea, Ephedra, Capparis*), forbs (*Obione, Allium*), and grasses (Papageorgiou, 1974).

Although conifer and oak stands give a well-vegetated appearance to the habitat of Kashmir markhor in the Chitral Gol, 97 plots in the winter range showed that nearly 95% of the ground consists of naked scree and earth, 3% of *Artemisia* and other shrubs, 2% of forbs, and less than 1% of grass. The markhor's main winter food is a forb (*Rumex hastatus*), a shrub (*Indigofera gerardiana*), and leaves from *Pistacia* and *Quercus* trees. The last-named is the most important forage species, the evergreen leaves being available even in deep snow. The goats readily climb up into the bushy oak trees to reach the foliage. It is an incongruous sight to see a tree full of markhor each balancing along swaying limbs and jumping from branch to branch. On one occasion 27 out of a herd of 40 markhor were in trees. The height above ground was estimated for 64 markhor, and of these 43 were fewer than 3 m above ground, 17 were up 3 to 6 m, and 4 up 6 to 9 m. The leaves on the lower branches (as well as all leaves on those oak that have been heavily lopped for livestock fodder) are spiny in contrast to those in the crowns which have smooth edges. Although markhor will eat spiny leaves they prefer the others. The two kinds are equally nutritious in winter with a fat content of 3.4 to 4.4%, a crude protein content of 8.0 to 9.5%, and a crude fiber content of 26.7 to 29.3%. In spite of this fairly high protein-level, markhor ate oak leaves mainly when little else was available, as in December 1970. But the animals seldom foraged on leaves during the winter of 1972–73 when the acorn crop in the Chitral Gol was exceptionally large. At that time some 46% of the markhors' feeding time was devoted to grasses and a few miscellaneous forbs, 34% to acorns, 15% to *Rumex*, 3% to oak leaves, and 2% to *Indigofera*. Animals often crowded beneath oak trees to pick up acorns. These acorns had high fat (8.7%), low fiber (10.7%), and low crude protein (4.0%) contents. In nearby Tushi the acorn crop was small that year, and the markhor ate mainly oak leaves. Of 4 rumens examined from Tushi, 3 were stuffed with oak leaves and 1 contained 40% *Pistacia* leaves, 30% grass, 25% oak leaves, and 5% acorns. I have no information on the summer food habits of markhor at high altitudes. One herd which remained on its winter range in the Chitral Gol foraged on *Rumex* and *Pistacia* in early August. The word "markhor" means "snake-eater" in Persian, but I lack evidence that reptiles feature in the diet of markhor.

According to Heptner et al. (1966), Asiatic ibex may in any one area consume some 80 plant species, mainly grasses, sedges, forbs, and the leaves of such shrubs as *Ribes*. This agrees with my observations (see table 23). Several plant transects revealed the kind of vegetation available to ibex (table 24). Two of the transects were made at 4000 m or below in ibex winter range heavily grazed by livestock. Grass was sparse but forbs were well represented. However, so many of the forbs were either spiny (*Cousinia, Astragulus, Arenaria, Acantholimon*) or aromatic (*Napeta*), that only 1 to 3% were of a kind usually preferred by herbivores. The aromatic shrub *Artemisia* was common but little eaten. The other four transects were made above 4300 m on alpine meadows with little or no livestock damage where ibex foraged in summer. Although the ground was almost bare, a few tufts of grass and such forbs as *Primula, Oxytropis,* and *Sedum* provided potential forage. Dagestan tur and Kuban ibex subsist mainly on grass (80–95%), on forbs, and on some browse (Heptner et al., 1966). Alpine ibex eat mainly grasses and forbs except in winter when they may also browse a fair amount (Couturier, 1962).

Pseudois

I visited bharal habitat only in winter. At that time the Lapche animals consumed mainly grass and a few dead forbs (*Thermopsis, Anaphalis, Polygonum*), and they sampled the branch tips and leaves of *Berberis, Ephedra* and other shrubs (table 23). One rumen contained 98% dry grass and traces of *Juniperus, Thermopis, Ephedra,* and *Berberis.* Shey bharal had a similar diet, most forage consisting of *Arundinella, Danthonia,* and other grasses. The tiny leaves of *Lonicera* and *Berberis* shrubs were also stripped and the branch tips of *Juniperus* and *Caragana* nibbled. Two rumens each contained nearly 100% grass and traces of *Lonicera* and forbs.

At Lapche, two vegetation transects (40 plots) on a slope below some cliffs at about 4500 m revealed 43% bare ground and a cover of 28% grass, 16% dead forbs, and 14% shrubs. *Juniperus* and *Rhododendron* shrubs, a meter or less tall, shaded about 40% of the slope except in a few places where fire had recently destroyed them. At Shey, three transects (150 plots) taken between an altitude of 4550 m and 4700 m showed 63% bare ground, 21% grass, 3% forbs, and 13% shrubs, principally *Juniperus* and *Lonicera* with a canopy coverage of 39%. Although the two areas were similar in their plant cover, the availability of bharal forage varied considerably. The slopes around Lapche were only moderately grazed by livestock and grass was readily accessible, especially on the broad ledges above the lower slopes. Shey was seriously overgrazed and much eroded. Most grasses that remained were protected beneath thorny shrubs. Forbs were scarce except for a species of *Polygonum* which neither livestock nor bharal liked.

Ovis

Punjab urial occasionally browse on *Acacia, Salvadora,* and other shrubs and trees, but their main food consists of grass (table 23). To test for grass preferences of urial, Z. Mirza and I presented 7 common species to 3 captive young at Kalabagh. The number of visits by the animals and the time they spent eating indicate that *Eleusine flagillifera, Digitaria bicornis,* and *Cenchrus pennisetiformis* were the most favored grasses and *Aristida depressa* and *Cymbopogon jawarancusa* the least. I analyzed the stomach contents of two dead animals: one was wholly filled with grass and the other contained 99% grass and 1% browse. Vegetation transects (196 plots) made in favored grazing areas during October showed that 81% of the ground was bare and 19% was covered with grass, forbs being present in negligible amounts. Only 7% of the grass cover consisted of preferred species. Palatable species were eaten to the ground by wildlife and livestock, in contrast to *Cymbopogon,* which grew in tall, dry swards. These were disdained by urial except for an occasional green blade or seed head.

I examined the rumens of two Ladak urial near the Baltoro Glacier in May. One contained 50% grass, the other 60%, and the remaining contents consisted of *Ephedra* twigs.

In Russia, urial also eat mainly grass with a supplement of forbs and browse. One population in Turkmenistan ingested beetles, scorpions, and other arthropods (Heptner et al., 1966). Mouflon in Corsica consume much browse, 35% of their diet consisting of leaves from shrubs and trees, 36% of forbs, and the rest of grasses, mosses, and lichens (Pfeffer, 1967). Marco Polo sheep prefer *Carex* and such forbs as *Primula* and *Delphinium* (Heptner et al., 1966). North American sheep are mostly grazers. Hoefs (1974) reported an average annual diet of 59% grass, 19% forbs, 17.5% browse, and 3.5% bark in Yukon Dall sheep, and Smith (1954) found a diet of 73% grasses, forbs, and lichens, and 27% browse in Idaho bighorns, to mention the results of just two studies.

Sheep and goats prefer grass when it is available. Both may also consume a wide variety of forbs and leaves from shrubs and trees. However, goats seem to browse more than do sheep in areas such as the Karchat Hills where during the dry season the grass may be dead but some trees remain green. Urial make no prominent seasonal switch to browse, this probably being one ecological factor enabling *Capra* and *Ovis* to inhabit the same desert ranges. Aoudad in California eat over 90% grass (Barrett, 1966) but in New Mexico they consume 42% grass, 50% oak and mountain mahogany (*Cercocarpus*), and a few forbs (Ogren, 1965), a diet typical of low-altitude goats. Possibly goats can detoxify secondary compounds better than sheep can. Browse is, of course, scarce above timberline, and Asiatic ibex and bharal subsist mainly on grasses and forbs.

Goats, including *Hemitragus,* show their adaptations to browsing by foraging in two ways not often used by sheep. They not only stand bolt upright on their hindlegs to reach low-hanging leaves (plate 34) but they may also climb trees. Although mouflon (Pfeffer, 1967) and Punjab urial may occasionally rear up briefly to reach a branch, goats often stand upright for minutes with one leg propped against the tree for balance and the other used to hook and bend in twigs. Interestingly, American sheep which resemble goats in appearance and habitat preferences are not known to feed in this manner.

The *Rupicaprini* and *Ovibovini* resemble goats somewhat more than sheep in the prevalence of browse in their diet. Chamois in the Caucasus eat primarily forbs (*Rumex, Pedicularis*) in summer, and grasses, acorns, and some browse in winter (Heptner et al., 1966). Mountain goats in Idaho consume mainly grasses and forbs and they browse particularly on spruce, juniper, and aspen in winter (Brandborg, 1955); in the Kenai area of Alaska they subsist on sedges, grasses, forbs (*Sedum, Epilobium*), and *Artemisia* during summer and on fern rhizomes and some browse during winter (Hjeljord, 1973). Japanese serow are mainly browsers (*Quercus, Acer, Sorbus, Tsuga, Viburnum*) and consumers of forbs (*Polygonum, Petasites*) according to Okada and Kakuta (1970) and Akasaka and Maruyama (in press). Goral in Russia prefer to feed on forbs and grasses, but in winter and early spring much of their available food consists of oak, ash, walnut and other browse (Bromlie, 1956). Goral in India have a similar diet (Dang, 1968b). Takin feed on birch, *Senecio,* and grass in summer and on bamboo and willow in winter (Wallace, 1913). The takins at the New York Zoological Park may rear up on their hindlegs to browse, plucking leaves as high as 3 m above ground with their prehensile lips. Muskox browse on low-growing shrubs (*Ledum, Betula, Salix, Vaccinium*) and they also eat grasses and sedges (Tener, 1965). The food habits of saiga vary seasonally, consisting of much grass in spring and summer and then increasingly more of forbs and shrubs (Bannikov et al., 1967).

To obtain forage in deep snow requires special techniques. Generally Asiatic ibex and markhor feed in areas where wind, sun, and avalanches have removed the snow, and this is also true of Marco Polo sheep (Heptner et al., 1966), Alpine ibex (Nievergelt, 1966a), and goral (Nasimovich, 1955), to mention a sample of species. Markhor sometimes seemed to sniff snow and then nuzzle it aside to reach acorns, apparently able to smell this food at a depth of 15 cm. One markhor female pawed snow aside on twelve occasions with 2 to 4 strokes of a foreleg while searching for acorns, my only such observation on this species. By contrast, Asiatic ibex paw snow often and vigorously in search of food. I observed ibex for 4 hours at Besti as they foraged in 10 to 30 cm of snow. During this time I tallied 274 instances of pawing, each bout ranging from 1 to 13 individual strokes with an average of 5. An animal faced uphill or parallel to the slope and with backward and sideways sweeps of its foreleg removed the snow over an area some 15 to 30 cm in diameter. After feeding a few seconds, it cleared another site. A herd soon fills a slope with such snow craters, some excavations being up to 3 m long. Similar behavior has been described in American sheep (Geist, 1971a). Muskox also paw snow aside to obtain food (Gray, 1973), as do Japanese serow (Akasaka, 1974), but not goral, according to Bromlei (1956). Feral goats in Wales may clear snow from vegetation with their horns (Crook, 1969).

Animals occasionally eat soil impregnated with natural salts. One salt lick in the Chitral Gol was used repeatedly by markhor. Once 21 animals scraped the soil with their incisors and licked it for 2 hours. A white substance seeped from cracks in some cliffs at Shey, and this bharal licked. They also licked limestone boulders and the urine-splattered rocks in the outdoor latrine of the local lama. I threw a little salt on a knoll at Lapche, and a herd of bharal visited the site for 3 consecutive days.

Wild and domestic animals often compete for the same forage. For example, domestic sheep and cattle at Kalabagh preferred the same grass species as the urial. Domestic goats are the black sheep among livestock in that they consume everything edible, thus competing directly with whatever wild Caprinae inhabit the area. Most mountain pastures within reach of livestock have been degraded, the thin soil trampled into trails, and grasses and other preferred food plants reduced to a small percentage of the ground cover. The vegetation composition has been changed into communities that are less preferred and less productive. Palatable species have given way to those with toxins and spines, as shown by the ibex range near Dorah Pass (table 24). In most terrestrial ecosystems the herbivores consume less than 60% of the primary production on an annual basis (Sinclair, 1975; Berwick, 1976), giving the vegetation an opportunity to recover. In some mountain valleys the percentage must be considerably higher: at Kilik Pass even *Artemisia* shrubs were trimmed to the base. In summer or during the rainy season when forage is nutritious and widely available, animals may not be affected by food shortages, but in winter or during dry seasons they may have to subsist on a low plane of nutrition. If snow blankets the ground, animals have to expend valuable extra energy in finding food, and then those plants available are usually not the preferred ones because livestock has already removed them. In general, wild goats are not as seriously affected as sheep for they often forage on cliffs which are inaccessible to livestock. Nevertheless, Asiatic ibex do on occasion become emaciated during deep winter snows (Nasimovich, 1955).

ACTIVITY PATTERNS

An animal must not only find enough to eat but it must also obtain its food without wasting energy and unnecessarily exposing itself to hazards such as predation. The daily cycles of feeding and resting and the seasonal cycles of moving are the subjects of this section.

Daily activity cycle

To find out if a species follows some diurnal pattern of activity, I noted how many animals in undisturbed herds were active or inactive at 5-minute intervals. The six points in each half-hour period were lumped and expressed in figures 25 and 26 as percent of animals active. Since social interactions and movements use little of an individual's day except during the peak of the rut, the activity curves represent feeding cycles. Taking the high-altitude species first, my data show that there are animals foraging at any time during daylight but that definite peaks of activity exist. The early morning and late afternoon are the main foraging periods in all species but no such constant pattern is evident at midday. In December and January the markhor in the Chitral Gol showed a drop in activity between 1100 and 1300 hours, and at that time fewer than 50% of the individuals were feeding. Himalayan tahr had a similar pattern in February and March except that the midday trough in activity began over an hour earlier than that of markhor. Barrett (1966) found that tahr in

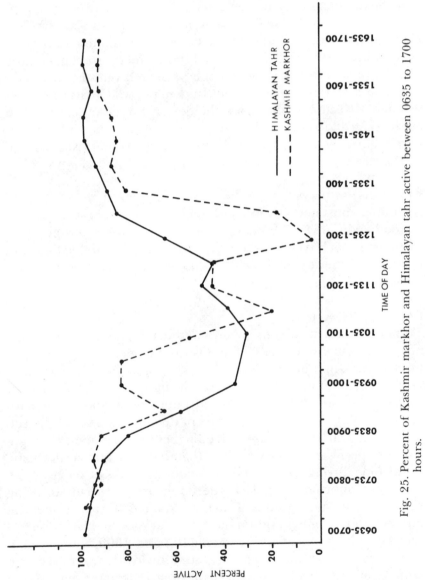

Fig. 25. Percent of Kashmir markhor and Himalayan tahr active between 0635 to 1700 hours.

California had three daytime feeding peaks—in the morning, at midday, and in the evening. I have few ibex observations. Between November and May most animals fed until around 0900 and after that some rested. By 1300 most were active again and they continued to forage until darkness.

The Lapche and Shey bharal have similar activity curves (fig. 26). However, the Lapche animals began their first inactive period over an hour earlier and they also had a more prolonged second inactive period than the Shey ones.

Nilgiri tahr resembled bharal in their activity cycles. Most fed until about 0800; there followed a sharp drop in activity until 1030. After that, and until 1430, at least half the tahr foraged. A second drop occurred between 1430 and 1530, and then the animals grazed actively until dusk (Schaller, 1971).

If the points on the curves are added and then divided by the 21 half-hour periods, the resulting figures reveal the average percent of time devoted by a species to foraging in daytime. The figures for Himalayan tahr, markhor, and the two bharal populations are 72 to 76%, a remarkable consistency considering the differences in the condition of the habitats. The figure for Nilgiri tahr was 55%. This species was sampled not in winter when forage is dead or dormant but shortly after the monsoon when there is ample nutritious food. These tahr averaged less than 6 hours of feeding between 0635 and 1700 hours as compared to about 8 hours in the other species.

I am unable to explain why markhor and Himalayan tahr have one major diurnal rest period and bharal and Nilgiri tahr have two. Temperature may have an effect on activity, with, for instance, Stone sheep and mountain goats feeding little on cold mornings (Geist, 1971a). Such a change was not noted in the Himalaya where temperatures are less extreme than those in Canada. However, temperature does influence the foraging behavior of Himalayan tahr. From mid-February to mid-March, when average daily minimum temperatures hovered around the freezing point, tahr tended to feed in the forest during the early morning hours. Not until sun reached the cliffs, usually around 0830, did they venture into the open. But during the second half of March, when the average minimum temperatures were 6° C, tahr were out at dawn and possibly did not retreat into the forest during the night. Tahr sometimes moved to and re-

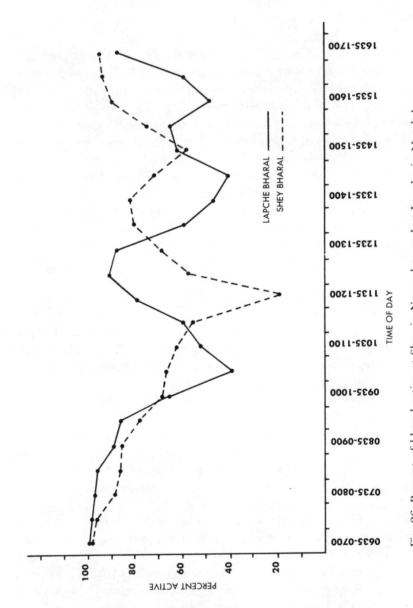

Fig. 26. Percent of bharal active at Shey in November and at Lapche in March between 0635 and 1700 hours.

mained in a patch of sun, something not observed in the other high-altitude species.

Stone sheep are thought to graze much at night (Geist, 1971a) in contrast to Dall sheep which feed only from sunrise to an hour after sunset (Hoefs, 1974). Both mountain goats and chamois are inactive at night (Geist, 1965; Christie, 1964). Himalayan tahr were often seen in the morning at the same place as the previous evening, indicating that they had moved little after dark. Markhor and bharal often fed into the night, but since they foraged for eight hours in daytime it seems unlikely that they would need many more hours at night to fulfill their daily requirements. Similarly, Grubb and Jewell (1974) found that feral Soay sheep spend about 9 out of 24 hours foraging, most of them in daytime.

Heat affects the activity of all animals in the lowlands. The Kalabagh urial and Karchat wild goats retreated beneath trees and into ravines as it became hot, and I was then unable to observe them well. In March and September the wild goats spent most of the night on cliffs and between 0600 and 0700 moved onto less precipitous terrain to forage. After feeding intensively for 2 to 3 hours some animals only nibbled at the vegetation in a cursory manner and others stood or reclined in the sun. By 1000, when the temperature hovered around 32° C, most animals had retreated into cover but occasionally a herd remained in the open until 1100. At around 1600, the animals again began to feed, first slowly around their rest sites and finally in the open. Sometime during the night they returned to the cliffs.

Some Kalabagh urial were active at first light between 0430 and 0600, depending on the season. Most animals sought shade by 0730 during the hottest weather and they remained in it until 1700 and later, indicating that they then foraged mostly at night. In October and November, when temperatures were moderate, urial fed in the open until around 0900 to 1000. At that time herds abandoned their feeding sites and moved at a fast walk to a shady ravine, not necessarily the one closest at hand. There they rested until about 1600 to 1630 hours. When shade temperatures reached 27° to 29° C urial avoided sun, revealing a lower tolerance for heat than did wild goats. In late December, when the daily maximum was only 18° C, some urial fed and rested on bare ridges throughout the day. Saiga show a similar switch in their feeding schedule during hot and cool seasons (Heptner et al., 1966).

Feeding and drinking behavior

In an arid environment or in one degraded by livestock, foraging animals must move constantly to satisfy their nutritional needs. Scattered loosely or aligned on a broad front, herd members tend to amble in the same general direction, nipping plants in their path. One feeding markhor herd traveled 300 m in 45 minutes. If a food source is localized and abundant, animals may concentrate there for a long time. One bharal herd fed for 3 hours on a 120×100 m plot as individuals cropped the short grass and poked their muzzles beneath spiny shrubs to extract the grass growing there. Markhor often spend an hour or more beneath or up in an oak tree. If the food is scattered, as on cliffs, herds may be dispersed over several hundred meters. Herds often give an impression of being restless. Feeding and inactive periods may follow each other at irregular intervals. Sometimes one animal in a resting herd begins to graze and all others soon join only to lie down agan, one at a time, after 30 to 45 minutes.

Most plants are simply nipped off, but some require special techniques. To bring a low-hanging branch within reach of its muzzle a markhor or wild goat may hook it, bend it, and then hold it down with a foreleg while eating the leaves. A female markhor bit into a branch and broke it with a twist of her head before eating the terminal part. Himalayan tahr and bharal sometimes pawed the ground as many as 21 times and ate some unidentified items afterwards. One bharal female kneeled on her forelegs to eat something, and a male wild goat used this stance to reach a shrub growing on a cliff face below him.

Goats must be good climbers to reach the vegetation on cliffs, and indeed they traverse ledges with amazing adeptness. Speaking of bharal, Shipton (1936) wrote that "never have I seen a more extraordinary display of rock climbing." I can make the same point with respect to Himalayan tahr. These animals balanced on ledges a few centimeters wide and leaped with precision onto grass tussocks growing from sheer cliff faces two meters below them. When confronted by a sloping rock face, an animal rocked back and forth and then propelled itself upward with a series of leaps, using the knee calluses and the flat chestpad rather than just the hooves to provide traction. When descending scree, snow, or a rock face, tahr may slide in a squatting position, the stiffly held forelegs and the hocks functioning as brakes. Such behavior was also observed in markhor. In spite of being adept moun-taineers, goats often dislodge rocks accidentally, and a good way to find herds is to listen for the clatter of falling stones. I discovered 11% of the markhor herds in the Chitral Gol by that method during December 1972. Markhor may find it difficult to cross cliffs after a fresh snowfall. Footholds are hidden, rocks are slippery, and cornices collapse. An animal sometimes tested a surface with a foreleg, and, if it judged the surface unsafe, it then leaped blindly ahead, behavior which surely caused occasional mishaps.

Hoofed animals differ in their ability to cope with deep snow. If the live weight of an animal is divided by its total track area, a measure of relative adaptation to life on top of snow is obtained (see Kelsall, 1969). The figure for one chamois male was 200 gm/cm^2, one goral male 365 gm/cm^2, twelve mouflon males 469 to 818 with a mean of 662 gm/cm^2, and ten Asiatic ibex of both sexes 551–996 with a mean of 848 gm/cm^2. Of these species the chamois is best able to walk on snow, but it is not as well suited as the muskdeer, which has a supporting area of only 80–120 gm/cm^2 (Nasimovich, 1955). Most Caprinae are not well adapted to snow, and they tend to seek out avalanche paths, wind-swept ridges, cliffs, and other terrain from which snow has been removed, a point well documented for Alpine ibex by Nievergelt (1966a) and for muskox by Lent (1971). Powdery snow more than 40 to 50 cm deep seriously impedes chamois, goral, and Dagestan tur, and snow of 25 to 30 cm hampers movements of mouflon and argalis, making them susceptible to predation and malnutrition (Nasimovich, 1955). Bharal walk slowly with small steps and with their legs spread more than normally on the fragile crusts of snowfields. When snow was more than knee-deep, they bounded with leaps of 150 to 175 cm. Markhor simply plowed through chest-deep snow in most instances, but at times they bounded downhill, using the snow to brake their momentum. Travel in deep snow uses up much energy, and it is significant that moving herds travel in single file, the animals following the path plowed by the leader. A female usually leads and adult males tend to be in the rear. I noted the leader of 22 mixed markhor herds containing at least one female and one male of class I or larger. A female was always in the lead. Similarly, a female led 28 out of 30 bharal herds, the exceptions being class III and IV males. With males using up so much energy during the winter rut, the effort saved in traveling may well help them to survive until spring.

Markhor, ibex, and Himalayan tahr were not observed to drink and they made no obvious treks to water. One bharal herd stopped for a drink at a stream on its way from one slope to another and so did two males. However, I observed these species mainly in winter when they commonly ate snow. Water sources in arid, hot environments are scarce and often seasonal. The Kalabagh urial have one stream and several rivulets available throughout the year and these they visit at intervals, especially toward evening in the hot season. Some individuals seemed to remain away from water for several days. Desert bighorns drink every 3 to 5 days on the average (Welles and Welles, 1961), as do aoudad (Brouin, 1950). The Karchat wild goat have no permanent water source except for a few widely scattered springs, some of them far from cliffs. A shower sometimes fills rocky depressions, but after these pools dry up herds may have no water within their range for months. Stockley (1936) noted that wild goat "never drink." While goats do drink when there is water, they can subsist for long periods on whatever moisture they obtain from their forage. Like some other desert animals, they probably decrease urine production and at the same time reprocess urea into protein. A flexible temperature-regulating mechanism possibly enables them to save water by decreasing perspiration rates (see Taylor, 1969). I noted that during the hot season fresh feces consisted of just dry pellets, little water thus being wasted through defecation.

Resting behavior
One seldom finds all members of a large herd resting at the same time for more than an hour except in bad weather. There are always a few individuals changing position or foraging in a desultory manner. As noted earlier, the choice of rest site depends in part on the weather. Punjab urial and wild goat withdrew into shade when temperatures climbed above 30° C, and some Himalayan tahr retreated at 15 to 20° C. Alpine ibex also avoid heat by seeking shade beneath overhangs and in caves (Nievergelt, 1966a). When temperatures were moderate the choice of rest site seemed haphazard as long as the ground was dry, visibility was fairly good, and, in the case of goats and bharal, a safe retreat was nearby. Favored sites were slopes, ridge tops,

and wide ledges, usually without reference to sun and wind. Alpine ibex respond to strong wind by seeking sheltered slopes in winter and by exposing themselves in summer (Nievergelt, 1966a). During sandstorms aoudad lie down with their rumps windward and their noses close to the chest (Ogren, 1965). Winds of up to 35 km per hour had no obvious effect on wild goats. Other species also ignored breezes, and it was not unusual to see a male Himalayan tahr or markhor rest on a promontory with his ruff whipping in the wind. Alpine ibex congregate on sunny slopes in winter according to Nievergelt (1966a). Such behavior was also observed in bharal and Asiatic ibex, but choice of exposure seemed to reflect the animals' need for food rather than sun: the snow had melted from the sunny sides of the mountains. Himalayan tahr appeared to be more responsive to cold than the other species I studied. On several occasions a herd sought a patch of sun and rested in it until it clouded over and the air became chilly, at which time the animals immediately began to forage. Resting animals may be crowded together or widely scattered depending in part on the terrain. On cliffs the individuals are typically dispersed, each finding a comfortable ledge or fragment of shade, whereas on ridges and slopes they often lie side by side.

Goats and sheep may paw the ground with a foreleg or alternately with both forelegs before lying down. This removes debris from the sites and creates resting platforms on steep slopes. Pawing depends to some extent on the substratum. Markhor, wild goat, and Asiatic ibex usually did not paw boulders and other solid surfaces whereas they readily scraped depressions in loose shale, soil, or snow with 1 to 14 sweeps of a foreleg. Himalayan tahr along the cliffs of the Bhota Kosi were not observed to paw but those in New Zealand, inhabiting less precipitous terrain, often scraped beds into scree with 3 to 10 sweeps of a foreleg. Twenty percent of a sample of 223 Shey bharal pawed the ground before lying down, using an average of 3.3 (1–11) scraping movements. Bharal used some rest sites repeatedly, as did Punjab urial, judging by the droppings of different ages that bordered their beds. Saiga may also dig depressions before lying down in them (Pohle, 1974). Most animals reclined with legs tucked beneath them, but some extended one foreleg, or more rarely both forelegs. Goats and bharal often rested with an extended foreleg and urial did so occasionally (plate 30). Of 49 resting wild goat sampled, 39% had a leg extended and of 298 bharal 42% had one extended. Sometimes an individual was wholly relaxed, lying on its side with all legs stretched out and its head resting on the ground. One yearling male markhor climbed 4 m into an oak and there reclined with two legs dangling down each side of a branch.

Animals often chew cud during rest periods. Although rates of chewing vary with such factors as size of individuals and coarseness of food, 3 chews per 2 seconds seem to be a rough average. Two subadult male Himalayan tahr were timed as they chewed a total of 25 boli. The average number of chews per bolus was 78 (65–87) and the time required to chew each bolus was 49 (40–65) seconds. A class III bharal chewed 18 boli an average of 62 (44–73) times in 43 (37–60) seconds, and a class I markhor chewed 7 boli 62 (50–82) times in 38 (35–43) seconds.

Movements

Food is a critical resource for most sheep and goats, but animals often circumvent the immediate effect of shortages by such means as storing fat and moving to favorable areas. Consequently many populations have specific seasonal ranges within their general home ranges and in them they may shift locally depending on the weather. The extent of movement varies considerably from population to population even within a species. For example, mouflon on Corsica may remain the whole year within an area of 1 to 2 sq km but elsewhere they sometimes travel seasonally as far as 25 km (Pfeffer, 1967). Thus my comments on this subject, as on others, apply only to my study populations, not necessarily to the species as a whole.

Species living on isolated mountains at low altitudes move little in the course of a year, especially if their existence revolves around cliffs. As described earlier, Karchat wild goat tended to move from the cliffs toward the plateau in the morning, spend the heat of day in some ravine, and return to the same precipices at night, having made a circuit on the order of 3 to 6 km. Sometimes the goats foraged only along the bases of their sleeping cliffs during the driest time of year. Many females probably spent their lives within an area of 20 sq km, but some males possibly used two to three times as much terrain. Nilgiri tahr showed a similarly restricted movement pattern. One feral goat herd had a home range of 1 × 8 km in size (Shank, 1972), such herds showing great fidelity to their ranges (Coblentz, 1976).

The Kalabagh urial reserve encompasses about 40 sq km. Those sheep that leave the area are shot by poachers and this rather circumscribes the range of the population. The animals are usually concentrated into 2 to 5 unstable aggregations with only a few individuals scattered over the remaining area. Such aggregations shift with the erratic availability of green grass, which for much of the year grows best on the low slopes rather than on the ridges. In July 1972, after several heavy rains, most sheep were along the base of the ridges, but a year later when a dry spell had left the plains dessicated the animals were high up. Urial segregate altitudinally to some extent by sex if food is widely distributed, male herds being found especially on the high ridges. Individuals may spend days within a few hectares only to shift suddenly to a different area a few kilometers away, using the many well-beaten trails along ravines and over cols. Some feral Soay sheep spend their whole lives within 35 ha (Grubb and Jewell, 1974). Domestic sheep in paddocks walk about 3 to 10 km in the course of a day depending on the condition of the pasture (see Squires, 1975).

The great altitudinal differences in the high Himalaya may involve species in considerable range shifts. But the factors that induce such shifts may vary from species to species. Some are possibly motivated internally to migrate, with external factors merely synchronizing the movements, whereas others respond only to local climatic conditions. The depth of the snow cover seems to have an influence on the travels of Asiatic ibex, temperature being of secondary importance. Markhor sometimes migrate to low altitudes over a month before their summer range is covered with snow, suggesting that a drop in temperature may stimulate them to move. Chamois seem to have little tolerance for temperature extremes (C. Clarke, pers. comm.) and this may cause them to shift to low altitudes during November in the Alps (Pfeffer and Settimo, 1973).

Having spent the winter in the valleys below 2500 m, markhor in the Chitral Gol begin their trek to the summer range in May, according to my local informants. Females and young are said to remain along the upper fringe of the forest but males may venture into the high crags. The total distance between winter and summer range in this population is 10 to 15 km and the total annual range is about 80 sq km (fig. 27). The downward trek to winter quarters occurs in November and most animals are at low altitudes by the end of that month. One herd of about 10 females and young remained on its winter range all summer in 1972. At nearby Tushi, markhor occupy the same range all year, only a few males wandering northward to higher altitudes. The Chitral Gol population is divided into fairly distinct units on its winter range, each confining itself to a certain part of the valley. While some individuals no doubt switch units, most stay in the same area if one judges by some easily recognizable animals. Units consist of up to about 35 females and subadults and the attending males. All members of a unit are seldom together, as singly and in groups they roam over the winter range. Some males are not attached to units, and these wander alone or in small male herds throughout the lower valley. In 1970 the population was divided into six units, as shown in figure 27; in 1972 I found five units but possibly overlooked a sixth one in a densely forested part of the valley. The distribution of units was similar in 1970 and 1972. I did not check the western half of the valley in 1974. The ranges of units were small, 2 sq km or so in size. The whole Tushi population of about 125 represented one unit, all members of which were sometimes crowded into 1 sq km or less. The markhor in the Chitral Gol had a conspicuous diurnal move-ment pattern probably related to the thermocline. In winter, mountain valleys show temperature inversions, with a layer of warm air above the cold in the bottom, a cline readily noticeable even to a casual hiker. Most animals were high on the slopes in the morning, often hidden in thickets of oak. At around 1300, after the bottoms of the valleys had warmed, animals descended and foraged low on the slopes until after dark. Markhor in the Haramosh Range also descend toward evening (Darrah, 1898).

Some Asiatic ibex descend to lower altitudes in winter, both by descending slopes and migrating down valleys, whereas others remain in the same general area all year. Ibex penetrate up the Baltoro Glacier as far as Concordia (G. Rowell, pers. comm.) but in May, when I visited the region, the animals were rare beyond Urdukass, which is over 15 km downstream from Concordia. I suspect that many ibex winter in the lower areas and move up the valley with the retreating snows in spring. The preferred feeding

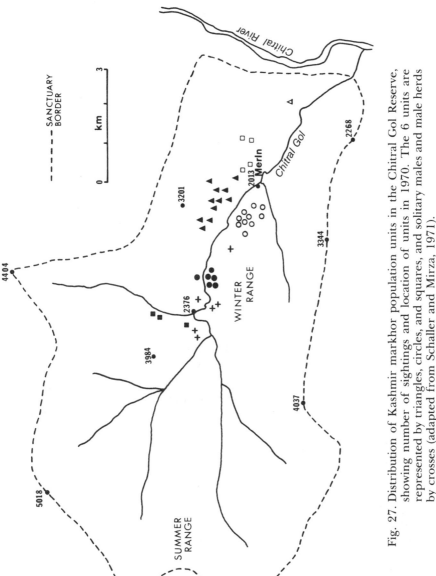

Fig. 27. Distribution of Kashmir markhor population units in the Chitral Gol Reserve, showing number of sightings and location of units in 1970. The 6 units are represented by triangles, circles, and squares, and solitary males and male herds by crosses (adapted from Schaller and Mirza, 1971).

sites of the animals were along the snow line in May. Many ibex fed on the precipices around the village of Besti in winter. Few are there in summer, most having moved 5 to 15 km up the valleys to different pastures but not necessarily to higher altitudes. Ibex were around Kilik Pass at 4500 m in late November, and villagers told me that they stayed there all winter. A downward movement may occur not in winter but in spring when young green grass first becomes available on the lower slopes (Stockley, 1936). A similar descent in spring was reported by Holloway and Jungius (1973) among Alpine ibex and chamois in Italy. Nievergelt (1966a), writing about Alpine ibex in Switzerland, noted that "in spring, the animals were in the lowest part of the area, in summer in the highest and in winter in the middle regions, which can be explained by the variations in temperature and vegetation."

Himalayan tahr on a cliff along the Bhota Kosi were divided into two population units separated by a vertical strip of forest. Females and subadults seemed to associate only with members of their respective units, but adult males were not as sedentary. These tahr remained for over a month on about 1 sq km of terrain, but in summer they possibly moved into the alpine zone. Caughley (1966) noted that female tahr in New Zealand have distinct home ranges about 1400 m in diameter where they are visited by the widely roaming males. Tahr descend with the winter snows "and seek the cover of rocky outcrops and other sheltered places in bad weather" (Christie and Andrews, 1964). In spring, the animals return to their summer pastures, which are seldom more than 1000 m above those used in winter. Female herds often move down from the cliffs in the late afternoon and retreat upward at dawn (Caughley, 1967).

Shey bharal were divided into four population units in late October, the largest one numbering 61 individuals. Each unit foraged on a certain slope, its pastures separated from those of its neighbors by deep valleys or snowfields. There was more movement between units in late November after the rut had begun and the snow had somewhat retreated. The total winter range of the Shey population was at most 20 sq km, more than half of it covered deeply with snow. Some units confined themselves for weeks to areas of 1 sq km or less, and the daily movements of animals probably did not exceed 2 km. Some herds remained in the same place for several days, something that was also true of those at Lapche. According to the villagers, some bharal disperse into the mountains during the summer. Bharal, like Asiatic ibex and markhor, tended to remain in the middle of the slopes in winter, avoiding the cold temperatures at high altitudes and in the valleys.

The Stone sheep studied by Geist (1971a) had more complex movement patterns than any species I observed: "The number of seasonal home ranges a sheep may occupy may be as high as six or seven if it is a ram or a maximum of four if it is a ewe. A minority of sheep have as few as two seasonal home ranges. The home ranges can be anywhere from half a mile to twenty miles or more removed from each other.... Sheep are loyal to their home ranges and return to them in the same season year after year." For rams these ranges may include a prerutting area in autumn, the rutting grounds, a midwinter range, a spring range, and a summer range.

Other Caprinae are as variable in their movements as those already discussed. The movements of goral in Russia seldom exceed a distance of 2 km, the animals retreating to patches of exposed cliff when the snow is deep (Bromlei, 1956). Dang (1968b) noted that goral and serow in the Himalaya do not change their ranges seasonally except for altitudinal shifts in severe winters. Japanese serow also tend to remain attached to one site (Okada and Kakuta, 1970). Takin move seasonally up and down slopes (Wallace, 1913; Bailey, 1915), as do mountain goat (Brandborg, 1955; Hjeljord, 1973). Among muskox, "Winter and summer ranges on the Arctic islands may be up to 50 miles apart, adjacent to each other, or even overlap" (Tener, 1965). Solitary bulls and bull herds may frequent areas different from those used by mixed herds (Spencer and Lensink, 1970). Chiru travel long distances to new grazing grounds (Bartz, 1935), and saiga may move at a rate of 100–120 km per day and roam seasonally over an area of 20,000 sq km in search of forage and water (Bannikov et al., 1967).

BODY-CARE PATTERNS

Sheep and goats use various comfort movements, especially during their rest periods. They yawn and they stretch, usually by tucking in chins and humping backs slightly or by extending the hindlegs far back. They defecate casually, walking or standing with their tails raised. Females may squat deeply when urinating, in contrast to males who barely lower their rumps. Most body-care patterns consist of grooming the pelage. Animals lick and nibble themselves on the sides, rump, legs, and groin. Unable to reach head or neck with the muzzle, they use a hindleg to scratch these parts. The horns are also useful tools for scratching, a point particularly evident among male ibex and wild goat, who can easily treat an itch on the rump with a tip of their long scimitar-shaped horns. Occasionally an animal shakes its body or rubs itself against a rock or tree. Table 25 shows how often several species groomed themselves by using the various techniques. Shaking was not quantified because its frequency is influenced too much by precipitation. *Capra,* bharal, and tahr licked more often than they scratched but among sheep the reverse was true. Geist (1971a) noted that mountain goats seldom licked and scratched and instead threw dry soil over themselves. He attributed this to the animal's thick coat. Captive aoudad take dustbaths by lying down and throwing sand over their bodies with their horn tips (Katz, 1949). Male Himalayan tahr have a voluminous and dense ruff, yet their comfort movements resemble in frequency those of the less hirsute goats. *Capra* females have much longer horns than do urial and bharal females and they use these proportionately 3 to 4 times as often to scratch themselves. Stone sheep scratch more with their horns than do urial (table 25), a difference for which I have no explanation. Stone sheep also rubbed themselves on objects more often than any species I studied. However, Geist (1971a) collected his data during the molt when wool peeled off the animals in mats.

The members of a courting pair and females with small young sometimes groomed each other, but otherwise comfort interactions between individuals were uncommon. Once two adult wild goat males licked each other, and twice markhor behaved similarly: a yearling male nuzzled the chest of a female and a class I male licked the forehead of a class II male. Punjab urial rams may rub each other's faces and bharal males each other's rumps, but such behavior was related to dominance, not grooming, as noted later, and it was seldom shown by females. Only Himalayan tahr exhibited grooming between adults in a way that seemed to be partly utilitarian. Social grooming with one tahr licking the neck and head of another was observed on 61 occasions. Interactions

may be cursory or last as long as 10 minutes. One animal sometimes invited licking by holding its head close to the face of
another. Most licking involved either two females (13 interactions)
or a female and a young with the former being licked on 8 occasions and the latter on 12. One yearling male licked the same
female 11 times and the female always reciprocated. An adult
female also licked an adult male and the reverse occurred once.
Muskox may groom each other mutually by rubbing their rumps
together (Gray, 1973).

Responses to Danger

Caprids occasionally meet other species of animals during their
daily routine. Most present no danger and are treated with indifference, passing interest, or casual avoidance. A tree pie may fly
onto the back of an urial or wild goat, and a yellow-billed chough
may perch on a bharal, eliciting at most a shake. Snowcock forage
casually among ibex and bharal. Markhor may watch low-soaring
griffon vultures. Chinkara may amble through urial herds without response, four sambar once passed within 2 m of Nilgiri tahr,
and the associations of chiru, kiang, and argali have already been
mentioned. Livestock is also tolerated. Lapche bharal grazed
within 10 m of yak, and Nievergelt (1966a) shows a photograph of
an Alpine ibex in a herd of domestic goats. Even potentially
dangerous creatures are not avoided as long as they remain at safe
distances. This distance depends on the past experiences of the
animals, as Krämer and Aeschbacher (1971) have shown for Alpine ibex. The average flight distance of herds in response to a
person on foot varied from 21 to 373 m. Lapche bharal slowly
retreated as villagers began to cut grass 120 to 150 m away. Shey
bharal permitted me to observe them from a distance of 60 to
80 m, and sometimes from as close as 40 m, without noticeably
altering their behavior. Several Punjab urial ignored an Indian
fox 5 m away, and several kiang barely looked up as a wolf passed
at 10 m (Stockley, 1928). However, man and the other large predators are usually dangerous, and the animals respond by fleeing.

Markhor and bharal retreated to cliffs when seriously threatened, as shown by the wolf incidents described earlier. When
disturbed in the open, Nilgiri tahr characteristically bunched up
and ran to the nearest cliffs, which in most instances were within
half a kilometer. On three occasions, however, tahr left the

protection of some precipice and fled across undulating terrain for two or more kilometers. Such behavior was also observed in Karchat wild goat, which at times retreated not to the closest cliffs but instead moved off across the plateau. Markhor, ibex, and bharal often just moved at a fast walk up or parallel to a slope in response to spotting a person, merely putting distance between themselves and the danger. As noted earlier, Eurasian sheep do not seek refuge in cliffs but American sheep do so.

The bunching up of Nilgiri tahr in times of danger is typical of many caprids, including wild goat, markhor, and bharal. Scattered urial immediately bunched when chased by dogs. Mountain sheep and mountain goat also crowd together when danger threatens (Geist, 1971a). Krämer and Aeschbacher (1971) twice saw eagles swoop over Alpine ibex herds that included small young. After bunching up, the adults reared on their hindlegs and lunged at the birds with their horns. A predator finds it difficult to select and pursue a victim in a crowded herd.

Muskox form a tight circular or semicircular defense with their horns facing the predator (Gray, 1974). Interestingly the yak, which in appearance and habitat resembles the muskox, uses a similar defense: "all rush together and remain thus with their heads toward the threatening danger" (Rawling, 1905).

"A sentinel is invariably posted to watch over the slumber of the herd," wrote Fletcher (1911) regarding Nilgiri tahr, and Markham (1854) made a similar statement about bharal, Anderson and Henderson (1961) about Himalayan tahr, and Macintyre (1891) about Asiatic ibex, to mention just a few examples that express a widely held belief. The sentinel is usually said to be a female. Indeed, one animal sometimes stands in a prominent spot. Females, being the most vigilant of the sexes as well as herd leaders, often halt on conspicuous places and scan the terrain. Rutting males often stop on pinnacles as if searching for females. If one animal responds to potential but not imminent danger, the others may seemingly ignore the situation, giving the impression that a sentinel is on duty. Once a female markhor watched me alertly for 10 minutes while the rest of the herd, equally aware of my presence, continued to browse. With respect to Nilgiri tahr, Schaller (1971) wrote:

> In 7 out of 24 resting mixed herds observed, one animal (5 females, 2 saddlebacks) stood or reclined conspicuously above the others and

would have fitted the popular definition of sentinel. However, such animals achieved their isolated position usually by accident rather than choice. On two occasions a female reclined while the herd continued to graze. The other animals passed and finally rested on the slope below leaving a "sentinel." Of course, an animal in a prominent position is more likely to spot a potential source of danger before the others and it thus functions as a sentinel without the need to imply that the behaviour is purposeful. The general restlessness of herds also helped them to detect danger. With one or another animal almost constantly shifting position or grazing, it would have been difficult for a predator to approach undetected.

While adept at spotting danger below them, animals often fail to glance up and it is sometimes possible to creep close to them from above.

The first animal to sense a critical situation flees immediately, drawing all others with it. Often the herd then stops briefly some 100 m away and looks back. Although sheep and goats have acute eyesight, able to detect a moving person at 1.5 km, they sometimes respond only after having scented the danger. Once 15 Nilgiri tahr came to within 20 m of me as I reclined on a slope. At first they whistled and stamped their feet, but then after several minutes they reclined and chewed cud. The wind shifted 30 minutes later and all bolted. Similarly, McDougall (1975) noted that feral domestic goats flee more quickly in response to human scent than to the appearance of a person. Bharal are particularly fearful of dark moving forms, even becoming alarmed at the sudden appearance of other bharal silhouetted against the sun. Perhaps these resemble wolves.

Caprids communicate the existence of danger by using several visual, auditory, and olfactory signals. Others often respond by facing the source of the disturbance and they may then signal too. All species use an alert stance with body held rigid, neck erect, ears raised. The tail may be vertical, a position which in bharal fans out the white rump hairs. A foreleg is sometimes stamped audibly. Air may be expelled forcefully through the nostrils to produce distinctive sounds. Punjab urial snort, a sound almost like a sneeze. Both Himalayan and Nilgiri tahr emit piercing whistles. One male Nilgiri tahr gave four whistling snorts that seemed intermediate between aggressive snorts and whistles. Wild goat grunt ö-ö-ö, a sound which can be heard for nearly one kilometer.

Markhor make a similar noise, and they also snort when startled. The Asiatic ibex has a high-pitched whistle or chirp reminiscent of a bird, as does the Dagestan tur (Petzsch and Witstruk, 1958). Surprised Alpine ibex snort (Krämer and Aeschbacher, 1971). Bharal make a sound resembling that of an annoyed red squirrel, a *chit-chit* or a slurred *chirrt*. Five out of 28 calls consisted of single chits, 19 of double notes, and 4 of triple ones. Other Caprinae also use alarm calls. Chamois whistle much like Alpine ibex, goral produce a combined hiss and cough (Stebbins, 1912), serow emit "a kind of sharp shriek" (Macintyre, 1891), muskox snort (P. Lent, pers. comm.) and takin give a hoarse cough (Wallace, 1913), to name just a few examples. A call by one herd member alerts all others and they respond by at least glancing around. When one Nilgiri tahr whistled, 33 others fled 30 m before halting and looking back.

Some species use distinctive gaits to communicate the presence of danger. Punjab urial hit the ground up to a dozen times with all four legs in unison before continuing their flight at a trot or gallop. This gait resembles the spronking of antelope, but it has less bounce and the emphasis is on thumping the forelegs. Such thumping is especially prevalent by animals in dense brush and by ewes when they have small lambs. Muskox may thump their forefeet when fleeing (A. Oeming, pers. comm.), and *Capra* also do so, but rarely. One markhor male made three exaggerated leaps during which he thumped his forelegs, and a wild goat female once behaved similarly. One Nilgiri tahr female repeatedly leaped into the air at the same spot and brought her forelegs down forcefully. A bharal male thumped the ground three times with all four feet as he fled. The distinctive gait alerts other herd members both by sight and sound, and it may also deposit concentrations of scent from the pedal glands as a delayed warning system to others passing later. The glands in Punjab urial exude a sticky substance which smells sweetly musky. It is significant that urial with pedal glands on all four feet use a spronking gait, whereas goats who have pedal glands only on their forefeet, if at all, thump only those two feet. Bharal, which may or may not have glands on all feet, bound like urial.

Falling rocks and avalanches are a constant source of danger in mountains. One markhor inadvertently loosened a boulder. Hearing it crashing down, the animals below scrambled aside without looking up. They then trotted back and watched a small rock avalanche tumble down a gulley. When on another occasion a boulder rolled toward a reclining markhor, she leaped into the air and sideways with one movement. Geist (1971a) noted that mountain goats cower against cliffs during rock falls, and similar behavior was observed in chamois (Krämer, 1969). The thunderous fall of a boulder in a canyon caused several Himalayan tahr to crowd against the cliff wall. K. Tustin wrote me that tahr "make very deliberate attempts to conceal themselves from helicopters, crouching under scrub, standing rock-still in crevices or under overhanging rocks and caves."

8 HERD DYNAMICS

The basic unit in caprid society consists of a female with her offspring. All other social systems represent permutations on this basic theme. With the exception of the rather solitary goral and serow, the Caprinae are all highly social animals living in herds consisting of a variable number of males, females, and young.

One problem in the field was to define what constitutes a herd. If the animals occupied the same piece of ground, were in sensory contact, and were more or less coordinated in their movements, I considered them to belong to one herd. While these criteria may sound vague, groups could in practice be delineated with little difficulty, only occasional peripheral individuals presenting a problem. Animals may restrict themselves to certain localities where they form fairly discrete population units composed of one or more herds of unstable composition. Solitary individuals aside, herds were of three types: (1) male herds, composed solely of males; (2) female herds, consisting of females with or without young and yearlings of both sexes; and (3) mixed herds, including adult males, females, and subadults. Herd compositions are here listed as they were at the first contact, to avoid inconsistencies due to the joining and parting of individuals. Herd size depends to some extent on population size. With wildlife so decimated, herds tend to be smaller than they normally would be, something to remember when trying to assess degrees of sociability among the various Caprinae.

HEMITRAGUS AND CAPRA

Himalayan tahr

I tallied 36 herds along the Bhota Kosi in March. The mean size of herds was 6.5 with a range of 2 to 23 (table 26). Female herds contained an average of 4.0 animals, mixed ones 10.2. Sixty-four

190

percent of the herds were mixed. Thirteen class III males were seen and of these 10 were in mixed herds and 3 were solitary. In addition, 4 other solitary males (1 yearling, 1 class I, 2 class II) and 4 male herds were seen. Three of the latter consisted of a yearling male and a class I male, and the fourth one contained 2 yearlings, one class I male, and one class II male.

At the beginning of the rut in May, I classified 14 herds in New Zealand. Average herd size was 8.8 (3–22), the maximum being considerably less than the herd of 170 that was once seen in the area (K. Tustin, pers. comm.). Thirteen of these herds were mixed, nine containing 1 class III male and two containing 2 class III males in addition to one or more other males. Of 24 class III males seen, 33% were solitary, 12.5% were with a class II male, and the rest were in mixed herds. Proportions were similar for class II males, but only 9% of the class I males were solitary, the rest being in mixed herds. All yearlings remained in mixed herds.

The segregation of sexes may at times be more marked than it was during my visits to the Bhota Kosi and New Zealand. Stockley (1928) once saw a herd of 24 male tahr, and Caughley (1967) counted 54 males together in New Zealand. Caughley (1966) wrote that "during the summer thar range in three main kinds of groups: one consists of females, juveniles and kids, a second consists of young males and the third of mature males." Male herds may further segregate into small groups of 4 to 8 immatures and into groups of adults (Caughley, 1967). C. Challies (pers. comm.) once counted 710 tahr in a New Zealand valley, only 5 of them adult males, and in another valley he saw 50 scattered males but no females, examples showing well how the sexes may segregate.

The tahr in my Bhota Kosi study area were divided into three population units, one with at least 10 females and 5 young and another with at least 12 females and 5 young as well as a variable number of males (table 27).

Nilgiri tahr

In the course of censusing Nilgir tahr, I counted 23 female and mixed herds and these varied in size from 6 to 104 with an average of 22.6. Sterndale (1884) mentioned herds with 60, 65, and 120 animals, and Kinloch (1926) with 60 and 90, indicating that the usual upper limit in herd size is around 100. Of the 22 herds whose composition was ascertained, 6 were female herds varying in size from 6 to 18 (mean 9.3.). Four of these herds contained a disproportionately large number of young. For example, one consisted of a yearling female and 5 young and another of 2 adult females and 6 young. One mixed herd of 15 contained a class I male, 4 yearling females and 10 young. These compositions indicate that young had left their mothers to associate with each other and with yearling females. A somewhat comparable situation occurs among bighorn sheep in Idaho in that yearlings and lambs separate from the ewes after the rut and remain on their own at least until late spring (Smith, 1954). Mixed tahr herds averaged 27.4 members in size and 12 of the 16 herds contained at least one class III male. During October, class II and III males were leaving mixed herds to become solitary or to join male herds. Early in the month, when I worked in the Nilgiris, half the males were in mixed herds, but by the end of the month, when I was in the High Range, only 19% were in mixed herds (table 28). All males remained in mixed herds until the age of at least 3 years, in contrast to Himalayan tahr males, which sometimes left mixed herds before the age of 2 years. Nine male herds varied in size from 2 to 12 with an average of 5.5. Some herds contained only class III males, others both class II and III males. Male herds in the High Range tended to congregate. Whereas mixed herds were scattered along the cliffs, 30 of the 69 large males were in the western corner of the reserve.

The large mixed herds seemed to be population units (table 27) in that each remained along the same cliffs for many days. Most or all members of a unit were often together, a feature in which they differed from Himalayan tahr.

Wild goat
Karchat wild goat herds ranged in size from 2 to 103 (fig. 28) with an average of 19 in March and 24 in September-October (table 26). In the Chiltan Range, where the density of animals was lower than at Karchat, 37 herds had an average size of 4.1 in August (Z. Mirza, pers. comm.) and 25 herds had an average size of 5.4 in November, the largest herd comprising 9 individuals. Heptner et al. (1966) noted many herds with up to 30 animals and a maximum of 90, and Danford (1875) and Stockley (1928) reported herds with up to 100 individuals, figures which correspond

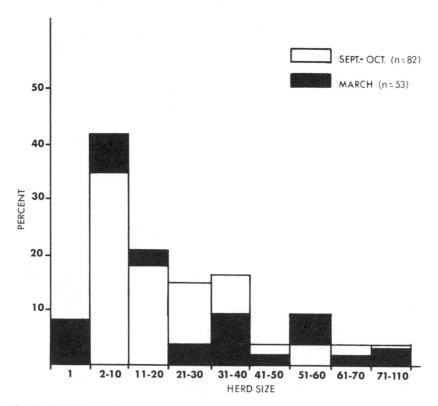

Fig 28. Herd sizes of wild goat at Karchat during the rut (Sept.–Oct.) and birth
season (March).

well with mine. Females are often solitary during the birth season.
After the newborn are mobile, females congregate into female
herds. Average female herd size at Karchat in March was 8.2 with
the largest herd comprising 52 animals; in September it was 4.0.

Most males are in mixed herds during the rut, only 8% being
solitary or in male herds at that time. Among these 8% were seven
herds of 2 males and two of 5 males, and the remaining males
were solitary animals of classes IV and V. Of the animals in male
herds, 54% belonged to classes IV and V, 13% each to classes III,
I, and yearlings, and 8% to class II, indicating that young males
tended to remain with the females. While males of similar horn
length tended to associate, occasional anomalous pairings also occurred, with for example one class IV male being followed by a

yearling. I saw males only in mixed herds during March (plate 33) probably because the male herds had moved outside of the study area. In the Chiltan Range some 95% of the males were in mixed herds in November and 66% in August (Z. Mirza, pers. comm.). In Iran, I saw several mixed herds in June as well as a male herd with about 60 individuals. These scattered observations indicate that some of the adult males may be found in mixed herds at all seasons.

Fifty-five percent of small herds, those with fewer than 20 individuals, were without fully adult (classes IV and V) males during the rut, whereas all large herds contained at least one such male (range 1–13, mean 6.4). The figures for March were 18% in small herds and 86% in large ones (range 2–10, mean 5.3). The largest herd at Karchat contained 103 individuals, including 10 class IV and V males, 6 III, 4 II, 2 I, 2 yearling males, 48 females, 5 yearling females, and 26 young. Adult males often formed a subgroup within or at the periphery of such a large herd.

An island population of 125 to 175 feral goats in Canada varied in herd size from 2 to over 100, with an average of about 10 to 15. The animals showed no obvious sexual aggregation at any season although occasional solitary males and small male herds did occur. Shank (1972) attributed this to the fact that the goats bred throughout the year.

Markhor

I observed markhor mainly between November and January. Mean group size in the Chitral Gol at that season was 9.0, excluding male herds (fig. 29); 63% of the herds were mixed and 37% consisted of females and subadults. Average mixed herd size was 12.4 (3–35) and female herd size 4.5 (2–10). Only one female was solitary. Thirteen percent of the males were alone or in male herds. Among the 33 such males tallied, 5 belonged to class I, 5 to II, 11 to III, and 12 to class IV. Yearling males remained with the females. The largest male herd comprised 3 individuals, but others (Darrah, 1898; Stockley, 1928) reported as many as 6 together. Only 44% of the small mixed herds, those with fewer than 10 individuals, contained a class III or IV male, whereas 65% of the large mixed herds had such males with them. Twenty-four percent of the large mixed herds had more than one class III or IV male, with a maximum of four.

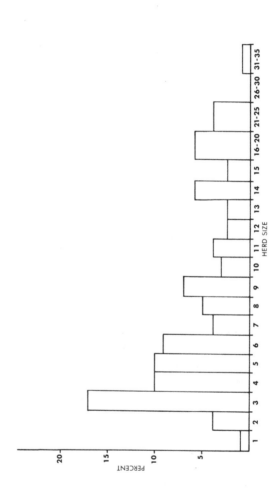

Fig. 29. Herd sizes of Kashmir markhor in the Chitral Gol, November to February.

Eighteen herds at Tushi had an average size of 16.1, the largest one comprising 92 animals. Ten of the 18 were female herds.

The females and subadults in the Chitral Gol were divided into population units as noted in Chapter 7 (fig. 27). One unit contained at least 38 individuals in 1972 but no more than 24 of these were seen together in one herd. Another unit consisted of 18 individuals in 1970 and of 24 in 1972. Table 27 shows the composition of the three largest herds seen in 1972–73, herds which contain most individuals in their respective units.

Ibex

Mixed Asiatic ibex herds varied in size from 2 to 26 and female herds from 2 to 15, with a combined average of 9.4. Heptner et al. (1966) noted that most ibex herds ranged in size from 3 to 40 with a few up to 100 and 200. In November and February, just before and after the rut, 71% of 16 herds were mixed, in May 80% of 4 herds were mixed, and in July-August 50% of 4 herds were mixed. Haughton (1913) reported a mixed herd of 29 in March, Darrah (1898) several small mixed ones in May, and Demidoff (1900) a mixed one of about 40 in August. Heptner et al. (1966) also noted that mixed herds occur at all seasons. Ibex form male herds too. Three such herds were seen, one of 4 in May and others of 2 and 33 in August. One male was solitary.

Grouping patterns of Alpine ibex and Walia ibex have been described in detail by Nievergelt (1974). This author noted that "ibex form all-female or all-male groups except during the rutting period when they join together in mixed groups." However, his data show that mixed herds may occur at all seasons. Walia ibex have an extended rut, and the percent of mixed herds in which at least one male was present was as follows: December to February, 56%; March to May, 54%; June to August, 48%; September to November, 39%. Male herds with up to 30 members—almost all the males in a colony—have been reported for Alpine ibex, and solitary males are also common, especially from November to March. Female Alpine ibex are alone mainly during the birth season. Walia ibex are so rare that large herds do not occur (mean herd size was 3.1). Consequently males and females are often alone.

PSEUDOIS AND AMMOTRAGUS

Bharal

Bharal may associate in herds of up to 200 (Stockley, 1928) and even 400 (Schäfer, 1937) individuals. The largest herd I saw numbered 61, and means ranged from 4.8 to 18.4 depending on area (table 26). The average size of female herds at Shey was 5.9 and of mixed herds 20.7. Males tend to separate from the females after the rut (Burrard, 1925) and become either solitary or form male groups which may contain as many as 40 members (Schäfer, 1937). However, a few males associate with the females throughout the year (Kinloch, 1892; Sheldon, 1975). Solitary males were seen in October, November, December, and March, but they were fairly common only during the height of the rut (table 29). A large percent of herds were mixed, 40% in the Sang Khola during October, 53% at Shey in November-December, and 83% at Lapche in March (plate 42). About a third of the class I to V males at Shey were in male herds with up to 15 members until the onset of the main rut in late November. Male herds then broke up, showing a significant drop in average size, and the animals then often traveled singly and in twos and threes or joined mixed herds (table 29). All solitary males at this time belonged to classes IV and V. Between November 1 and 9, 52% of the males in male herds belonged to classes IV and V, 36% to class III, and 12% to classes I and II, but after that date 96% of the males in male herds belonged to classes IV and V, showing that most young males had joined mixed herds. Yearling males remained with the females.

Schäfer (1937) noted that herds have preferred ranges. The bharal at Lapche were divided into three population units and those at Shey consisted of four units at least while deep snow confined each to a separate slope (see table 27).

Aoudad

In their native habitats, male and female aoudad are said to be quite solitary (Brehm, 1918) or they "associate in small family groups" (Walker, 1968). The sexes separate after the rut (Brouin, 1950), behavior which also occurs among the introduced aoudad in California where males may roam up to 65 km from the nearest female herds. The 530 aoudad studied by Barrett (1966) in

California were divided into at least two population units, one of which, the Red Rock herd, contained 258 animals including 180 adult females.

Ovis

Punjab urial at Kalabagh sometimes collected into herds of from 30 to 40 individuals and in one instance two disturbed herds joined to form a herd of 85. Table 30 presents mean herd sizes at various times of year. Females herds were usually small, averaging 4 to 6 members, but up to 50 individuals congregated briefly on occasion. The average size of mixed herds was 9 to 13, with the least number of animals remaining together in October-November when rutting rams disrupted herds. Some rams associated with mixed herds at all times of the year but the frequency with which they did so varied with the seasons as well as the age class. Figure 30 expresses this in terms of percent rams in male herds. The most rams were in male herds during the April birth season and the fewest in December just after the rut. Yearling rams seldom left the ewes, at most 16% joining male herds. Class I rams were more venturesome than yearlings, 3% being in male herds in December and 41% in April. Class II rams resembled young rams in that most were in mixed herds in December but differed in that by April 57% had entered male herds. Adult rams (III and IV) preferred to associate with their own sex. Over a third of these rams were in male herds even during the rut, and by April the figure had risen to two-thirds. Generally, the older the ram the less time he spends with ewes. While this sounds paradoxical in view of the fact that adults do most mating, the results can be explained by noting that adult rams roam widely in search of estrous ewes during the rut without tarrying in herds that have none available. It is also noteworthy that more young males are in mixed herds after the rut in December than during the rut. Possibly the aggressiveness of rutting adult rams deters some young rams from remaining near the ewes. The largest male herd contained 24 rams, and the average varied with the seasons; it was lowest (3) during the rut and highest between December and April (7–8). Male herds broke up progressively at the beginning of the rut as the average sizes from 1970 illustrate: October 6–22, 4.6; October 23–31, 2.3; and November 1–9, 1.7 (Schaller and Mirza, 1974). Rams of all ages associated in male herds

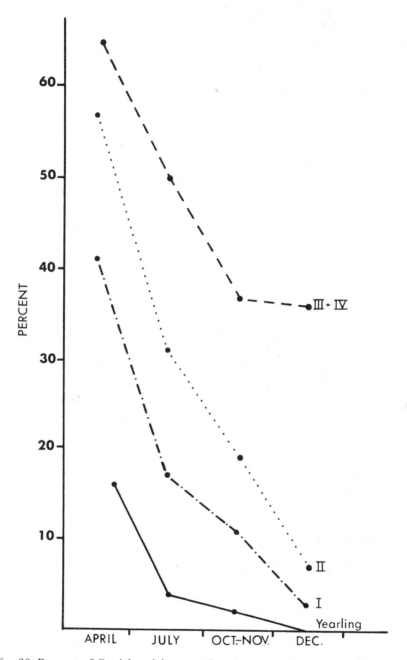

Fig. 30. Percent of Punjab urial rams of various age classes in male herds at different seasons.

throughout the year, but, as table 31 shows, an animal preferred another of about its own size for company. Some rams become solitary for brief periods. Of 157 solitary individuals tallied, 9% were yearlings, 17% class I, 19% class II, and 56% classes III and IV, figures not too different from the composition of the ram population as a whole (see table 17). Thirty-seven out of 40 lone ewes were seen during the birth season in April.

Although urial form population units which superficially resemble those of goats, these units consist of unstable and shifting aggregations at temporary food sources. This, of course, does not preclude the presence of small stable units in areas of low urial density. Grubb (1974a) described sedentary and persistent female herds among feral Soay sheep, and rams too formed close associations although they readily switched their allegiance to other rams, especially after the rut. One herd of 7 desert bighorn sheep remained together for at least 5 weeks (Welles and Welles, 1961).

Other *Ovis* resemble urial in certain aspects of their herd dynamics. Mouflon live in small female and mixed herds between October and December, the time of rut, and many rams are at that time solitary. Average herd size increased after the rut, reaching over 25 individuals, but by March most rams had separated from the ewes (Pfeffer, 1967). With respect to Stone sheep, Geist (1971a) wrote: "The rutting season in late November and December is preceded by a gathering of rams from late September to mid-October. In early November the rams disband and move to their respective rutting areas .... The rams remain with the females for about two weeks after the females have been bred, and then the larger and older rams move to their own wintering areas." Male herds contained up to 10 individuals and female herds up to 19 in this population. Stray notes on argalis also point to a segregation of the sexes during much of the year. Cobbold (1900) noted that Marco Polo rams remain at higher altitudes than ewes until the rut in December. In September, Dunmore (1893) saw many Marco Polo herds, and he complained of "very few rams being amongst them, and those that were with them were young ones." However, he did see several small ram herds and later in October one with 32 individuals. Demidoff (1900) encountered a herd of about 60 rams in July in the Altai, and many authors (Macintyre, 1891; Darrah, 1898) met Tibetan argali in separate ram and ewe herds in Ladak.

OTHER CAPRINAE

A few comments on herd structure in the other Caprinae help to place my observations on sheep and goats into a broader perspective. Rutting saiga males collect harems of from 5 to 15 females, and young males form herds with up to 100 members at that time. The sexes segregate after rut, with male herds then averaging 5 to 10 and female herds up to 50 individuals, with some exceeding 1000 (Heptner et al., 1966). Rawling (1905) wrote about chiru: "Anywhere and everywhere above 15,000 it is to be found, either singly, in twos or threes, or in tens of thousands ... ;" males and females move to different areas in summer, and "as far as the eye could reach the whole country seemed covered with does and their young." Muskox may be found in herds of up to 100, but more typically in herds of 5 to 20 (Tener, 1965). Mean herd size on Bathurst Island was 4 to 7 in summer and 21 to 22 in winter. Herds there were unstable in composition, with individuals commonly switching from one to another, but a few herds retained their cohesion for a month or two (Gray, 1973). Some bulls are solitary or in small bull herds (Spencer and Lensink, 1970). Little is known about the herd structure of takin. Ward (1921) observed a herd of 14, Cooper (1923) of 29, and herds with up to 300 individuals may be seen in early summer (Yin, 1967). Old bulls either live alone (Allen, 1940) or in small herds with up to 4 members (West, 1926). ·

As for the Rupicaprini, the serow is essentially solitary but as many as 5 may be together (Prater, 1965). According to Dang (1962) they "dwell in devoted pairs." I observed two lone animals and a group consisting of 2 males, a female, and a young. Akasaka and Maruyama (in press) studied the social organization of Japanese serow:

> Through the observation of 17 identified serows, 4 social units were classified. The pair, consisting of a male and a female, was formed from late autumn to early winter and maintained till parturition in late May and early June of the next year. The family, consisting of a pair and its young, appeared after parturition and was maintained till early autumn when the male left. Then, the mother-child group appeared, and lasted till the next summer. Members of those units did not necessarily act together. In a family and a pair, a male took a role as a leader. And solitaries were also observed.

Male goral are often solitary but pairs form during the rut

(Heptner et al., 1966). "Small parties of from four to eight in number" also occur (Stebbins, 1912). I saw 4 single goral and two pairs. Mountain goat herds seldom contain more than 30 to 40 individuals. "Females leave the bands to seek isolation before the kids are born, only to rejoin them a few days later with the new offspring. The males are usually seen singly or in small groups during the spring and summer. During the rut they join the females, yearling and kids . . . . Aggregations grow smaller during March and April as the males gradually move away and the gravid females begin to split from the groups" (Brandborg, 1955). Mountain goats also tend to aggregate in January and February when snow is deep and good feeding sites are few in number (Holroyd, 1967). C. Clarke (pers. comm.) observed herds of 200–300 chamois in the Tatra Mountains. But such concentrations break up when during the rut some males become territorial and others join drifting female herds. At any time of the year males may be solitary, in male herds, or in mixed herds, and during the rut they tend to avoid contact with others of their sex. Chamois live in an open society without personal bonds except those between a mother and her offspring (Krämer, 1969).

DISCUSSION
The various herd systems in the Caprinae show some common features but also some important differences. Except for the rather solitary goral and serow, all species may aggregate into herds numbering at least 50 and often 100 or more individuals. Male, female, and mixed herds occur in all these species too. Some female herds have characteristic compositions at certain seasons. Chamois females with small young tend to segregate from females without young (Krämer, 1969), behavior also observable in wild goat and Punjab urial. Yearlings and large young may form their own groups in some populations of mountain sheep, Nilgiri tahr, and bharal. Himalayan tahr yearlings often establish temporary groups during the birth season and then return to their mothers (Caughley, 1967). Males may become solitary, but females seldom do so except shortly before and after giving birth. Unstable as herd members, adult males often switch from mixed herds or male herds to a lone life. Although males tend to separate from females at certain seasons, the degree of sexual segregation varies with the species. In Punjab urial, wild goat, Asiatic ibex, and

chamois, to mention just four, a fairly large number of adult males are with the females all year, whereas in chiru, mountain goat, argali, bighorn, and Himalayan tahr the sexes appear to separate more completely. The formation of male herds after the rut seems to proceed much more rapidly in sheep than in goats.

Most Caprinae herds are open, in that animals join and part, often when population density is great and not as often when it is sparse. However, the members of such unstable herds may belong to a fairly cohesive population unit which is isolated from its neighbors either by fragmented habitat or by a preference of individuals for their own unit. Such units seem to exist in tahr, *Capra*, and bharal, and they have been described by Grubb and Jewell (1974) in feral Soay sheep. Their study population consisted of ten ewe home-range groups and several ram groups. "Although the range of groups overlapped, it was evident that a group of sheep occupied its home range in some way that deterred intrusion by other sheep. This restriction on movement operated between interacting ewe home-range groups and between ram home-range groups but not between ewe and ram groups . . . . Often two groups grazed close together, but very little reaction was shown between them." If there was an interaction, "the stranger would be nosed by one or two ewes or one or two ewes would show a more positive reaction and run up to the stranger, nose her and perhaps butt her."

Such behavior implies that members of a group can recognize and avoid each other; and it raises the interesting question of whether territorial behavior exists among the Caprini. Territories in the sense of an actively defended area have not been well documented among sheep and goats, and I found no evidence for their existence among the animals I studied. Pfeffer (1967), with respect to mouflon, wrote: "It is only at the moment of reproduction that certain males choose a territory which they defend against the approach of rivals. The other males practice a hierarchical promiscuity." It thus seems possible that in certain situations sheep exhibit territorial behavior, especially if the concept of territory is broadened from a defended area to an area of exclusive use regardless of the mechanism employed to maintain it.

Rutting saiga males collect up to 50 females and patrol the periphery of their herd, fighting off intruding males, but it is not quite clear from the account of Heptner et al. (1966) whether males maintain and defend plots of ground or only their harems of females. Chiru males also reveal behavior which suggests territoriality: they dig shallow pits and lie down in them. Their inguinal glands exude a musky-smelling yellow substance which may be used to mark these pits. The glands are so large that one can put a fist inside them (Rawling, 1905). However, as noted earlier, saiga and chiru may be Antilopinae not Caprinae.

Tener (1965) found no evidence for territorial behavior in muskox, nor does the literature provide positive information on takin. Mountain goats are not territorial (Geist, 1965). But territorial behavior has been proven in one rupicaprid and it possibly occurs in two others. Chamois males may lay claim to certain plots of land on open hillsides during the rut and from these they chase other males and try to retain females by herding them. Territory centers are about 300 m apart (Krämer, 1969). Nothing is known about land tenure among goral except that they deposit their feces at certain spots on ledges and in rocky niches.

The land tenure system of Japanese serow is complex and to date only partially understood. According to Akasaka (1974) and Akasaka and Maruyama (in press), solitary males, pairs, mother-young groups, and families consisting of a male, a female, and her young each tend to restrict their movements to small, discrete home ranges for much of the year, although temporary feeding aggregations with up to 7 animals may occur at favored sites in winter. Serow remain closely attached to their home ranges except in summer and autumn when social units tend to be changeable, with males roaming and grown young leaving their mothers. Home range sizes vary from about 3 ha (1.3 to 4.4) among solitary individuals to as many as 15.5 ha (9.7 to 21.7) among family units. A few observations suggest that these ranges constitute territories. On two occasions Akasaka and Maruyama (in press) watched as a male in a family unit chased an intruding solitary male out of the unit's home range, and both males and females mark the range peripheries with scent from their preorbital glands.

I observed marking behavior of a male and female serow in the Bangkok zoo. Both sexes had enlarged suborbital glands which exuded a clear fluid smelling like acetic acid. The male walked around the enclosure and repeatedly wiped the opening of each gland against fence posts while his tongue flicked in and out of his mouth. Later the female marked the same sites. One part of the enclosure was wet with urine. The male visited that spot, extended his hindlegs and urinated; ten minutes later he repeated this. The female also came to the site twice but she merely stood on it. In another part of the enclosure was a fecal mound spread over an area 5 m long and 1 m wide. The male ambled to it once, sniffed at several places, and then defecated while slightly squatting. Serow are also said to paw up the earth and horn shrubs (Allen, 1940).

9 AGGRESSIVE BEHAVIOR

Every society represents an adaptive unit, an evolutionary compromise between selection pressures tending both to maintain and disrupt it. Individuals in herds must compete for forage, mates, rest sites, and other resources. Those that circumvent the constraints of society by becoming solitary may expose themselves to a disproportionately high level of predation. Those that remain must apportion resources in such a way that a minimum amount of energy is wasted in competition for them. Many societies have established rules of hierarchical behavior. By creating predictability in the interactions between herd members, a rank order reduces levels of aggression and concomitantly raises the biological fitness of the species. Since individuals of high status have priority of access to mates and food, as many studies have shown, a subtle and not so subtle striving for dominance is one of the most pervasive aspects of many societies, especially among the male members. Not surprisingly, many vertebrates have developed a rich repertoire of olfactory, auditory, and visual signals to denote status. In fact aggression is the most common form of overt interaction among the Caprinae. This chapter describes aggressive and submissive behavior. Its purpose is not to test hypotheses and formulate new ones but to present data on the frequency and kind of interactions with the aim of clarifying how sheep and goat societies are organized. Most of the space is devoted to males because they interact more often than females and their behavior is more complex.

Two basic kinds of male rank order can be recognized in ungulate societies. In one of these, the relative rank order, males are dominant mainly within their own territories. Such a system is found in many African antelopes (see Estes, 1974) but it is rare in the Caprinae, the chamois being one well-documented example. A second system is based on absolute rank, one in which members form a more or less linear hierarchy. All Caprini have this type of rank order. Individual recognition, a memory of past encounters, is often thought to be the basis for such a hierarchy (Nievergelt, 1967). Indeed, all males in small populations or cohesive social units probably know each other, but another type of absolute hierarchy exists in large fluid populations where most meetings are between strangers: animals appraise each other's physical characters and establish rank on this basis almost instantaneously without overt interaction. For example, such a system has been described in axis deer, gaur (Schaller, 1967), and mountain sheep (Geist, 1966), and it seems to be found in all Caprini. Of course the two forms of absolute rank are not mutually exclusive, there being both acquaintances and strangers in any large population. For such a hierarchy to function, not only are prominent rank symbols essential but there must also be a gradation in the size of these symbols. The various Caprini males continue to grow in bulk and horn length throughout life, and in addition may have beards, manes, and other adornments which can be used to impress opponents. By contrast, in territorial species growth almost ceases in adults (Estes, 1974). While horn length seems to be the main status criterion among all Caprini except the tahr, other features also contribute importantly. Among urial and bharal I have several times seen a decidedly dominant male who had shorter horns but larger body than his opponent. Sparring matches and clashes teach each male to associate horn and body size with fighting potential, and in that way the animals create for themselves a social milieu in which serious altercations are few. Although I observed many hundreds of aggressive interactions not one ended in a damaging fight.

A plethora of names have been applied to various aggressive patterns, some German in origin (*Laufschlag*), others functional in derivation (threat jump). In naming the various kinds of behavior, I have used simple descriptive terms. Two main types of aggressive behavior are generally recognized (Walther, 1974): (1) direct aggression that involves overt threats—warnings of imminent attack—or actual physical contact; and (2) indirect aggression during which the animal attempts to assert or achieve dominance not by a test of strength but by intimidating the opponent solely through use of its rank symbols. Some semantic maneuvering is always necessary when making categories, and there are a few patterns such as horning vegetation which do not fit neatly into one or the other type.

Although much of this chapter is devoted to visual signals, the probable importance of olfactory ones also needs comment. As Müller-Schwarze (1974) has shown for black-tailed deer, glandular odors may have a great influence on the interactions between individuals. Sheep have preorbital glands and goats subcaudal ones, to mention just two, and males no doubt use the odors of these to enhance their physical attributes. For example, muskox often rub their suborbital glands against an extended forelimb prior to an attack (Tener, 1965), and goats urinate on their pelage. Sounds are of minor importance among interacting males, being confined to occasional grunts and snorts.

Most individuals respond to a dominant herd member by ignoring or avoiding him, but on occasion one assumes a submissive posture to signify lack of aggressive intent. Each successful assertion of dominance by one animal implies submission in the other, and at the end of this chapter I discuss this topic.

DIRECT AGGRESSION

All Caprini may use several gestures to threaten opponents overtly and these are listed here in a roughly increasing order of intensity. Some actions are subtle, a lingering look or casual approach, too subtle to be quantified with precision. However, there are also actions that are quite unambiguous. An animal may lower or *jerk* its head downward or sideways, pointing its horns toward an opponent in an implied threat to butt. Sometimes it *lunges* a meter or more, or *chases* the retreating animal for as much as half a kilometer. Urial, however, sometimes chase in what seems to be indirect rather than direct aggression. One ram may merely follow another at a run or trot. If the ram in the lead stops, the other kicks him or otherwise induces him to flee again. Walther (1961) noted similar behavior in Alpine ibex and Geist (1965) in mountain goats. Chases during courtship are not included here. Body contact is made in the form of a *butt*, the animal being bashed in the neck, side, or rump either with the blunt part of the horns or with the tips. On rare occasions one animal *bites* another.

An individual may rear or *jump* up on its hindlegs in front of an opponent, a gesture which seems to represent an intention to clash with a downward thrust of the horns. The *clash* is the most conspicuous behavior among Caprini. Unlike most other aggressive patterns, a clash requires the synchronization of both combatants if the horns are to meet with precision. Animals clash by sparring gently, by pushing and twisting with their horns, and more violently by rushing at each other from 20 m or more away to meet with a crash. Frequently one combatant clashes while the other catches the blow. In quantifying clashes each horn contact was tallied as a distinct encounter even when several followed in rapid succession.

Another form of physical contact is the *head-to-tail*. The two opponents stand in reverse parallel position as they push with their shoulders and hook each other in the abdomen and sides with their horns, sometimes circling as they do so. Minor forms of contact include the *shoulder-push* in which two animals stand side by side and shove with the shoulders, the *horn-pull* in which a male hooks his curving horn through the horn of another male after which the two pull sideways, and the *neck-fight* in which one animal places the ventral part of its neck over the neck or shoulders of another and pushes downward (fig. 33). As with the clash, both participants in a head-to-tail, shoulder-push, and horn-pull behaved aggressively (see tables 32 to 37).

Hemitragus

Himalayan tahr. Anderson and Henderson (1961) felt that tahr were extremely placid animals, that "a more docile assembly would be hard to visualize." The populations I watched were quite aggressive. As shown in Table 32, the jerk was seen 15 times, directed mainly by females at courting males (6 times) and at young (4 times). Females lunged on 3 occasions, twice at another female and once at a yearling male. The butt was common only in females who responded to persistently courting males. Attacks were directed at the neck 8 times, at the shoulders and sides 8 times, and at the thigh and rump also 8 times. The jump was rare, seen twice in young and once in New Zealand in a female.

Caughley (1967), after studying tahr for several years, noted that "sexual fighting between males has never, to my knowledge, been observed." Similarly, K. Tustin (pers. comm.) has not seen combat between adult males after many field seasons in New Zealand, but a sheep rancher told him of two males standing parallel, then stepping back several paces before rushing forward to clash in the manner of domestic rams. I observed 18 mild clashes, most often between yearling males and females, between yearling males and yearling females, and between young. Sometimes one animal took the initiative, the opponent merely absorbing the blow, and at other times both clashed in unison. Shoving and twisting with locked horns followed some clashes.

Roberts (1971) watched as "two bulls confronted each other, whistled sharply, and began to wrestle like domestic cattle. The tactics appeared to be to try to put the opponent off balance, for after a period of pushing, twisting, and sliding downhill one bull was heaved off balance and the victor immediately shot his horn under him and ripped him in the belly." However, fighting injuries are rare. K. Tustin wrote me that "I've never seen any body injury on any thar which resembled one that an animal could have obtained in a fight, and I have personally inspected the carcasses of at least 2500 thar." Interestingly, Himalayan tahr have as yet not been observed to rear on their hindlegs and lunge down to clash in the manner of Nilgiri tahr and *Capra*.

On two occasions, once in Nepal and once in New Zealand, two young stood head-to-tail and jabbed at each other's sides. On two other occasions a female and a yearling male assumed a similar stance and circled rapidly with heads cocked as if to jab. After standing head-to-tail one young placed its neck briefly over the neck of the other as if intending to neck-fight. I observed one neck-fight between two young at the New York Zoological Park.

Table 32 shows frequencies of aggression in the various age and sex classes. I observed the animals during a socially placid time of year when they were neither rutting nor giving birth. However, one female came into estrus unseasonably late, and the courting animals became markedly more aggressive. When compared to their actual composition in the population, most tahr interacted roughly as often or slightly less often as expected, but the yearling males were more than twice as aggressive as expected. Another and more precise method of comparing frequencies of aggression is to convert the interactions into number per animal-hours of observation, one animal observed for one hour being an animal-hour. This measure proved useful for such species as Himalayan tahr, markhor, and bharal in which small herds may remain visible for long periods, but not for urial and wild goat, species in which scattered herd members constantly moved out of sight in the broken terrain. Yearling male tahr were most aggressive, followed by females and by young (table 32). Butting, clashing and other forms of aggression in young often seemed playful, involving exaggerated gestures. However, even playful aggressiveness may serve to establish and maintain rank.

Nilgiri tahr. I did not quantify jerks, lunges, and butts. One female snorted and lunged at another female with both forefeet off the ground, behavior similar to that described for mountain goat by Geist (1965). Butts were common, especially when a disturbed herd was bunched. On 6 occasions a tahr walked to a resting animal, jabbed it in the side or thigh with a horntip and then appropriated the site. Five of these instances involved a female displacing a young, and one a class I male crowding out a yearling male. When a yearling male poked both another yearling male and a class I male, neither budged from his rest site. Tahr clashed horns on 50 occasions, most interactions involving subadult males and females (table 33). The animals clashed in two ways: in one they stood parallel and in unison jerked their heads sideways once or twice to make horn contact—the side-clash; in the other they faced each other a meter or less apart, then lowered their heads and met with a crash. Clashes were seldom preceded or followed by other interactions. Nilgiri tahr side-clashed 36 times, a striking difference from Himalayan tahr which were not observed to fight in that manner. According to Hutton (1947), "two animals almost simultaneously reared up on their hindlegs and seemed to 'dance' in front of each other, while keeping their distance and circling. Suddenly they would close in and bring their heads together with a resounding crack." After half an hour one male ended the fight by ramming the other in the shoulder, and he then mated with a female. I did not witness this method of fighting. However, once a class I male reared up before another class I male, his chin tucked in and turned slightly to one side and with forelegs drawn close to the body.

Two tahr stood head-to-tail on 13 occasions and pushed each other shoulder to shoulder, often circling rapidly while doing so (table 33). Once two males kneeled as they circled. At the same time the animals may jab each other in the side, abdomen, or thigh with their horn tips. Tahr occasionally paw the ground with a foreleg before or after such fights.

Adult males interacted only as often as expected, but yearlings, especially yearling males, were aggressive over twice as often as expected and class I males some four times as often. Females and young clashed little (table 33).

Capra

Kashmir markhor. The data in table 34 were collected in the Chitral Gol between November and February, a period which includes the rut. Jerks involved mainly adult males threatening those herd members that intruded on their courting and females that claimed rights to food, rest sites, and trails. Lunges and chases, the latter covering up to 50m, were mostly used by males to drive other males from the vicinity of estrous females. Butts were rare, those by females being employed to deter ardent suitors. Markhor seldom jumped but when they did so it was in the typical *Capra* manner, with body upright while facing or standing sideways to the opponent, with chin tucked in and turned to one side, and with forelegs bent and held close to the body (see fig. 33D). Markhor clashed often, sometimes two of them doing so several times in succession. Of the 168 animals involved in clashing, 15% reared up on their hindlegs before lunging downward. Just before contact the animal turned its head in such a way that the waiting opponent could catch the blow either between its horns or rarely with horns crossed. The ears were laid back, out of harm's way. Occasionally both combatants reared in unison. Walther (1961) noted that a captive male of unspecified race braced his hindlegs against a rock while absorbing the clashes of an opponent. Markhor often clashed downhill on an opponent, apparently a means of hitting harder. For example, once a young reared up and clashed 8 times in succession on another young and on 7 of these it first maneuvered itself up the slope. Class III and IV males never initiated a clash and their participation was limited to catching attacks by females during courtship. Females interacted mostly with other females and with yearlings and young. Young often fought playfully among themselves, as this 10-minute extract from my field notes at Tushi illustrates:

> Two young bash heads 3 times in succession from about .3 m apart, each time locking their stubby horns while pushing and twisting. The larger of the two then rears up, sometimes directly before the smaller and at other times a few meters away after retreating with

exaggerated leaps. Taking several steps or hops forward it then clashes on the other who just catches the blows. This happens 14 times. The large young then mounts, a gesture to which the small one responds by rearing and clashing 3 times. Both tussle. The smaller jumps and clashes and the larger then does so twice. Another sparring bout follows. The small youngster rears again, as usual with its body sideways to its opponent, and clashes. The other reciprocates first with a mount and after that with 11 jumps and clashes. A nearby female rises and both dash to her and suckle.

The head-to-tail was observed only once between two young. Equally rare was the shoulder-push. Two youngsters in the Chitral Gol rammed each other on one occasion, and a yearling male and a class I male fought near Tushi:

> After tussling head to head, one male swivels so that his body is parallel to the other. They continue to spar but now also push hard with their shoulders. The large male breaks away, rears up 3 times and clashes, once delivering such a hard blow that the yearling goes to his knees. They spar 4 times more, one or the other initiating contact. When the yearling turns away, the class I follows until the former turns and spars.

Most markhor behaved aggressively only as often as one would expect from the composition of the population, but the yearling males were an exception in that they fought about three times as much as expected (table 34). Converted to number of aggressions per animal-hour, yearlings still remained the most active age and sex class, followed by the males of classes IV, II, and III, in that order. Class I males seldom interacted, and females and young did so moderately often.

Wild goat. Wild goat are more aggressive during the rut (7.2 aggressive individuals per hour of observation) than during the birth season (1.4 aggressive individuals) even if the fact that there were fewer males in mixed herds during the latter period is taken into account. Males jerk, lunge, and chase vigorously and often during the rut (table 35), sometimes pursuing rivals as far as 50 m in their effort to keep them from the vicinity of estrous females. But most attacks consisted of a mere jerk, with the head occasionally lowered so far that the horn tips pointed at the opponent. Females jerked most often at other females and at yearlings and young, usually if these came too close. Some females also threatened courting males. If, as happened now and then, one male was not nimble enough to escape the lunge of another, he was butted. One class IV male had mounted a female when a class V male rushed up and hit him so hard in the neck that he almost fell. Females butted on 14 occasions, and courting males elicited twelve of these butts. In contrast to the usual butts, which are directed at the sides, thighs, and rump, these consisted of jabs at the face and neck. The jump was relatively uncommon.

Wild goat clashed often, using several techniques. On 9 occasions two animals stood or walked parallel and clashed horns with a sideways jerk. They also sparred with much pushing and head-twisting, and they delivered quick, hard blows. One goat may rear up and clash, and occasionally both do so. Their horns hit either before or just after their forelegs touch the ground. Eleven percent of the animals reared before clashing. Males seldom clashed while defending estrous females. Out of a total of 81 clashing encounters between males, some involving repeated horn contacts, only 6 were the result of direct competition over a female. Females clashed on males on 6 occasions and 5 of these involved a yearling male and 1 a class II male. Females usually fought with each other and sometimes playfully with their young; yearling males also sparred with young.

Neck-fights twice involved class II males. On one occasion the smaller of two males low-stretched behind the other, who whirled around and clashed. The former then placed his neck over the neck of the other and pushed with his chest whereupon the larger jumped and clashed and was in turn mounted. On 5 occasions two young males walked parallel, tussled with their horns, and pushed with their shoulders.

Class V males were proportionately over twice as aggressive as expected and class IV males were not quite twice as much (table 35). However, the two classes differed strikingly in their mode of aggression. Class V males mainly jerked, lunged, and chased whereas class IV ones mainly clashed. The other age and sex classes interacted with about the expected frequencies except the females, who were aggressive only half as often as expected.

Other goats. My observations on Asiatic ibex include jerks, lunges, chases, jumps, butts, and clashes, the last-named either frontally or standing side by side. Once two young males shoulder-pushed while sparring. On several occasions a male in the New York

Zoological Park climbed on a boulder before rearing and clashing on his opponent below. Males may show an extreme horn threat in that they sometimes lower their heads almost to the ground, horn tips pointing at the opponent (Walther, 1961). Fighting in Alpine ibex has been described by Couturier (1962) and Nievergelt (1967) and in captive Kuban ibex by Walther (1961). Shank's (1972) account of direct aggression in domestic goats closely resembles mine for wild goat and markhor. He defined two main categories of direct aggression. "Rush association behaviour is delivered by larger individuals to smaller ones, while clash association behaviour is restricted almost entirely to equal-sized individuals." Most rushes were directed by large males at small ones, in contrast to clashes which involved mainly small males, at least from July to November, when Shank's study was conducted. On 11 occasions a goat placed its head under the other animal's abdomen and lifted, behavior which resembles head-to-tail fighting. Shoulder-pushing also occurs in domestic goats (V. Geist, pers. comm.).

Ovis

Punjab urial. Urial were aggressive throughout the year but most so during the rut. Since my sample of interactions is large only for the rut, I limit this discussion to October and November. Yearling rams, ewes, and young threatened or clashed so seldom that the following notes must by necessity confine themselves to rams of classes I to IV (table 36). Jerks were quite common, especially when a ram kept trespassers from the vicinity of an estrous ewe, but this form of threat seldom led to lunges and chases. The jump was rare. The animals did not stand upright like goats but leaned forward with forelegs hanging loosely. However one day in July I watched 7 class I and II rams fight in a rather playful manner for 20 minutes and during this time 6 of them jumped almost as upright as goats. The horn-pull was seen twice. Butts were uncommon perhaps because rams reacted quickly enough to catch attacks with their horns. Most butts were light, delivered toward the side or rump, but a few could have caused injury. Once, while a class III ram sparred with a class IV ram, a second class IV animal crashed into the rump of the unsuspecting class III, who stumbled sideways. Nearby males quite often disrupted a fight, something also observed in wild goat and bharal.

Rams clash often (table 36). Usually two animals jerk down their heads and either crack horns or first move apart before clashing from a distance of a meter or so, sometimes twisting and shoving with their heads briefly afterwards. Two or more such clashes may follow in rapid succession. Sometimes one or both rams walk or trot up to 25 m apart and run at each other, often starting their charge with a slight leap. If the charge is from a short distance the ram may simply back up a few steps before rushing at his opponent. Just before impact the ram tucks in his chin, turns his head slightly sideways, and occasionally leaps a little so that his forelegs are off the ground on contact. The two collide with precision, one catching the blow adeptly between his horns. The hindlegs may be flung into the air on contact, either from the force of the impact, or, as Geist (1971a) suggests, on purpose, to increase the strength of the blow. The force of the clash may throw the recipient backward as far as 5 m even though he may have braced himself by crouching slightly. Occasionally a ram trots away from several others and turns as if to charge—but if none of the others lower their horns to receive the blow the attack aborts. Sometimes, too, a ram turns aside after an attack has begun, thus terminating the interaction. In 6% of the clashes from 3 m or more apart rams failed to make proper contact, the glancing blows causing them to stumble, jackknife, or in one instance to fall over each other. Most vigorous clashes were on relatively level terrain, but occasionally a ram charged down a slope, thereby increasing the force of his attack. Urial run at each other on all fours before clashing. However, once 2 class III rams both reared and fell forward into a clash.

Although class IV rams comprised only some 9% of the population they were involved in nearly half of all aggressive acts (table 36). Rams of classes III and II were over twice as aggressive as expected. The other age and sex classes seldom threatened and clashed.

Other sheep. Schaller and Mirza (1974) summarize aggressive patterns of mountain sheep, Soay sheep, Marco Polo sheep, and mouflon. I mention here only a few points of comparative interest. The jump is common in Marco Polo sheep and in mountain sheep. Geist (1969) tallied 329 instances of clashing, butting, jerking, and jumping among bighorn rams, and of these 27% consisted of jumps, as compared to 1% in the Punjab urial observations. Grubb (pers. comm.) saw a Soay ram rear only once in a playful manner. The behavior also seems to be uncommon in mouflon, but my observations in the New York Zoological Park suggest that it is more common than in urial. Ewes in particular often jumped up slightly during aggressive encounters. "A fight between [Soay] rams began when they marched backwards away from each other, goose-stepping and raising the legs high and smartly as they did so" (Grubb, 1974b). Similar behavior occurs in other sheep, though usually not in as accentuated a form except in mouflon (V. Geist, pers. comm). Mountain sheep and Marco Polo sheep often rear up and race at each other on their hindlegs with body at an angle leaning forward and forelegs hanging loosely before jumping into a clash, in contrast to urial and Soay sheep, which charge on all fours. Judging by my observations in the New York Zoological Park, mouflon are somewhat intermediate between urial and Marco Polo sheep in their form of charging. Rams often reared slightly before running into the clash on all fours (fig. 33C), and during some attacks they essentially moved on their hindlegs, touching the ground with their forelegs only intermittently. Mountain sheep may face each other with raised and slightly averted heads for as long as a minute after a clash, whereas urial either part immediately or halt only a second or two.

The shoulder-push is prominent in Soay sheep, with rams "leaning heavily on each other" (Grubb, 1974b); it is often seen in mouflon (fig. 33F) and mountain sheep too, but urial seldom use this behavior. Vestigial neck-fighting with one animal placing its chin over the shoulders of another, the *cérémonial d'allégeance* as Pfeffer (1967) termed it, is conspicuous in mouflon and it has also been described in mountain sheep, especially in lambs (Shackleton, 1973). Grubb (pers. comm.) saw no unequivocal instances of neck-fighting in Soay sheep. In Chitral I several times observed domestic rams as they placed their necks over the necks or backs of others and in that position twisted and kicked. The head-to-tail occurs occasionally in bighorn sheep (Geist, 1971a; Shackleton, 1973) and in Soay sheep (Grubb, 1974b).

Pseudois
During the rut at Shey, bharal commonly jerked and lunged at opponents, females usually threatening other females and young, and males other males (table 37). The jump was common too,

bharal rearing up in the manner of goats either standing sideways to or facing their opponents (plates 47 and 48). Females usually jumped before other females but they also threatened males in this manner after having had their rumps sniffed. Males sometimes reared up when other males mounted them. A male may walk or hop as far as 5 m toward another male while balancing on his hindlegs. Out of 90 jumps among males some 28% were directed by a larger at a smaller animal, 21% by a smaller at a larger, and in 51% the participants were in the same horn-size class. Twenty percent of the jumps by males were directed at a group of males rather than at a specific individual. Typically, a male would dash with exaggerated bounds some 5 to 10 m from a group, rear as if inviting someone to catch a blow, then return at a trot.

Animals tended to butt members of their own sex, usually in the rump or side. Males often butted when an opponent was off guard. One male saw the charge of another but he was unable to swivel around in time and received such a hard butt in the side that he stumbled. On another occasion two males sparred and a third entered the fray by ramming one of the combatants in the rump. The shoulder-push was observed only once. Bharal females bit at others on several occasions, the only caprid in which I observed this behavior.

Bharal may clash in several ways. The animals sometimes twist, push, and circle with locked horns. Usually they face each other, but they may also tussle while standing side by side. Contact tends to be brief, a lunge and clash or several quick blows in succession. On 3 occasions a male walked some 10 m from his opponent and then rushed at him and clashed on all fours like an urial. However, during most fights one combatant rears up, tucks in his chin, twists his head, perhaps takes a few hops, and then lunges down on the other, who catches the blow between his horns, sometimes bracing himself for the impact by bending his hindlegs slightly. Contact may occur before, during, or after the forelegs touch ground, depending on the distance between the participants. Sometimes a male bounds 10 m or so from his opponent, rears, then runs on his hindlegs several steps with body leaning forward and forelegs hanging down like those of a mountain sheep before ending the attack on all fours. The charging animal often increases the force of the blow by clashing downhill. Two bharal may also take several steps backward or trot some 10 m apart and rear in unison, standing upright with forelegs hanging loosely or sharply flexed. Hesitating a moment, they then fall into a clash, sometimes first giving a little leap as if to hit harder. In a variation of this method, two animals stand head-to-tail, then jump up, and with horns cocked toward each other meet in a clash. In 45% of 179 clashes observed both bharal reared, in 34% one reared, and in 21% neither did so. One percent of the clashes were inept in that the animals did not make proper contact. The animals parted immediately after a clash except on one occasion when two males faced each other for several seconds with heads averted.

Males usually clashed with males, females with females (but 11 times also with yearling males), and young clashed only with young.

Most animals were aggressive only about as often as expected except class V males which were somewhat more so and young which were much less so (table 37). In terms of aggressive acts per animal-hour of observation, class V males were most active, class IV and III males, yearling males, and females showed about the same intermediate levels, and the others interacted little. When comparing the figures for Shey with those for Lapche (table 37) it is clear that rutting animals are more aggressive, something most evident in class V males and in females. However, when I presented the Lapche bharal with a competitive situation, in this instance a small salt lick, aggression in adult males increased promptly to levels even higher than those during the rut.

Class V males became more aggressive with the onset of the main rut on November 29 as they began to assert their rights to estrous females, an increase that was based on a greater propensity to threaten rather than to clash (table 38). Most other males were quite aggressive early in the rut but became increasingly less so as time went on, a trend conspicuous in yearling males and males of classes I and IV. As class V males became less tolerant of competitors, these males grew more passive: they seemed to have become inhibited. Class II males showed little aggression at any time. Possibly they are at an awkward age, neither large enough to provide competition for adults nor small enough to be accepted as innocuous partners by young males. Class III males were most aggressive during the first and fourth weeks, a pattern for which I have no explanation except to suggest that they are large enough to compete for females but, unlike the class IV males, not so large that they pose a serious threat to the class V males. Females were generally more aggressive prior to the main rut not only toward each other but also toward importunate males. Grubb (1974b) noted a similar trend in Soay sheep. Although bharal females generally behaved calmly, they erupted at times and in the ensuing melee jumped, butted, and clashed often and vigorously. One such affray on November 10 lasted ten minutes, a second one on November 21 persisted for thirty minutes. Some yearling males joined the commotion. The number of interactions during these melees has raised some figures in table 38. For example, with melees excluded, the figures for November 2–10 in the adult female column would be .18 instead of .46, and for November 20–28 .11 instead of .20. Young seldom threatened or fought.

Ammotragus

Katz (1949) and Haas (1958) studied aoudad behavior in captivity, and I supplemented their observations with some of my own. Animals jerked, lunged, and butted, but neither Haas (1958) nor I noted chasing and jumping. Katz (1949) described clashing: "The sheep walked rapidly toward each other, gradually picking up speed and breaking into a run shortly before they collided. Just before impact their heads were lowered and turned slightly to opposite sides. They attempted to meet squarely with their noses crossed . . . . The second type of fighting consisted of close butting, and locking and twisting horns . . . . Each tried to twist the head of the other by pulling downward and away. Also, attempts were made to hook the belly or the flank." One animal may also run into a clash while the other waits with lowered head, catching the blow between its horns. Clashing with a sideways jerk of the head when two animals walk parallel is common too. The shoulder-push is a conspicuous pattern, and sometimes one or both combatants hook a horn over each other's neck or shoulders and pull as well as shove. Aoudad also push while standing head-to-tail. By placing its head and neck on the rump of the other, an aoudad may force its opponent onto its knees (Haas, 1958). Horn-pulling is often seen with the animals standing parallel or head-to-tail and sometimes pushing with the bodies too. The neck-fight is a complex and prominent pattern (fig. 33E). At its simplest, one aoudad places its head and neck over the body of the other and presses down, sometimes until the animal sinks to its knees. The full weight of the body may also be used, with the attacker draping his whole forequarters so far over the shoulders of the other that his head hangs down the other side.

Ovibovini and Rupicaprini
Takin in the New York Zoological Park hook the air with their horns, butt, and occasionally clash head-on from a meter or so. Muskox behave similarly (Tener, 1965), and in addition bulls may go over 30 m apart before they turn to charge and clash, absorbing the violent impact on their broad bosses (A. Oeming, pers. comm.).

The rupicaprids with their sharp, slender horns cannot clash vigorously without injuring themselves, and their fighting methods differ from those of the Caprini. "Mountain goats do not fight head to head," but instead "they keep fighting side by side while moving about one another. Goats strike up and sideways with their head, driving the horns into the opponent's ventral body region" (Geist, 1965). Mountain goats also threaten with lowered horns, lunge, and butt. Chamois clash rarely. Krämer (1969) reported that an occasional animal may rear up and clash like a goat. Jerks, lunges, chases, and butts also occur. The chase is conspicuous in this species with the pursuer trying to hook the fleeing animal with its horns. Chamois also perform a "threat bound" (Krämer, 1969), as do mountain goats (Geist, 1965), a gesture perhaps analogous to the jump. I observed two chamois, probably two-year-old males, circle rapidly in a head-to-tail position. After that, one stood broadside to the other with his head averted. Young may also circle head-to-tail (fig. 33G), and they also neck-fight (Krämer, 1969). Akasaka (1974) noted that fighting Japanese serow chase each other and may kill each other with stabs in the bowels.

Discussion
It might be assumed a priori that direct aggression is the prerogative of the larger and hence the dominant of two individuals. This is indeed so among males with respect to jerking, lunging, chasing, and butting, as shown in figure 31 for urial, wild goat, and bharal, three species for which I have large samples of data from the same phase of the animals' reproductive cycle, namely, the

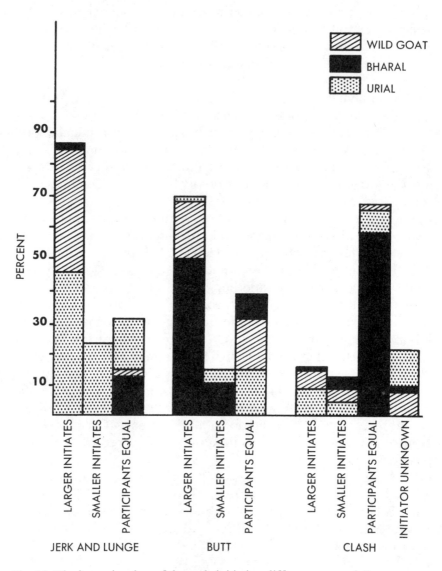

Fig. 31. The horn-size class of the male initiating different types of direct aggression in relation to the size of his partner, based on data for wild goat,
Punjab urial, and bharal.

rut. Even if the participants are in the same age class a size difference in horns and body is often apparent and the larger tends to
take the initiative. Occasionally a small individual attacks a large
one, either in retaliation or surreptitiously while the other is inat-

tentive. Such reversals were most common among urial. Females may jab males with impunity during courtship.

Clashing shows a somewhat different pattern. It should be remembered that clashing requires the participation and coordination of both animals, dominant as well as subordinate. Either the larger or smaller of two individuals may initiate a clash, the former perhaps to assert its position and the latter to test it. The fact that a large male tolerates the attacks of a small one seems to be an expression of dominance in itself. Most clashes are between males in the same horn-size class. Not only are such males not likely to be intimidated by their opponents but also they may use a battle to resolve questions of status. Sometimes two males clashed seemingly in unison, preventing me from ascertaining the aggressor.

The jump shows a pattern similar to that of the clash, something one would expect since the gesture represents an intention to clash. In this context it is significant that those species such as Himalayan tahr and urial, which seldom rear up before clashing, also fail to jump much. *Capra* and bharal which commonly rear up to clash also jump often.

The various forms of direct aggression fall into two main categories: (1) assertions of rank in the form of jerks, lunges, and butts, all gestures which are used to maintain the hierarchy; and (2) testing situations in which one or both participants clash to examine or settle their respective rank. All species I studied behaved similarly in these respects, as did domestic goats (Shank, 1972) and probably ibex (Walther, 1961; Nievergelt, 1967). Geist (1971a) studied aggression in mountain sheep, but it is difficult to compare our respective data because of differences in quantification and analysis. He found that "a large-horned ram treats a small-horned ram in a conspicuously different manner than a small-horned ram treats a large-horned one." This generalization cannot be extended easily to any species I studied. A clearly subordinate individual, judging by his size, sometimes treated an opponent as if he were the one of low rank, and in some interactions, especially in clashes, there was no obvious difference in behavior between dominants and subordinates. Geist (1971a) noted that jerks and lunges were mainly the prerogative of dominant individuals, a finding similar to mine, but he further wrote: "Thus the opponent with the heaviest horns, heaviest armor, and greatest capacity to deal out crushing blows is the recipient, not the donor of most blows." Subordinate mountain sheep initiated forceful clashes in 72% of the instances. My data do not fit this pattern (fig. 31), even taking differences in quantification into account and realizing that Geist referred to what he considered dominance fights. As noted earlier, I observed no long or damaging fights, and during many interactions it remained unclear whether two animals were merely testing their strength or striving for dominance.

Males do not threaten and clash randomly, as is evident from this discussion. Males interact mainly with members of their own age class or the next smaller or larger one (fig. 32), but some differences exist in the type of aggression. A jerk, lunge, or butt is most often directed by a large male at a small male, sometimes one which is two to five sizes smaller, whereas a jump or clash usually involves participants of the same horn-size class. Domestic goats (Shank, 1972), Alpine and Walia ibexes (Nievergelt, 1974), markhor, and tahr show similar patterns of behavior.

Females interact mainly with other females, young, and sub-adult males, animals with whom they presumably have not wholly settled their rank. Since rank is based on size, adult females tend to be dominant over yearling males but over only some class I males. Courtship presents a special situation, a time when reversals are tolerated. For example, courting wild goat females jerked at, lunged at, or butted class II to V males 18 times and only twice were they threatened in return.

Several factors have an influence on the frequency of direct aggression in a population. These include the age and sex of the animal, the season, the size and composition of the herd, the occurrence of competitive situations such as a salt lick, and possibly the level of nutrition. As can be expected, the presence of any limited resource such as an estrous female causes more aggressive interactions: aggression must be profitable if it is to have selective advantages. However, the kind of aggression shown by animals depends to some extent on the situation. For example, Petocz (1973b) found that deep snow affects the kinds and frequencies of aggressive interactions in bighorn rams, the animals competing during severe winters for limited food mainly by threatening with their horns and other such overt actions rather than expending energy on kicking, twisting, and clashing hard. Among the species

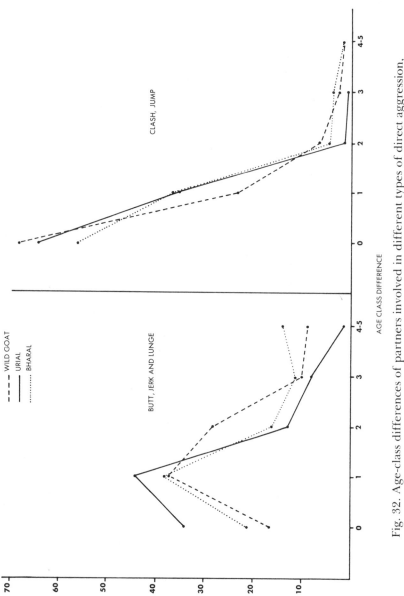

Fig. 32. Age-class differences of partners involved in different types of direct aggression, based on data for wild goat, Punjab urial, and bharal.

Fig. 33. Types of direct aggression among Caprinae: (A) aoudad spar (redrawn
      from Haas, 1958); (B) a Marco Polo sheep jumps (from Walther, 1974);
      (C) mouflon clash (from Pfeffer, 1967); (D) a Nubian ibex rears up be-
      fore lunging downward into a clash (from photograph in Reed and
      Schaffer, 1972); (E) aoudad neck-fight (from Haas, 1958); (F) mouflon
      shoulder-push (from Pfeffer, 1967); and (G) head-to-tail in chamois
      young (from Krämer, 1969).

I studied, males defended estrous females from other males
mainly by jerking and lunging, clashing being uncommon in this
context. Only 7% of the recorded wild goat clashes and 6% of the
bharal clashes occurred in direct competition over a female in
estrus, and the percentages for the other species were even lower.
Nievergelt (1967) wrote that "from spring to autumn ibexes of
similar age indulge in frequent fighting or rather play-fighting.
The behavior described might explain that during the rut hardly

any fights were observed." Rutting Alpine ibex usually do not tolerate rivals in a herd, with the result that the dominant individual has little need to assert himself. The same was true of those markhor herds which contained just one adult male. Wild goat and bharal herds were usually large and included many males of all ages. It seems likely that animals which are acquainted with each other and therefore already know each other's strength interact less often than do those which are strangers (see Shackleton, 1973); in other words, males which are fluid members of herds might interact more often than those which remain together. Among urial 78% of all aggressive individuals interacted by clashing as compared to 61% in wild goat, 51% in bharal, and 20% in markhor. These figures generally confirm expectations except the one for wild goat, which is rather high. But group size may also affect the frequency of aggressive behavior, friction increasing with numbers. Markhor had the smallest average herds, followed by bharal and wild goat. However, total levels of aggression were higher among markhor than among bharal (see tables 34 and 37). This could be due to a basic species difference, the level of nutrition, or other factors.

In general, adult males display a disproportionate amount of aggression during the pre-rut and rut. One might assume that they would fight less often after the rut when the size of the testes and with it, presumably, the testosterone level decreases. That hormone levels have an effect on aggression has been shown in many studies. However, various data presented earlier show that the aggressive potential remains even if it is not expressed. During July, I observed urial rams clash only once until one day several rams bashed horns 25 times in a few minutes, showing how the vagaries of field observations could bias ideas.

Yearling male Himalayan tahr and markhor were markedly aggressive but those of other species were not, a difference for which I have no explanation except to note that perhaps good nutrition was a contributing factor. Yearlings are sometimes more aggressive than males of intermediate size, especially in the frequency of clashing, possibly because their age still gives them a certain freedom of expression without eliciting retaliation.

Female markhor, Himalayan tahr, bharal, wild goat, and Nilgiri tahr were quite aggressive, the first three species in roughly the expected proportions and the rest in about half the expected proportions. Urial ewes, as well as urial lambs, differed from the others in being almost passive. At first glance horn size seems to be the deciding factor. *Capra* and *Hemitragus* have conspicuously pointed weapons some 15 cm long, whereas Punjab urial have blunt stumps. However, a lack of horns does not deter tahr, bharal, and goat young from bashing foreheads, and bharal females with their puny horns clash often and hard. A genetic difference might be postulated.

Being similar in physical appearance, the various Caprini threaten visually and fight in basically similar ways with only slight modifications in technique and differences in frequency. Table 39 compares several species as to their techniques. Some gaps in the chart are no doubt the result of too little observation time rather than the absence of a trait. For example, the shoulder-push, head-to-tail, and neck-fight are rare in many species, given in some instances only by the young. The jerk, lunge, chase, butt, clash, and jump seem to be universally present in the Caprini except that aoudad and possibly the Soay sheep may lack the jump. The species differ greatly in the frequency of jumping, with, for instance, tahr doing so rarely, urial seldom, and mountain sheep, wild goat, and bharal often. Clashing consists of sparring and of delivering sharp blows. This behavior will be discussed more fully later, but I want to make one point now. Although the Caprini use various means of increasing the force of a blow, from rushing at each other on all fours and running bipedally to rearing bolt upright before lunging downward, the basic technique of making horn contact with precision remains similar: the recipient generally neutralizes the blow by catching it between his horns. The rupicaprids, with their pointed spikes, clash rarely or not at all and use instead such fighting postures as the head-to-tail which do not require a dangerous face-to-face encounter. Clashing in Saigini has not been adequately described, but Heptner et al. (1966) mention that during the rut many saiga males have bloody stab wounds in their bodies and that deaths due to fighting are common, suggesting rather haphazard brawls.

INDIRECT AGGRESSION
Both participants risk being injured during a direct confrontation if they fail to parry blows by catching them with their horns, or, in the case of some species such as the mountain goat, on their shields of tough hide (Geist, 1965). There is little selective advan-

tage for an animal to expose itself to injury if the same results can be achieved in other ways. Instead of attacking an unknown opponent it is best to first test his qualities, to try to intimidate him without eliciting either a defensive gesture or a retaliatory one. By expressing their aggression indirectly animals save energy and perhaps their lives.

This section describes the various indirect means of threatening. The displays are by necessity treated as isolated events, obscuring the dynamic interplay between aggression and submission that characterizes a herd of sheep and goats. A society whose males are in effect perambulatory status symbols revolves around dominance so subtly that many interactions do not include obvious displays. This can best be illustrated by a few observations on bharal.

Seeing a large male approach, a small one may begin to graze or amble away. Sometimes a dominant animal detours and casually walks past the other which just as casually turns aside. Or one male may stand either parallel or at right angles to the other until he moves away or at least averts his head. Two males often forage side by side in what seems to be a peaceful association. But if the smaller of the two steps ahead, the larger will quickly bring his muzzle even again. As many as a dozen males may graze on a broad· front. Should one move ahead, the others immediately catch up and this in turn may prompt some male to move even faster and so on until all hurry along with their heads still lowered as if grazing. To take the lead is obviously a minor assertion of dominance among males. It partly explains why the largest animal in a male herd tends to be the leader.

Perhaps the most difficult aspect of dealing with indirect aggression is this subtlety of interactions. Lunges and clashes are unambiguous events, blatant in their purpose as compared to the many indirect forms which may be so delicately expressed that a human observer is uncertain whether an interaction has occurred. This obviously makes it difficult to quantify some displays.

Direct aggression is commonly observed in all age and sex classes. In contrast, indirect aggression is so seldom seen among females and young that any instance is a notable event. Hence this section deals almost exclusively with males.

Most displays fit into one of two general categories. (1) By standing broadside to his opponent a male shows off his profile, his large size, as well as his special adornments such as horns and ruff. There are various permutations to this basic lateral posture but all make the male appear impressive without displaying his weapons in such a way that they seem ready for imminent use. (2) Several displays are so much a part of the male's courtship repertoire that they tend to be thought of mainly in that context. Yet such gestures as exposing the penis, mounting, and low-stretching are also forms of dominance. (3) In addition a few displays, among them huddling and horning vegetation, do not fit neatly into the above categories.

Lateral displays
Head-up. By raising his muzzle and standing stiffly erect a male enhances his appearance. All his physical features are displayed to best advantage, and he may even raise his tail and hold his ears sideways as if to make himself even larger (fig. 34). A slow approach or motionless stance in front, beside, or behind an opponent is usually enough to displace him. This display is widespread among the Caprini (table 39), but only in urial did I see it fairly often. In this species a quick head-up may precede and follow a clash, and rams may also follow each other or walk parallel in that posture, sometimes with their heads pulled so far back that the lower neck bulges conspicuously. Once a markhor stood parallel to another in a head-up with his white incisors exposed, and another walked slowly through a herd with his head raised and a leafy branch balanced between his horns. One class IV bharal gave a head-up and a tongue-flick when a class I male walked past him and then inserted his penis into his mouth. In New Zealand I observed two chamois stand side by side with muzzles raised high, exposing their white throat patches. Gray (1973) describes the head-up in muskox.

A mountain sheep ram may face an opponent in a head-up, a gesture which Geist (1971a) termed the "present." However, the present is not a lateral display.

Head-down (hunch). In previous publications I treated the head-down and hunch separately, but I now think that these represent one display at different intensities. A wild goat may pass broadside or walk parallel to another with head lowered, muzzle pointing toward the ground or tucked close to the neck, and back humped. His steps are slow and stiff, his tail is up, and the crest on his neck bristles (fig. 34). This posture was seen about 10 times in males and once in a female. Punjab urial occasionally behave similarly. The head-down is common in bharal. A male points his muzzle obliquely down, tucks in his rump, thereby humping his back, and with his tail often partially raised and legs in a slight crouch cuts in front of or parallel to another male (plate 48). Sometimes he flicks his tongue or stops to unsheath his penis. Bharal often roam among herds in a postural hybrid with muzzle forward and back humped. Geist (pers. comm.) noted a similar posture in mountain sheep on rare occasions. From this position they can switch either to a low-stretch when a female is near or to a head-down. On one occasion a female bharal displayed the head-down to another female. I am not certain whether the head-down exists in markhor; my only observation concerns a female who humped her back while facing a yearling male.

Tahr have a conspicuous head-down. The Nilgiri tahr hunches its back, arches its neck down, sometimes so far that its nose almost touches the ground or points back between the legs, and moves in a stiff gait with its legs bunched beneath its body. This display was seen 16 times, females giving it two of these times. A male occasionally displays in no particular context, one male doing so 60 m from the herd. I observed the head-down in Himalayan tahr on 5 occasions. Walking broadside to his opponent, a male erects his ruff more than usual so that it stands up like a keel above his back. The tail is vertical and the anus puckers out.

Among mountain goats "the male erects itself by stretching front and hindlegs until they are straight and stiltlike. The back is arched upwards; the belly is drawn in, the neck is arched down" (fig. 35) and he may emit harsh roars (Geist, 1965). As Walther (1974) has pointed out, the head-down in mountain goat may function more as a direct than indirect threat. Chamois present a humped broadside with lowered or raised head showing off their dorsal ridge of hair (Krämer, 1969). The male and the two female Burmese takin in the New York Zoological Park display by arching their neck down, tucking their chin in, retracting their ears, and adducting their tail (plate 24). They move jerkily past or sidle toward their opponent, giving harsh staccato grunts while holding their heads averted. I also observed a bull muskox display in this

Fig. 34. (Left) head-down display of wild goat; and (right) head-up display of courting Himalayan tahr.

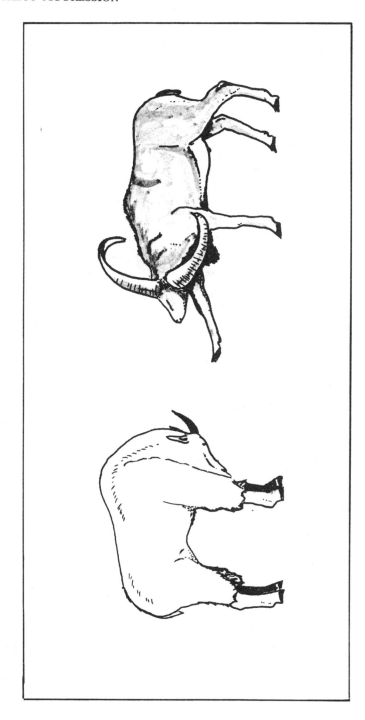

Fig. 35. (Left) head-down display of mountain goat (redrawn from Geist, 1965); and (right) twist and kick by a urial (from Walther, 1974).

fashion at the Calgary zoo. He moved stiffly with his head turned away, and his tongue was partly extended. Although his head was not markedly lowered, the hairy hump on his shoulders gave his profile an arched look (plate 23). He grunted *brrooo*. Tener (1965) found "no evidence of a musky odour of the muskox," but this displaying bull smelled strongly, a muskily sweet scent resembling that of gorilla.

Broadside. Wild goat and markhor sometimes stand broadside to an opponent with body erect and with chin tucked in. After a few seconds one or both animals may circle each other. Walther (1961) noted that ibex generally lack broadside displays. I observed only one broadside in Asiatic ibex: as one adult male walked toward another though smaller male, the latter turned broadside and the two stood motionless for 15 seconds some 3 m apart; the larger then cut in front of the other, almost touching him, and stood parallel for another 10 seconds before departing. An urial ram often walks slowly past or stands beside another ram with his neck more erect than usual. On about 10 occasions an urial ram stood motionless for as long as a minute facing the side of another ram in a stance resembling the broadside. Bharal usually lower their neck somewhat during the broadside. Two animals may stand side by side with heads slightly averted, tails horizontal, and their tongues flicking.

Himalayan tahr have a conspicuous broadside, as this description from the Bhota Kosi illustrates:

> One tahr may stand close to another with its neck stretched some-what forward, and with its muzzle held horizontally or tipped slightly up or down. The displaying animal either stands parallel to the other and facing in the same direction, a position which shows off the size of the ruff, or it halts in front or behind the other. A conspicuous feature of the display is that the muzzle is seldom pointed directly at the opponent but is averted to show a partial to complete profile. The threatened individual often responds by assuming the same posture with the result that the two display side by side or face to face, a metre or two apart, each with its muzzle turned away. I observed this display 7 times, once between 2 yearling males, 4 times between a subadult [class I] and a yearling male, once between an adult male and a yearling male, and once between an adult and yearling female. Most contacts were brief, lasting less than

a minute, but on one occasion a subadult and a yearling male displayed to each other 3 times within a period of 50 minutes, each interaction lasting some 5 to 10 minutes. An animal sometimes terminated such an interaction by licking or scratching itself before turning away (Schaller, 1973a).

Once, in New Zealand, two class III males walked in single file. The one in the lead stopped and turned broadside to the other, which then displayed too. Abruptly the lead animal continued on, followed by the other, but both halted after a few meters to display once more. A third broadside followed. Suddenly the lead male trotted off across a cliff face.

Parallel walk. Two males may walk or trot parallel .5 to 4 m apart, their horns tipped toward each other. Sometimes their posture is normal or only slightly erect and at other times they display, a head-up or head-down among urial, and only a head-down among bharal. These two species also walk parallel while grazing. Soay sheep and mouflon use this display too (see Schaller and Mirza, 1974) as do mountain sheep (V. Geist, pers. comm.). The parallel walk is uncommon in most *Capra*. I saw the behavior twice in markhor and once in domestic goat. On one occasion two yearling Nilgiri tahr walked side by side for 5 m in a head-down. Two Alpine ibex males may run shoulder to shoulder if they are of equal rank but the subordinate one takes the lead if they are not (Walther, 1961). I also observed the parallel walk in chamois (see below), and Gray (1973) reported it in muskox.

Block. A special form of broadside is the block, which I noted 13 times in urial, 11 times in bharal, and 6 times in wild goat. To halt the advance of one male another and larger male typically moved in front of him and stood broadside. One urial ram prevented another from following a ewe on 7 occasions; one ram chased another, overtook him, and halted his advance with a block on 5 occasions; and once a ram left a huddle but another hurried after him, blocked, and waited until he returned to the others. A blocked ram may turn broadside and the two then continue in a parallel walk. Bharal males mainly blocked others from access to females, using a variety of displays including the head-up, low-stretch, and extended penis, and wild goat also blocked in such situations. The block is common in mountain sheep (Geist, 1971a).

Body-shake. Chamois have a unique display during which they stand slightly crouched with neck parallel to the ground as they shake their back, sides, and belly. As he shakes, a male scatters urine over himself. He also grunts (Krämer, 1969). An excerpt from my field notes describes an interaction between an adult and subadult male:

> The adult stops parallel to the young male and shakes, a vigorous shake that makes his long dorsal hairs conspicuous. He then walks parallel to the young one with his neck lowered obliquely. The subadult shakes and immediately afterwards the adult does too. Going to a shrub, the latter horns it, holding his head sideways to bring twigs into contact with his post-cornual glands. He then urinates, not in a normal male posture but in an exaggerated squat like a female. (Krämer, 1969, noted that this posture is used mainly by subordinates). As the two walk parallel, the larger head-ups and the other shakes.

Humped approach. The head-down is basically a lateral display, but some species increase the intimidatory value of the posture by facing their opponent, sometimes so much so that the posture becomes a form of direct aggression. Krämer (1969) noted this in chamois. Himalayan tahr approach opponents in a head-down, looking like a huge grass tussock with a black muzzle peering from it. When approaching or being approached, a male bharal may face the other animal with his back humped and neck lowered but with his muzzle pointing upward and with his tongue flicking. Sometimes bharal combine the head-down and humped approach, standing broadside but with the head facing the opponent.

Aggression-courtship displays

Spraying urine and touching penis with mouth. Chamois scatter urine on themselves by shaking their hide as noted above. The male takin at the New York Zoological Park occasionally lowered his head, raised a hindleg, and squirted urine against his forelegs and chest. Female takin urinate with tails clamped close to the body thereby soaking the long tail hairs. *Capra* use another technique.

On 4 occasions a wild goat lowered his rump and with unsheathed penis urinated against the back of his forelegs. A more common gesture, tallied 61 times, was for a male to tuck in his rump and twist his muzzle toward his bared penis. Sometimes he nuzzled the penis and at other times he inserted it into his mouth. The mouth, face, and beard may then be doused with urine. Markhor squirt urine over their face and forelegs in similar fashion, as do Asiatic ibex (Walther, 1961), Kuban ibex (Steinhauf, 1958), and domestic goat (Shank, 1972). Alpine ibex may ejaculate after nuzzling the penis (Aeschbacher, in press). In New Zealand, Himalayan tahr were twice seen to mouth their penis. By examining males shot in cropping operations at the beginning of the rut, I was able to check for evidence of urine on the hide. The hair on the belly is dense and soft, a spongy mat perfect for absorbing urine. Males up to an age of 2½ years have white bellies and they do not smell musky. When males are 3½ years old their belly hairs are some-what rust-colored, slightly damp, and faintly musky. Males 4½ years old and older have rust-colored hairs heavily soaked with urine and their odor is powerful. Some rest sites of Nilgiri tahr smelled strongly, suggesting that this species also sprays urine. However, I have not noticed such strong odors when tracking wild goat and markor even though these species urinate on themselves.

The penis may be extended in three situations: when courting, during aggressive contacts between males, and in no particular context. For wild goat the percentages were 46% when courting, 2% during aggression, and 52% in no special context. Six of my 7 observations in markhor were made on courting males. The function of the gesture in *Capra* seemed to be mainly utilitarian, to enhance the attractiveness of the male by perfuming his pelage.

The situation was quite different in bharal. Males extended their penises 39 times (38%) in no particular context, 4 times (4%) when courting, and 60 times (58%) while interacting with another male (plate 46). On 6 of these occasions a male merely extended his penis and in the others he licked it, nibbled it, or poked it into his mouth. I have no evidence that males urinate on themselves. This, together with the fact that the gesture is used during domi-nance interactions, suggests that the extended penis has a threatening function in itself. In 43% of the interactions a male displayed to an animal of a smaller horn-size class and in 50% to one of the same class (though usually with smaller horns), indicating that the behavior is mainly the prerogative of a high-ranking individual. Once, for instance, three males walked in single file with the largest in the lead and the smallest in the rear. When the one in the rear tried to pass, the middle one stopped and exposed his penis. The small male halted too. As they continued he was again tagging behind. Two further points suggest that the penis has display value. In 58% of the instances the male unsheathed his penis while standing broadside to another male who then sometimes averted his head or turned away (plate 46). When not in rut males have a pinkish penis but during the rut it is crimson. Backhaus (1959) reported that domestic goats can distinguish red, and possibly bharal can too.

An aoudad may jerk his extended penis against his belly and he may insert it into his mouth (Haas, 1958). Aoudad have also been observed as they lay "on their backs and sucked their penises for short periods" (Katz, 1949). Sheep do not spray themselves with urine nor do they mouth their penises except on very rare occasions. A sheep may, however, extend his penis and whip it against his belly, such behavior having been reported in mountain sheep, urial, and Soay sheep, as well as in saiga (Pohle, 1974).

Mount. Mounting as an expression of dominance was uncommon in mountain sheep (Geist, 1971a), Soay sheep (Grubb, 1974b) and in most of the species I studied. Wild goat mounted on 5 occasions and markhor once. Urial mounted 9 times during October and November, but one day in July several class I rams mounted each other 29 times within a few minutes. One yearling urial ewe mounted another. Bharal mounted often when compared to these species. The 115 mountings included one yearling, 3 class I, 2 class II, 4 class III, 22 class IV, and 83 class V males, indicating that the behavior was mainly expressed by adults. A male mounted another of a smaller horn-size class on 24% of the occasions, of a larger class on 6%, and of the same class on 70%. Among the last-named a difference in horn size was often apparent and in two-thirds of such instances the larger animal mounted a smaller one. Among mountain sheep the dominant ram generally mounts subordinates (Geist, 1971a) and in Soay sheep the reverse is usually true (Grubb, 1974b). The behavior of bharal cannot be categorized quite as clearly. Sometimes a male used the gesture to test his rank, the result of the interaction becoming apparent from the response of the mounted animal. If a small-horned male had the temerity to mount, the act was either ignored or retaliation followed swiftly in the form of a jump, clash, kick, or butt. If a large-horned male mounted, the other usually trotted away, grazed, or otherwise behaved as a subordinate. However, in 25% of the interactions a low-ranking male threatened in return with a jump or clash. An example illustrates a typical interaction:

> Two class IV males trot parallel and 15 m apart. They stop, rear in unison, and then the larger of the two mounts. The smaller jerks his head in response. After running 6 m uphill the larger rushes and the other first catches his blow and then paws the ground. The larger jumps as if to clash again but this time the other walks off. Following closely, the large one cuts in front, blocks, and hooks the air with a horn. Then he runs 13 m ahead and rears up on his hindlegs. The small male merely turns aside, ignoring the challenge. A mounting follows but he trots ahead. Noting the commotion, another class IV male runs up and mounts the smaller one. Three class V and another class IV arrive, all joining to form a huddle.

This sequence also illustrates two other aspects. One male may goad another, he may pressure him to respond aggressively. Such behavior was also observed in urial, and Geist (1971a) reported it in mountain sheep. When two males fight, others gang around and often join the brawl, a response apparently common to all Caprini as well as to many other ungulates.

Young urial, markhor, and wild goat mounted other young or females on occasion, but I do not know if such behavior is related to dominance.

Low-stretch, twist, and kick. These displays may be given separately or in association, a low-stretch followed by the twist and an almost simultaneous kick being a common sequence. Grubb (1974b,d) termed this combination of patterns a nudge. In quantifying the displays I tallied every occurrence even if an animal repeated itself during an interaction. A low-stretch immediately preceded by a twist was not counted because I found it difficult to distinguish a slight low-stretch from the dip of the head that normally accompanies a twist. A kicking male sometimes jerked his foreleg upward 2 to 5 times without lowering it to the ground. Such jerks were not counted as kicks. Tongue-flicking sometimes accompanied the displays, as did grunts.

A low-stretching male holds his neck low and horizontal, strains his muzzle forward or has it slightly raised, and he may lower his body in a slight crouch (plate 44). I saw this posture twice in wild goat, displayed once by a class I male to a yearling and once by one class II male to another. Punjab urial low-stretched 5 times, these displays always involving class III and IV rams. Grubb (1974b) noted that Soay sheep do not low-stretch during aggressive encounters, but he then goes on to describe a ram as "he lowered and extended his neck along the side of the other ram. He then turned his head . . . ." This would seem to be a low-stretch followed by a twist. Mountain sheep commonly low-stretch, much more so than do urial. Geist (1969) found that 29.2% out of 1602 patterns in bighorn were low-stretches as compared to .4% out of 669 patterns in urial (Schaller and Mirza, 1974). Our methods of collecting data were not wholly comparable, but a striking difference obviously exists. The low-stretch was observed 21 times among bharal, 17 of these displays being given by class IV and V males (plate 44). Twelve times a male displayed to one in a smaller horn-size class and the reverse occurred 3 times, indicating that dominant individuals low-stretched most often. A low-stretching animal approached from the front 4 times and from behind or the sides 17 times.

When twisting, a male lowers his neck into a position that resembles the low-stretch and then rotates his head up to 90° so that the horns face away from the opponent. He may twist slowly or with a jerk. The display was seen once in markhor, twice each in wild goat and Himalayan tahr, and 5 times in bharal. Twisting is common in all *Ovis* (fig. 5). The display was observed 205 times in Punjab urial: 11 times by class I rams, 23 times by class II, 61 times by class III, and 110 times by class IV. Of these rams, 60% displayed to an opponent in a smaller horn-size class and 32% displayed to one in the same class. About 66% of the twists were given behind an opponent, 13% while facing him, 10% while walking parallel, and 11% toward his side. Occasionally a ram shoved with his chest while twisting.

In the third display, the kick, a male lifts his foreleg some 5 to 30 cm off the ground, holding it stiffly or bending it at the carpal joint. The gesture usually fails to touch the opponent. One class V wild goat male kicked another 3 times while sparring, and a class IV male kicked a reclining male before usurping his rest site, my only such observations in this species. Bharal kicked 24 times, all except 3 kicks being given by class V males. Six of the kicks were delivered by the smaller of two males, the reverse occurred 3 times, and in the rest the combatants were in the same horn-size class. A small male immediately rubbed his face on the rump of a large male after kicking him, as if to reaffirm his subordinate position. Most kicks were directed at the hindquarters, but twice a bharal merely flailed the air while another animal rubbed his rump. The kicks were variable: the leg was straight in most of them, limply bent at the carpal joint in some, and once an animal made pawing movements.

The kick was the most common display of urial rams (fig. 34). Twenty-six kicks were delivered by class I rams, 43 by class II, 194 by class III, and 302 by class IV, a total of 565. Rams may kick repeatedly during an interaction and sometimes they did so reciprocally, either in unison or succession. Two rams once kicked each other 44 times in one minute. The blow sometimes hit the opponent between the hindlegs, on the thigh, rump, or other part of the body. The leg was usually held straight, but one ram made pawing movements with his foreleg. Rams kicked each other from the front (33%), from behind (40%), from the side (18%), or while parallel (7%). In the remaining 2% of the cases they kicked while facing away from each other, and twice rams kicked while thrashing a shrub with their horns. Some 45% of the kicks were directed at a ram in a smaller horn-size class, 46% were between rams of similar size, and 9% were reversals, with a small ram kicking a large one. In the last category were mainly rams who responded to a kick by returning the gesture. When two rams were in the same class they readily kicked each other even if one was larger: dominance could be inferred from horn size but not behavior, especially since both tended to part casually. One ewe twice kicked another. Mountain sheep commonly kick, though bighorns do so less often than Stone sheep (Geist, 1971a). Kicking is also prominent in Soay sheep (Grubb, 1974d), and I have seen mouflon in zoos kick each other as well.

The kick appears to feature in the fighting repertoire of Japanese serow too: "Serows are also known to chase one another intensely, and when the pursuer catches up with its quarry it kicks the quarry in the femoral region. While chasing the serow utters the *shu shu shu* call" (Akasaka, 1974).

Tongue-flick. A wild goat male flicked his tongue at another male on 3 occasions. *Ovis* rams also used this display. Bharal commonly flicked their tongue in and out of the mouth while showing a broadside, head-down, low-stretch, or other such forms of behavior. Muskox also tongue-flick (P. Lent, pers. comm.).

Lip-curl. Males usually lip-curl after having tested the urine of a female. However, chamois may lip-curl while displaying the head-down. The species seems to have incorporated a sexual gesture into its threat display (Krämer, 1969).

Poke. A Punjab urial ram may poke another once or twice with the muzzle in the rump, thigh, or side. The poke was usually given in conjunction with a kick or twist and was observed 31 times. Mountain sheep also use this pattern (Geist, 1971a).

Miscellaneous displays

Dig rutting pit. Mountain goats dig pits some 30 cm wide and 45 cm long during the rut. "Prior to pawing, a billy would sit on the ground and assume a posture not unlike a sitting dog. Arching the neck and looking towards the ground, he would paw quickly and vigorously with one front leg, thereby throwing snow, sand and dirt at his belly, hindlegs, and flanks" (Geist, 1965). Males possibly urinate and defecate in the pits, for they smell strongly after they have been used. Takin may also sit like dogs though I have not seen them excavate anything. Saiga (Pohle, 1974) and chiru (Rawling, 1905) paw depressions too, perhaps as manifestations of territorial behavior (see Chapter 8). Muskox dig "shallow pits" (Tener, 1965) with their horns but a function has not been ascribed to these. P. Lent (pers. comm.) noted that these pits smell strongly of the bulls.

Head-shake. Walther (1961) observed that markhor of unspecified race, Afghan urial, Asiatic ibex, and Marco Polo sheep shake their heads sideways during aggressive encounters, and such behavior has also been noted in mouflon and mountain sheep (see table 39). I noted this pattern only once in a bharal male which shook his head slowly before lunging at his opponent. Muskox swing their heads from side to side before clashing (Gray, 1973).

Paw. Individuals may stamp or paw with a foreleg before lying down, when digging for food, and when sensing danger, but only some species use the behavior in aggressive situations other than as a kick. One urial ram pawed as he horned a shrub and another did so as he threatened a ram. Bharal males also pawed in these circumstances. Mouflon at the New York Zoological Park sometimes pawed before clashing; Shank (1972) noted similar behavior in domestic goats and Walther (1961) in Marco Polo sheep. One wild goat male stamped hard as he confronted another male, and chamois also stamp when threatening (Krämer, 1969).

Huddle. On 15 occasions some 3 to 6 Punjab urial rams faced each other in a circle with lowered heads and clashed, poked, kicked, grunted, and rubbed faces almost continuously and seemingly at random. Occasionally a pair broke away from the huddle for a brief parallel walk. Huddles lasted as long as 10 minutes and usually included only adult rams. I made no attempt to quantify the many rapid interactions. Mountain sheep and mouflon behave similarly and Soay sheep congregate and interact rather casually (Grubb, 1974b), as if huddling at low intensity. I have no records of huddles in *Capra,* but bharal display the behavior. As many as a dozen bharal males may cluster and mill around for as long as 10 minutes while mounting each other, horning the ground, jumping, and clashing. Occasionally a male dashes from the huddle, rears up, and, if no male lowers his head to receive a clash, trots back. All may move along in a tight cluster and if one steps ahead the others hurry to surround him. Bharal huddles differ from those of urial in being less cohesive: males do not face each other in a circle, probably because some of the main displays are oriented not at the heads of other animals as in urial but toward the rumps. Huddles seem to be means by which males can test rank informally without the risk of retaliation, and this possibly helps to reduce tension among these inveterate status seekers.

Horn vegetation. Horning is probably practiced by all Caprinae. Using upward, downward, and lateral motions, an animal may spend a minute or more in vigorously belaboring shrubs, saplings, and grass tufts with its horns or in wiping its face on leaves and branches. The behavior often occurs as if in vacuo. The sight of one thrashing animal may stimulate others to horn nearby or even at the same spot. Some shrubs in an area are horned repeatedly, no doubt deposits of odor and frayed branches making them particularly inviting. Horning differs from the other indirect forms of aggression in that animals of all ages and both sexes participate. I observed horning in Himalayan tahr (a total of 15 times), Nilgiri tahr (8 times), Asiatic ibex (once), markhor (14 times), wild goat (11 times), and Punjab urial (10 times). A Nilgiri tahr twice kneeled on its forelegs while rubbing the ground. Mountain goats horn vegetation, and they slide grass stems and twigs over their post-cornual glands (Geist, 1965), behavior also typical of chamois as described earlier. Muskox horn vegetation, and they mark it, as well as their forelegs, with their preorbital glands (see Gray, 1973).

Bharal commonly horn vegetation. I tallied 147 instances at Shey, 24 of them by females, one by a yearling male, 5 by a class I male, 2 by a class II, 6 by a class III, 20 by a class IV, and 89 by a class V, or .04 hornings per animal-hour of observation. (Markhor horned at the rate of .01). As many as 4 males horned at the same time. Bharal used two techniques to horn: they thrashed shrubs in a typical manner, and they placed the base of their horns on the ground and twisted from side to side, leaving a semi-lunar mark in the dirt. The horned objects included shrubs (46%), junipers (20.5%), the ground (29%), snow (3.5%), and rocks (1%). In some situations horning seemed to function as an intimidation display, as the following example suggests:

> A class IV male mounts another of equal size. The mounted one lurches ahead, rears, and clashes. Briefly the two stand side by side with averted heads. The same male is mounted again and he responds by sparring. The male who had mounted turns aside and thrashes a shrub. He mounts once more whereupon both rear and clash. Then one male horns a juniper while the other, the one who had been mounted, rubs his rump. After that both hook shrubs.

Discussion

A lateral display of one form or another exists in all species listed in table 39 and in some others as well. The head-down is particularly conspicuous and frequently displayed in chamois, mountain goat, tahr, and takin—all animals with small horns. In these species the strikingly humped back and attendant pelage adornments seem to have the same intimidatory function as do the large horns of Caprini. Furthermore, in the short-horned forms the head-down is more of a direct threat than it is in the sheep and *Capra* goats. Speaking of mountain goat, Geist (1965) found that "the lack of rushing and chasing on the part of adult males" was striking. Instead they displayed a head-down, thereby threatening seriously without wholly unsheathing their dangerous weapons. Himalayan tahr males behaved similarly in that they seldom clashed, even during the rut, and used lateral displays instead. In this context it is significant that female rupicaprids, tahr, and takin readily present the head-down, in contrast to the sheep and *Capra* goats, which seldom do so. As noted earlier, females often threaten directly but seldom indirectly. If the females of a species use the head-down frequently it seems to indicate that the gesture functions to some extent as a direct threat.

A number of displays occur both in the contexts of courtship and aggression, a subject that Grubb (1974b) discusses thoroughly for sheep. Mounting is obviously a sexual gesture which secondarily has been incorporated into the aggressive repertoire of several species. An extended penis has a number of useful functions. *Capra,* Himalayan tahr, and takin, among others, spray urine over their faces and other parts of their bodies and bharal threaten with their penises. Several primate species also mount and erect the penis as forms of threat (Wickler, 1967). The lip-curl, normally employed to test the estrous state of a female, is a threatening gesture in chamois. The derivation of the low-stretch, twist, and kick remains obscure, and it is not my intention here to enter into an interpretative fray with ethologists. Walther (1974) felt that the patterns represent phylogenetically old fighting forms. By assuming that the low-stretch is an intention to bite, the kick an exaggerated way of pawing (Walther, 1974), and the twist a presentation of horns to an opponent (Geist, 1971a), then such an intepretation is plausible. However, other explanations seem just as plausible. The low-stretch could be the antithesis of the head-up, a gesture indicating lack of aggressive intent during a direct approach; the kick may be a preliminary to mounting (see Shank, 1972); and perhaps the twist was originally a means of preventing flaring horns from hitting an animal accidentally when moving up behind and close to it, as I observed occasionally in bharal. (In the small-horned Caprinae the twist is rare or absent.) Of interest here is not the derivation but the function of these displays. All are relatively mild expressions of dominance, so mild that a subordinate may use them to initiate contact with a dominant or actually to retaliate. To court successfully a male must remain near a female without scaring her away, and this he can do best by deemphasizing his physical appearance somewhat (low-stretch, twist) and asserting himself with rather gentle gestures (kick). On rare occasions these gestures may also be directed at females in contexts other than courtship. For example, one urial ram kicked a ewe to make her move from a rest site.

The degree to which different species have incorporated the low-stretch, twist, kick, mount, and the extended penis into the threat repertoire varies greatly. Figure 36 shows the relative frequencies with which 4 species use the five displays in interactions between males. The low-stretch and twist are uncommon in all these species except in the urial. The Punjab urial and other *Ovis* kick often and in this they differ from the goats. *Capra* and bharal often mouth their penises but sheep do not. Mounting appears to be a common expression of dominance only in bharal.

Most interactions between males involve relatively few displays. Taking only the low-stretch, twist, kick, and mount, I tallied an average of 1.8 displays per bout in Punjab urial during the 1972 and 1974 ruts. Some 52% of the interactions consisted of one display, 30% of two, 9% of 3, and 9% of 4 to 10. Interactions with 2 displays were either repetitive, two kicks or twists, or consisted of a twist and kick in that order; interactions with 3 or more displays followed a similar pattern. The only other species for which I have a large sample of displays is the bharal. This animal averaged 1.2 displays per interaction, with 69% being single displays—usually a mount—18% double displays, and the rest mostly repetitive kicks and mounts. The frequency of mouthing the penis increased as the rut progressed but that of mounting did not (table 40). In general, *Capra,* bharal, and tahr seldom extended their penises except in the rutting period.

Adult males are most active in displaying indirect threats. Figure 37 shows this for urial, wild goat, and bharal, using the sum of all low-stretches, twists, kicks, mounts, and penis mouthings

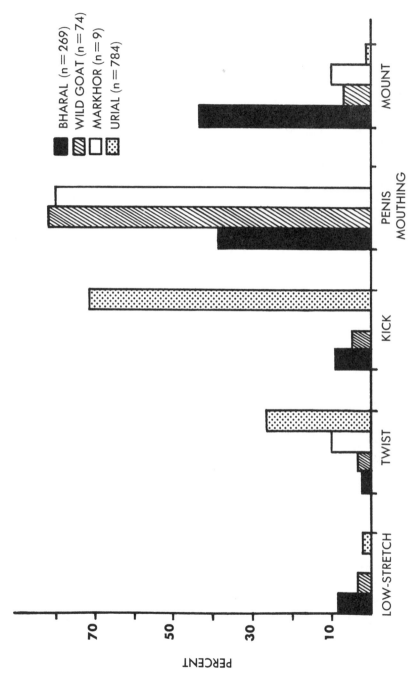

Fig. 36. Relative frequency (in percent) of several indirect forms of aggression, comparing bharal, wild goat, Kashmir markhor, and Punjab urial.

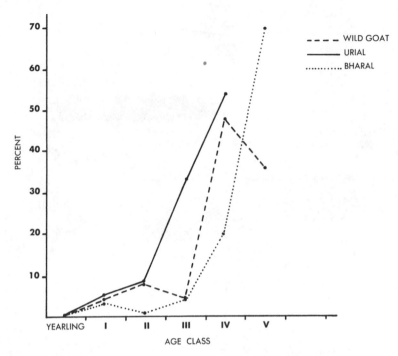

Fig. 37. Frequency of use of indirect forms of aggression (low-stretch, twist, kick, penis mouthing, and mount, combined) by different male age classes in wild goat, Punjab urial, and bharal.

observed. Males tend to indirectly threaten others of their own horn-size class or the next larger or smaller one: taking only the low-stretch, kick, twist, and mount, 100% of the wild goat interactions, 90% of the bharal, and 83% of the urial fit this pattern. Although species differ in their displays, the circumstances in which indirect aggression is expressed remain rather similar.

Young urial, markhor, and wild goat mounted other young and females on occasion, the only behavior of young which could possibly be considered a form of indirect aggression. I observed markhor young mount each other 4 times, urial once, and wild goat once; a wild goat young also mounted a yearling female on one occasion and an adult female on two. Mounting by young is also prevalent in aoudad (Haas, 1958), mountain sheep (Geist, 1971a), and muskox (P. Lent, pers. comm.). Grubb (1974d) observed the behavior among Soay lambs, males doing it more often than females. "In a minority of cases (31 out of 179 observed) mounting is preceded by a sequence of 'nudging', (a 'twist' frequently accompanied by a 'kick' or 'laufschlag') but only in the males .... Much lamb activity can be envisaged as an ontogenic stage in behaviour and since individual behaviour patterns (mounting, nudging, butting) are modified in form, function and context on attaining adulthood, these can be regarded individually as stages in a developmental sequence." However, it should be pointed out that development in domestic sheep is highly accelerated, with animals often reaching sexual maturity as lambs.

SUBMISSION

Considering that most species use a dozen or more forms of behavior to intimidate an adversary, one might ipso facto expect several distinct appeasement gestures too. However, an animal usually conveys submission either by retreating casually or indulging in a clearly innocuous activity. Certain types of direct aggression are of course so unambiguous that evasive action is the most timely response. Subordinates often begin to graze when confronted by a dominant individual, behavior which obviously signals lack of aggressive intent. This was observed in all species I studied and seems to be typical of Caprinae in general. Himalayan tahr groom their pelage to terminate a confrontation or at least to reduce tension. To face someone is a form of threat, and animals may use the antithesis of this position—they avert the head. For instance when a large-horned bharal walks past one with smaller horns the latter may do nothing more than turn his muzzle away. In a more extreme response the subordinate individual swivels completely around and presents its rear, thereby concealing its weapons. Some species have prominent rump patches, and one function of these may be to signal a submissive attitude. It is perhaps significant that this probable appeasement structure is located farthest away from the weapons on the head. There is thought to be an interrelation between threat and appeasement structures (Guthrie, 1971). Geist (1971a) noted that snow sheep have smaller rump patches than Stone sheep, which in turn have smaller ones than bighorn sheep, and he concluded that the sheep which have the heaviest horns and probably deliver the most forceful blows also have the largest rump patches. There may indeed be a correlation between forcefulness of clashing, as measured by weight and not length of horns, for Asiatic ibexes have seasonally larger rump patches than Alpine ibexes and Altai argali have larger patches than Tibetan ones. However, it is not clear why eastern and western urials differ in rump-patch size and distinctiveness, and why *Capra* goats have relatively much smaller rump patches than do *Ovis* even though the former use proportionately more direct than indirect aggression.

Males sometimes behave like estrous females and thereby presumably convey their submission. Subordinate chamois and mountain sheep males may urinate like females when approached or sniffed by a dominant animal, and subordinate rams may even perform lordosis when mounted (Krämer, 1969; Geist, 1971a). "In the interactions of dominant and subordinate rams it is typical that the subordinate remains and only rarely withdraws" (Geist, 1971a), behavior characteristic of an estrous but not an anestrous ewe. Urial may behave similarly on occasion and subordinate bharal males also tend to permit mounting. I did not observe such seeming female mimicry in *Capra* and *Hemitragus,* but Shank (1972) reported it for domestic goat.

Subordinate urial sometimes stretch their necks diagonally downward in a neck-low posture. A ram may also stand in this position at right angles to another ram and lick or nuzzle his penis sheath. Both the neck-low and the penis-nuzzle have also been observed in Soay sheep (Grubb, 1974b). Mouflon emphasize their neck-low posture by kneeling at times on their forelegs (Pfeffer, 1967). Himalayan tahr and bharal use the neck-low too, but I have not seen it in markhor and wild goat. Mountain sheep apparently lack this submissive display too (Geist, 1971a). A submissive chamois lowers its body, stretches its neck and muzzle forward, and raises its tail (Krämer, 1969), and a mountain goat sometimes crouches so low that it is crawling (Geist, 1965). On one occasion a class I urial ram chased a ewe until she crouched with neck extended on the ground for 20 minutes. Such crouching has also been observed in chamois (Krämer, 1969) and saiga (Pohle, 1974), and it seems to represent an extreme form of submission.

Urial rams often make gentle contact in two different ways. One ram may rub his horns against the head or body of another, and on a few occasions two rams horned in unison while at the same time flicking their tongues. Such horning was observed 36 times. It was initiated 17 times by the smaller of the two rams and in the remaining instances the animals were in the same horn-size class. On one occasion a ewe horned a yearling male. A ram also rubbed his face on the face, horns, neck, and shoulders of another, and sometimes he licked or nibbled rather than rubbed. Of the 47 instances of rubbing tallied, 70% were initiated by the smaller of the two rams, 3% by the larger, and the rest involved rams in the same size class. Rams of all ages participated in horning and rubbing. Obviously these two gestures are mainly displayed by subordinates. They enable a low-ranking animal to make contact with and remain near a high-ranking one. Young rams possibly derive some status by rubbing themselves against the suborbital glands of adults, thereby acquiring the odor of dominant rams. Rubbing and horning have also been reported in mouflon, Soay sheep, and mountain sheep (see Schaller and Mirza, 1974).

Bharal rub rumps, not heads. One bharal typically approaches another from the rear and rubs his face and forehead up and down on the rump (plates 45 and 46). Sometimes he nuzzles the anal area, behavior reminiscent of that used to test urine in estrous females. The recipient often raises his tail horizontally; if he does not, the other may push it aside with his muzzle. The eyes of rubbing animals tend to be closed, as they are in sheep. One male may rub two or three others in succession. Rubbing stimulates others to join, and on 11 occasions 3 males and twice 4 males stood in a row each rubbing the rump of the one in front. Several bouts lasted as long as 4 minutes. I observed the behavior 332 times between males, 10 times between females, and 7 times a male rubbed a female. Yearling males rubbed at the rate of .03 per animal-hour of observation, class I at .10, class II at .23, class III at .12, class IV at .21, and class V at .29, showing that males of class V rubbed most often. The frequency of rubbing increased in early November and then maintained the same level into December (table 40). A male of a smaller horn-size class initiated the contact in 49% of the interactions and the reverse occurred in 3%. The remaining participants belonged in the same size class, but 63% of these rubbing individuals had noticeably smaller horns. Thus, at least 79% of all rubbing was initiated by subordinates. Three examples illustrate typical interactions:

A lone class V male walks and trots toward a herd containing 7 animals including a class IV male. He enters the herd and in a low-stretch approaches a female. She squats and he sniffs her rump then follows her slowly. The class IV male approaches the new-comer from the rear and rubs him. Ignoring the gesture, the male sniffs a second female and then a third. Still tagging behind him is the other male who rubs once more.

A large class V male follows a female in low-stretch some 20 m until she squats. He lipcurls then waits behind her. Another but smaller class V approaches and rubs the male. And a class IV hurries over and rubs the smaller class V, making a string of three. The big male disregards the others and sniffs the female once more.

A class V male stands. A larger male moves parallel to him but the only response to him is an averted head. The newcomer swivels his rump in front of the other's muzzle and after this emphatic invitation receives a rubbing.

Bharal often use rubbing as a means of approaching a dominant individual, especially if he is near an estrous female. The recipient usually ignores the gesture but sometimes he reacts aggressively by jumping (seen 4 times), jerking (7 times), horning vegetation (7 times), or unsheathing his penis (6 times). On 5 occasions a subordinate animal both kicked and rubbed one of high rank, the two actions presumably nullifying each other. A dominant twice rubbed a subordinate and then mounted him as if to reassert himself after having provided reassurance.

Capra and *Hemitragus* do not rub, and other friendly gestures between males are rare too. Once two adult wild goat nuzzled and licked each other, one markhor male once licked another male, and on three occasions a yearling Himalayan tahr licked a class I male, my only such observations in these genera.

Definite submissive displays are obviously rare among the Caprinae, and those species which have one in their repertoire seldom exhibit it. As noted earlier, submission is usually conveyed by behaving in an unaggressive manner. Some genera such as *Capra* seem to regulate conflict by simple avoidance and by keeping a low profile, whereas others use the neck-down or similar display. In addition, subordinate sheep and bharal, but not other species, display an overt pattern—the rub—to make contact with dominant individuals and in effect to assert their low rank. *Ovis* and *Pseudois* herds are more fluid than those of goats, and a friendly gesture signaling low status might help reduce aggression among strangers.

CONCLUSION

Two main points need to be emphasized in the conclusion of this brief overview of direct and indirect aggression. Most forms of direct aggression are widespread in the Caprinae and with minor variations are similar in execution. All age and sex classes exhibit these forms: animals seem to be born with them ready for use. By contrast, the indirect forms are sometimes species-specific, and those in general use are often modified in various ways. Most displays are given mainly by males and not at all by small young, indicating that they are under maturational control. Adult males, whether sheep, goats, bharal, or tahr, exhibit a disproportionately large amount of both direct and indirect aggression during the rut.

When perusing the long list of aggressive patterns, one is struck by their seeming redundancy. For example, table 39 lists 22 threatening gestures for Punjab urial and 17 for chamois. Why should each of these species use 5 variations on a lateral display? The advantages of such diversity can best be understood if aggression is viewed as a graded system similar to that often found in the vocal repertoire of some animals. Most forms of aggression fall into three categories: the direct threats which are offensive, direct challenges; the lateral displays which are strongly intimidatory but usually without the immediate intent to attack; and low-level aggressive displays which represent relatively mild assertions of dominance. The use of the different types of aggression depends largely on the situation, and consequently a threatening individual tends to limit itself to one category although it readily switches to others as the circumstances change. Some forms of threat within each category have a rough hierarchical sequence of intensity, with, for example, the jerk, lunge, and chase making up one such sequence, and the low-stretch, kick, and mount another. Individual displays may be modulated on the basis of intensity too, the head-down according to the amount of arch in the back and the clash according to force, to give two examples. By varying categories, displays, and intensity of displays a male can convey his changing levels of aggression with great subtlety. Social life depends on communication, and the Caprinae, being relatively nonvocal animals, have evolved an intricate system of visual signals which serve to reduce overt aggression and provide a predictable social milieu in the herd. When to this system the vocal and olfactory systems are added, signals may indeed become highly refined.

10 COURTSHIP BEHAVIOR

Males of high rank have priority rights to estrous females, but most Caprini males do not develop the status symbols and strength that go with such rank until they are at least 4 years old. As few individuals live longer than 10 years, a male has perhaps 5 seasons in which to mate and pass on his genes. The main rut lasts at most a few weeks. Thus a male's entire reproductive life, his whole raison d'être, may be crowded into less than half a year of his existence. It is no wonder that the males seem to seek their goal with such urgency and that the rut is the most hectic time of year in caprid society.

The onset of the rut is gradual, generally characterized by a pre-rut during which males become restless and begin to show interest in the females. Soay rams start to investigate ewes about two months before mating (Grubb, 1974b), and among urial, wild goat, and markhor this period is at least 3 weeks long. Large male herds break up during the pre-rut, and individuals either join mixed herds or roam alone or in small groups.

Species differ in the percent of males that associate with females in mixed herds during the main rut. From November 1–9, 1970, some 13% of all urial rams of class I to IV were solitary and 28% were in male herds (fig. 30). Trotting or walking rapidly, rams hurried across the terrain and often halted on ridge tops as if scanning the slopes for ewes. Having spotted some, they approached and entered the herd. If other rams were already present, the newcomers might first interact with them; if not, they checked for estrus in one or more ewes. Finding none in heat, the rams either tarried or resumed their wanderings. Other *Ovis* behave similarly (Geist, 1971a; Grubb, 1974b), but wild goat revealed a somewhat different pattern. Between September 20 and October 3, 1972, a period during which the wild goat were at a similar stage of rut as the urial, only 5% of the males were alone or

256   Courtship Behavior

in male herds and none were so after September 27. Males and
females tended to remain together once they had joined. Markhor
males also wandered little during the main rut. Some individuals
stayed with certain herds for weeks, but in late January, a month
after the peak of rutting, their fidelity became less marked. About
10% remained nomadic, possibly because the mixed herds were
too small to accommodate them. The domestic goats studied by
Shank (1972) showed no marked sexual segregation. Bharal were
intermediate between urial and wild goat in their proclivity to
rove (table 29). During the first week of the main rut about 20% of
the males were solitary or in male herds. One herd of 12 males
usually remained above the Shey monastery, but on November 25
at 0840 hours it broke up. The rams crossed an icy stream and
mingled with and displayed to some females. Although about 1 to
4% of the males continued to wander between mixed herds, the
animals lacked the tension and haste that were so conspicuous in
the movements of urial.

The amount of time that males spend roving can be correlated
with herd cohesion. With the rut being of short duration, a male
has little time to discover females in heat. For an urial ram to
remain with a few nonestrous females would lower his fertilization
rate, and this would place him at a selective disadvantage (Parker,
1974). Urial rams have to wander widely in search of scattered
ewes, whereas wild goat males have no need to behave similarly,
most females being found in a few large herds.

Various hormonal and environmental factors help to induce
estrus. Important among the latter is the social stimulation provided by the odor, sight, sound, and touch of males, as has been
noted in domestic animals (Fraser, 1968). Much of the nuzzling
and displaying of courting males must be viewed as having this
effect. The behavior of males can be divided into four general
categories: (1) that which primarily involves spreading odor;
(2) that which emphasizes physical attributes; (3) that which is
used mainly to test the estrous state of the female; and (4) that
which is used actively to court.

SPREADING OF ODOR
Animals raise their tails when excited—whether in play, in response to danger, in combat, or during courtship. A male's tail
position may change constantly during an interaction. For example, a wild goat male foraged with his tail down but raised it

horizontally when a female ambled past and vertically as he approached her. The raised tail itself is a conspicuous signal, and it also makes the rump patch prominent and presumably helps to dissipate odor from the anal glands in those species that have them. Urial and other sheep seldom raise their tails far above the horizontal, and this is also true of chamois and mountain goat. Himalayan tahr and bharal often hold their broad tails vertically. *Capra* males not only carry tails straight up but also folded over against the rump. Indeed one characteristic of rutting Alpine ibex (Nievergelt, 1967), wild goat, and markhor is the raised tail. Adult males have their tails up more often than do young males. For instance, 83% of the class V and 68% of the class IV wild goat had their tails above the horizontal during the rut as compared to 12 to 44% of the young males who had their tails in that position (see Schaller and Laurie, 1974). Markhor often walked around with their tails raised during December, but within a few days after the main rut, by January 11, they rarely lifted them. Domestic goats differ from wild species in that both males and females often carry their tails up even when they are not sexually active.

Chamois, takin, Himalayan tahr, and *Capra* spray themselves with urine, and mountain goats possibly urinate into their rutting pits. Only adult males douse themselves, except that perhaps female takin do so too, the urine with its rather rank odor probably being a rank symbol in itself (see Shank, 1972). Impregnated with urine, the beards of goats may act as fragrant sponges in addition to being prominent visual signals.

LATERAL DISPLAYS

Males may make themselves look impressive by assuming lateral displays while following or standing near a female. A wild goat male often holds himself erect or displays a head-up while behind, beside, or even on a ledge above a female. Occasionally he walks by her side in a head-down display (fig. 35). Markhor, too, may court in a head-up, as do chamois (Krämer, 1969). Urial rams tend to remain parallel or slightly behind ewes, either in a normal though slightly rigid posture or in a head-up. Marco Polo sheep and on rare occasions ibex (Walther, 1961, 1974) may also show themselves off with lateral displays. Bharal often tag along beside or behind females in a head-down posture. *Capra* and bharal sometimes mouth their penises.

Himalayan tahr display the head-up as well as another distinctive posture. A male may face the female with his neck lowered and muzzle slightly raised. Slowly he lifts his nose until it points almost straight up and at the same time he retracts his neck, transforming his shoulders into a hump (fig. 35). Standing rigidly, and perhaps flicking his tongue, he exposes his incisors and gums, whose whitish color contrasts with his black face. Such tooth-baring may be an exaggerated lip-curl. The male may interrupt his display to nod his head up and down, all the while shaking it vigorously from side to side.

TESTING FOR ESTRUS

Because most ewes are rather passive even during estrus, the approach of heat is difficult to detect behaviorally, the main sign of it being that they begin to tolerate the advances of rams (see Geist, 1971a). However, female goats in heat may show such symptoms as bleating and tail-waving (Fraser, 1968). From an evolutionary point of view it is imperative that males in a polygynous society waste as little time as possible on nonestrous females. By stimulating the female to urinate, a male can determine the stage of her heat by the odor of her pheromones. After approaching in a normal posture or a low-stretch, often with his tongue flicking, the male sniffs, nuzzles, or licks her anal area (plate 44). More often than not she responds by moving away. But he may persist in his attentions by sniffing and displaying until she either urinates or he recognizes his attempts as futile. If she squats and urinates, he sniffs the fluid then raises his muzzle high with his upper lip retracted, a gesture termed *Flehmen* in the German literature (plate 35). The raised lip seems to open the palatal passages to the vomeronasal organ where odors are probably analyzed (Estes, 1972). This sequence of events is widespread among bovids in general. If the female is not in heat, the male ignores her and perhaps checks others, as this example of a roving bharal male illustrates.

> A class IV male approaches a female in low-stretch, sniffs, and walks on. He flicks his tongue at a class II male in passing. A second female urinates when she perceives him and he lip-curls after sniffing the ground. Ambling past a third female, he gives her rump a cursory sniff then ignores her. Nearby is a fourth female. He follows her in a low-stretch until she urinates, but after sniffing, he just stands beside her. She squats again. He places his chin on her

rump as if to mount, and she urinates a third time, only a few drops. He lip-curls. A fifth female passes and he follows in low-stretch. She urinates and he lip-curls.

Males of various species test for urine in slightly different ways. Sniffing is usually rather cursory, but bharal males may nuzzle for some 15 seconds, and in one instance the gesture turned into a rub. Possibly rump rubbing in this species was derived from such sniffing and nuzzling. A wild goat male rubbed his forehead around the vulva of a female for 30 seconds, my only observation of this kind in species other than the bharal. Most males wait until the female has urinated before sniffing the ground, though on rare occasions they may place their open mouth into the stream of urine. The domestic goats studied by Shank (1972) usually let the females urinate directly into their mouths. Himalayan tahr sniffed females surprisingly seldom during the pre-rut in New Zealand. Among 37 courtship interactions there was only one instance of sniffing. Himalayan tahr males direct many of their displays toward the heads of females in contrast to males of other caprid species who orient toward the hindquarters of females. Possibly male tahr judge the state of estrus more by behavioral than olfactory means. When lip-curling, wild goat males tend to keep their heads almost motionless, in contrast to mountain sheep (Geist, 1971a), ibex, Punjab urial, bharal, and markhor, which often wave them from side to side or on rare occasions jerk them up and down. Lip-curling muskox also remain still (P. Lent, pers. comm.).

When a male shows interest in a female, others may be attracted and either hover around or join in the sniffing and lip-curling. Even the sight of a squatting female may induce males to hurry over to check the site from as far as 30 m away. Two or three urial, wild goat, or bharal may then amicably lip-curl side by side.

After displaying to and following a female persistently, a yearling bharal finally prevailed upon her to squat. He sniffed but turned his muzzle abruptly aside. She had defecated.

Females ignore many sniffs or respond to the propositioning by merely raising and wiggling their tails. Bharal males sniffed twice as often as they lip-curled and urial half as often again. In wild goat the incidence of urinating in response to a sniff increased with the rut, rising from 18% between September 6 and 19 to 38% between September 20 and October 3. In bharal the percentages dropped from 54% between November 2 and 19 to 47% between

November 20 and December 5, possibly because bharal males sniffed so often that females frequently seemed to have no urine left. Urial seldom sniffed during the prerut—only 6 times between October 6 and 22. But between October 23 and November 6 we counted 131 instances, of which 45% induced ewes to urinate. Geist (1971a) noted that mountain sheep rams seldom lip-curled when with estrous ewes, and I observed the same in bharal and others. At such times males can presumably discern the female's state from her behavior without need to confirm it.

The various age and sex classes sniff and lip-curl at different frequencies (figs. 38 and 39). Among urial and wild goat, all males through class III sniff and lip-curl less often than expected and fully adult males do so more often. Markhor behave similarly, if

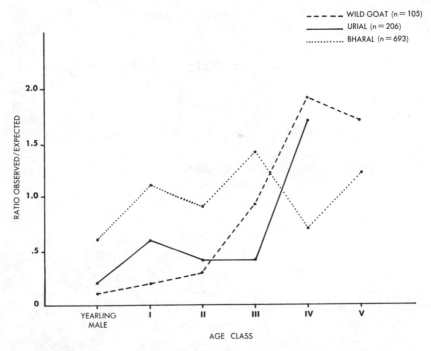

Fig. 38. Frequency of sniffing the anal area of females by males of various age classes during the rut, comparing wild goat, Punjab urial, and bharal. The frequency in figures 38 to 41 is expressed as the ratio of the percent observed/expected. In wild goat and urial the expected number of males is based on the percent in the population and in bharal on male-hours of observation.

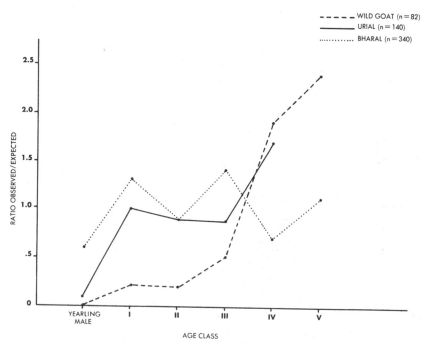

Fig. 39. Frequency of lip-curling by wild goat, Punjab urial, and bharal males of
different age classes after sniffing females during the rut (see fig. 38
caption for explanation of the ratio).

one judges by my small sample of 17 sniffs and 12 lip-curls. Bharal differ from these species in that young males test females as actively as do adults. A reason for this difference became apparent when I plotted the frequency of sniffing in bharal separately for each age class as the rut progressed (fig. 40). Class I males were highly active during the pre-rut, a time when adult males showed little interest in females, but during the main rut the positions were reversed. Thus the rut passed through three main phases during November: (1) mainly young males tested females until November 19, (2) adult males became increasingly interested in females between November 20 and 28, and (3) adult males tested and courted vigorously from November 29 onward. Among bighorns, too, young rams showed the most interest in ewes early in the rut (Shackleton, 1973).

The sniff is only one of several actions that may induce a female to urinate. The mere approach of a male may have this effect, as

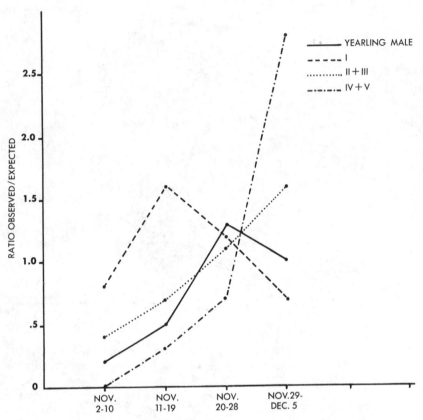

Fig. 40. Frequency of sniffing the anal area of bharal females by males of different age classes during various phases of the rut (see fig. 38 caption for explanation of the ratio).

may a mount and some displays, such as the low-stretch. How effective these methods are varies somewhat between species (table 41). A sniff is the most common gesture stimulating females to squat. Wild goat females frequently urinate at the approach of males, something that urial and bharal females seldom do. Urial rams commonly induce ewes to urinate by displaying, whereas bharal males usually are successful only after presenting a medley of sniffs and displays. A characteristic of all these forms of contact is that they are gentle, delicate. The urination response is generally thought to be a means by which the female can reduce the male's harassment or at least deflect his interest (Walther, 1974). While he lip-curls, she can move off unmolested. Sometimes a male mounts a female without preliminaries. Such unrefined behavior seldom stimulates the female to urinate. Usually females avoid crass males or threaten them. Bharal females reared up threateningly in response to an importunate male on 7 occasions. When a class III markhor male sniffed a female, she whirled around and twice jumped up and crashed her horns against his. On another occasion a class II male chased a female until she turned and jabbed him so forcefully in the lower neck that his forelegs were lifted off the ground. Similar aggression in wild goat was described earlier. A urial ewe only once butted a ram, a yearling.

Occasionally a male sniffs or low-stretches to another male in what seems to be a case of mistaken identity. Large urial rams sniffed yearlings on 4 occasions, and this also happened once in bharal. An urial ram once low-stretched to a yearling, and a bharal male did so too. The scrotum of yearlings is still so small that the animal resembles a female from the rear. Older males have progressively larger scrota, graded rank symbols like horns, which other males are unlikely to overlook. Young are about 6 months old at the time of the rut and they resemble small females. Bharal males sniffed or low-stretched to young on 5 occasions, and markhor and urial did so once each. Domestic sheep and goat young often take part in the rut and become pregnant (Grubb, 1974b; Shank, 1972), something that does not occur in wild species except among the saiga (Bannikov et al., 1967).

Sniffing and other such behavior is seldom seen outside of the rut. A wild goat male once sniffed in March, my only such record in a month of field work at that time. During the second half of January, shortly after the rut, markhor males sniffed only twice. Bharal rarely displayed in March (Schaller, 1973b). Urial remained more active than the other species after the rut. Rams sniffed or displayed to ewes on 13 occasions in late December and on 5 occasions in July. Mountain sheep rams may court ewes at any season (Geist, 1971a).

COURTSHIP

Males display the low-stretch, twist, and kick to females in the same general manner as they do to males (see Chapter 9). By walking with body held low and averting his horns in a twist a male probably reduces the probability of chasing the female away.

A male typically approaches a female in a low-stretch from behind—85% of the bharal and 93% of the urial did so—and when his head is level with her thigh he may give a twist or two and follow this with some kicks (fig. 34). Tongue-flicking and grunting may accompany the displays. There is much individual variation. A low-stretch may consist of a cursory dipping or a tense crouching walk, a twist can be slow or fast, and the kick is either flaccid or high and vigorous. The kick of a mountain sheep may propel the ewe forward (Geist, 1971a), and this sheep, as well as the urial, also sometimes pushes with its chest as it kicks. Species differences exist too. Ibex kick by pawing the air with the carpal hanging limply; markhor, wild goat, and urial usually kick with a straight foreleg; and bharal kick with rather variable leg positions. Himalayan tahr raise their legs limply and bent at the carpal joint, in this resembling the domestic goats studied by Shank (1972). Wild goats execute the low-stretch quickly; markhor, urial, and bharal with moderate haste; and ibex very slowly, the male often just standing by the female in that position. For example, one 3½-year-old Asiatic ibex in Chitral stood behind a female in a low-stretch, almost in a crouch, with his forequarters lowered. He jerked forward while flicking his tongue and he pawed the air with a foreleg. When the female wheeled around and lowered her horns, he froze in the low-stretch with his forequarters even lower than before.

The frequency of displaying varies with the species and time of rut (tables 42 to 44). Taking wild goat first, the low-stretch was the most prominent display early in the rut, but it was rare later on. As courting grew more intense the low-stretch was seemingly discarded in favor of more assertive behavior. Twists and kicks increased significantly during the rut with the former being the most common display. Twists, low-stretches, and to a lesser extent kicks are prominent during the pre-rut and early main rut in Punjab urial. Then the last two are almost discarded in favor of long series of rapid twists as rams chase ewes. Kicks decline in importance during the rut among urial whereas they increase in importance among wild goat. The low-stretch is the most important display of bharal, but neither it nor the twist and kick show marked changes in frequency during the pre-rut and early rut. I observed 17 tahr displays during the pre-rut in New Zealand and all consisted of single low-stretches; several kicks and 2 twists were also seen in Nepal. A total of 7 low-stretches, 38 twists, and 16 kicks were observed in markhor, proportions similar to those of wild goat. Aeschbacher (in press) recorded 3786 displays among captive Alpine ibex, of which 11% were low-stretches, 45% twists, and 44% kicks. Wild goat and markhor twisted more and kicked less than ibex.

A male often presents a sequence of displays, either several of the same ones or a combination of two or three. Animals tend to limit themselves to a single display early in the rut, but later they often give several (tables 42 to 44), behavior also reported for Alpine ibex (Aeschbacher, in press). The averages in tables 42 to 44 represent minima because courting pairs often moved out of sight. One wild goat male displayed rapidly 75 times in succession and one urial ram 58 times. Class II and younger wild goat males always courted in sequences of fewer than 8 displays, and the same applied to yearling and class I urial rams. Small bighorn rams also displayed in shorter sequences than did large ones (Shackleton, 1973). Bharal differed from the other species in that their sequences remained short at all times, usually just a display or two.

Males of all ages display to females, but yearlings and class I animals do so infrequently (fig. 41). Shackleton (1973) showed that yearling bighorn rams participate in the rut according to their nutritional condition: robust yearlings courted more than did those from a population of low quality. Class II wild goat and bharal males also displayed little in contrast to urial rams, which showed a striking peak at that age. This peak can partly be explained by a sampling bias: of 197 displays observed, 102 were given by 3 courting animals. Among wild goats of classes III to V the older the male the more he displayed. Other species did not follow such a tidy pattern. Class V and class III bharal displayed more than did class IV animals, and in urial too the largest and third largest rams courted more than the second largest ones. Class II and class III markhor were equally active (ratios of 1.7 and 1.8) and both were more so than class IV males (ratio of .6). However, most of my observations on this species were made in 1970, when the population contained few large males and this gave the others an opportunity to court.

The low-stretch, twist, and kick are also found in some other Caprinae. Domestic goats (Shank, 1972), Alpine ibex, Marco Polo

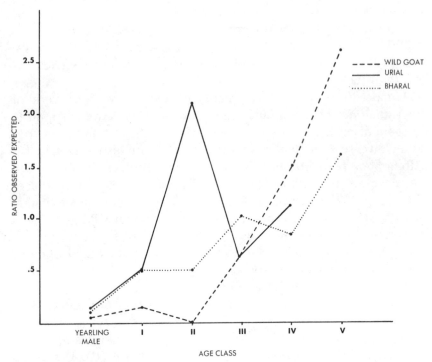

Fig. 41. Frequency of displaying courtship patterns (low-stretch, kick, and twist, combined) by males of various age classes, based on data for wild goat, Punjab urial, and bharal (see fig. 38 caption for explanation of the ratio).

sheep (Walther, 1961), Soay sheep (Grubb, 1974b, d), and mountain sheep (Geist, 1971a) behave much like the species I studied. Aoudad low-stretch and twist but do not kick. Mountain goat low-stretch and kick (Geist, 1965). A chamois male approaches the female in a low-stretch and then "the courting billy stands behind the nanny with his head erected, thereby displaying his white throat" (Krämer, 1969). This species does not kick, twist, or grunt, and its tongue-flicking, if such it is, consists only of extending the tongue. A male may stamp his foreleg which possibly is this species' equivalent of a kick. Japanese serow males may kick females during courtship (Akasaka and Maruyama, in press). I observed courting muskox in the Calgary zoo. A bull approached a cow and stood almost parallel to her with his head facing her thigh. He kicked her repeatedly but displayed neither low-stretch nor twist. He then stood by her side, their bodies almost touching.

Once he stepped behind the cow while audibly blowing air through his nostrils. Raising his leg as if to kick, he suddenly mounted. Muskox also twist (Gray, 1973), emitting low, husky sounds while doing so (P. Lent, pers. comm.). Rutting bulls may also bellow (Gray, 1973). Courting bulls occasionally circle cows slowly, according to P. Lent (pers. comm.). A captive saiga male sniffed females and followed them in a slight crouch with neck held horizontally, tail raised almost vertically, and nose inflated (Pohle, 1974).

Courting males sometimes jerk or wave their heads at females. A chamois may bob his head up and down (Krämer, 1969) and a mountain goat may jerk "to and fro" (Geist, 1965). A markhor, wild goat, and bharal male occasionally jerks his head at a female, causing her to walk or run, and he then follows. The distinctive nodding and shaking of Himalayan tahr have already been described. It is unlikely that these gestures all have a common origin and function. Among bharal, markhor, and wild goat the gestures seem aggressive.

Several species conduct at least some of their courtship at night (see Sambraus, 1973; Pohle, 1974), a time when displays would seem to lose much of their importance.

Courtship does not just consist of energetic displaying. As the female approaches estrus, a male often tends her gently, remaining close but not too close, as if being careful not to frighten her away. A wild goat, markhor, or urial male may remain up to 20 m from a female, moving and halting whenever she does. But usually he stays nearby, standing beside her or behind her with his head almost touching her rump. Himalayan tahr seem particularly discreet in pressing their attentions. One male remained motionless as he faced a reclining female for 12 minutes. Occasionally a male nuzzles a female discreetly. Bharal, too, tended females but they seldom did so for long, their displaying seemed to lack finesse, and their courtship appeared to be conducted with greater impatience than that of the other species. Tending bonds have also been reported for most other Caprinae. The mountain goat is the only species so far reported in which adult males may tend females before they are in heat (Geist, 1965).

Although there would seem to be selective advantage in courting with restraint on cliffs, a female sometimes rushes away from the male, either on her own accord or stimulated by his actions.

He then follows and both race over slopes, in and out of ravines, along ledges, around trees. Some chases are dangerous. For example, one Himalayan tahr couple descended wildly over precipices and rock falls for over 300 m. After a while the female halts and the male may display again, not in the gentle fashion of the early rut but with intense, jerky movements to which she may respond with aggression. Bighorn sheep (Geist, 1971a) and *Capra* females may be quite aggressive in such situations, whereas Soay sheep (Grubb, 1974b), urial, and bharal are not easily provoked. Wild goat and bharal females sometimes deterred insistent males by retreating to cliffs. Chases occurred in all the species I studied. Here is a typical example from urial.

> At 0905 a class II ram chases an ewe for 5 minutes around a peacefully grazing herd. He twists 36 times as they race along and finally dip out of sight into a ravine. They soon reappear, he still twisting. She halts. He places his chin on her rump and mounts but she steps ahead, causing him to fall off. Another twist and mount. She grazes while he stands behind her with his neck diagonally downward in a seemingly submissive posture. As they move slowly he mounts her 3 more times. Both drift into the shade of an acacia and stand awhile. At 0945 he twists 3 times, mounts briefly, twists some more, kicks, and mounts again. She walks, wiggling her tail rapidly. After twisting twice, he mounts her but as usual steps down after a second or two. Another kick follows and another mount, this one apparently successful. They then stand one meter apart.

Chasing also was a prelude to copulation in bharal, but, in contrast to the other species, males seldom bothered to display. In 27 chases only one twist and 4 kicks were observed. Two examples illustrate the behavior.

> A class V male and a female circle back and forth through a herd, running, trotting and walking. Both are tired: they pant with tongues hanging out. Twice he mounts during his pursuit but each time falls off as she continues to move. She leads him to a cliff, along a ledge, and into a rocky niche where he mounts her again. Restlessly she rushes off, cutting through and around the herd, the male still behind her, and then she returns to the same ledge. The male mounts. A class II male ventures near. The female bounds off pursued by the young male—but not the adult one. He tries to mount while running at high speed, but then further attempts are ruled out by a class IV male who hurries over to block him yet ignores the female as she drifts away.

A class V male chases a female, mounts her while going full tilt and then follows her to a ledge where she halts. Mounting again, he thrusts rapidly but departs hurriedly when a larger class V male trots up. The female races off, now followed by the new male who tries to mount twice. Circling uphill, the female clambers onto a ledge and stands there with her chest pressed against the cliff while the male mounts. He leaves her abruptly, and a class IV male who has been tagging along mounts her. Then he too leaves.

These examples illustrate several points in bharal courtship. Although a male may induce a female to run, she usually flees on her own volition but makes no concerted effort to escape. Bharal retreat to cliffs mainly in times of danger. But females also seek lofty ledges and platforms during the rut. By choosing such precarious retreats, females reduce competition among males for there is seldom room for more than one to maneuver. Several males often hover around an estrous female. Should the largest one lose interest the one next in rank is ready to take over. In the two examples above the lack of continuing interest by the males suggests that the females were not in full heat. Mountain sheep ewes also retreat into difficult terrain when pursued by rams (Geist, 1971a).

Several domestic goat males may chase and mount a female indiscriminately (Shank, 1972) and several Alpine ibex males may display communally to a female (Aeschbacher, in press). Similar behavior has been reported in mountain sheep at high densities (Geist, 1971a). At such times there are obvious breakdowns in rank order. I observed one instance in bharal where a dominant individual lost control over subordinates but then regained it.

Ten males are together when a female suddenly appears among them. As all converge on her she flees to a nearby cliff. Running along a ledge she reaches a platform. She stops there facing the wall while 6 males pile on her and on each other until she disappears beneath them. After much hooking and lunging only 3 males remain—two class V and one class IV. While the largest attacks a competitor, the female slips past him and escapes. She is pursued immediately by all males. A fleet class III mounts her on the run but a class IV butts him in the rump. Having lagged behind, the largest class V now arrives and displaces the rest as he follows the female back to the same rocky platform. He first mounts her, then lunges at any suitor that ventures along the ledge. Suddenly the female departs, closely trailed by the males. The largest is right behind the female. He rears and clashes on his closest pursuer, the second largest male in the herd. The latter catches the blow and before the other can recover he dashes to the female. But the dominant male races after him and blocks. A class IV male trots carelessly close and gets rammed in the side. The female climbs up a cliff face and into a cave still followed by the largest male. Standing nearby on a ledge are two males, a class IV and a class II. The former lunges, and to escape the blow the latter leaps down. After barely touching a thin ledge 3 m below, he plunges over 6 m through the air to land with a thud at the base of the cliff. The courting pair stands side by side for 25 minutes. A class IV approaches them along the ledge. The class V lunges, clashes, and backs up to charge again, thereby giving the other male an opportunity to slip into the cave. While the large one tries to evict the intruder, two others crowd in—and the female runs out. Another chase ensues....

While a dominant male may fail to impose social restraints at all times, he can and does claim his prerogatives if the reward—a female fully in heat—justifies the effort. Courtship in urial also included much chasing, as did that of Soay sheep (Grubb, 1974b, d) and mountain sheep. However, the amount of chasing is dependent on several factors. Geist (1971a) noted that the bighorns he studied chased mainly during the first week of the rut and that later courting couples were dispersed over the terrain. Shackleton (1973) studied two bighorn populations in which chasing was common in one but not the other. He attributed the lack of chasing to the presence of only a few rams with stable dominance relationships.

Mating often proceeds peacefully. A receptive wild goat female may walk with her back slightly humped, giving occasional wiggles to her tail. Schaller and Laurie (1974) describe the further behavior:

> The class V male twists and kicks and twists twice more as he stands beside the female. Gently he licks her neck. She takes a step forward. Twisting and rearing up with the same movement, he mounts her briefly. He dismounts and twists twice. The female rubs her face on his. He mounts again, gets off, twists vigorously, and mounts once more, just for a second or two. Then, standing parallel to the female, he twists and kicks. She nuzzles his face and neck and he reciprocates. Stepping back, he twists and kicks behind her, moves up beside her again, and they mutually nuzzle each other's face. He mounts her and she rubs her face against his. Twice he twists beside

her. After another cursory mounting, he stands by her, his penis unsheathed, and twists three times. A class III male walks up, but, when the adult turns toward him, he scampers away. Slowly, walking side by side, the courting pair disappears from view into a ravine after having been in sight 15 minutes.

Frequent licking and nuzzling were typical of markhor and wild goat, and they have been reported in some other species too. In fact, females may actively solicit a disinterested male by nudging and licking him and bounding around him, as has been noted in domestic goats and sheep (Shank, 1972; Grubb, 1974b) and mountain sheep (Shackleton, 1973).

Himalayan tahr may court gently too as one observation illustrates:

At 0815, a female is attended by a class III male and two class II males, all somewhat separated from the main group. When a class II male approaches the female, the class III male by her side advances toward him in a hunch display [humped approach]. The smaller male turns aside, joins the other class II male, and both rest at least 10 m from the courting pair. The class III male assumes the low-stretch, but the female ignores it and he reclines. At 0910 hours the approach of a class II male brings him to his feet, and a hunch display causes the interloper to veer off. Once again he faces the female in a low-stretch. For 15 minutes they stand, he with muzzle raised, she with head averted. After that both feed and rest. At 1105 hours, the female approaches the male who lifts his muzzle so high that the underside of his jaw faces her. She licks herself, advances, licks again. Whenver she moves, he adjusts his position so that his muzzle points at her. Suddenly he steps behind her, his shoulder by her rump. He gives a low-stretch coupled with a twist, then shakes his head and kicks. Twice more he shakes and kicks before moving around to face the motionless female. There he alternately low-stretches with teeth bared and shakes a total of 9 times. Occasionally he nudges the female with his nose as if to get her attention, for when she looks at him he intensifies his low-stretch. The female begins to feed at 1200 hours. Slowly the male steps behind her and rears on his hindlegs, mounting her. He thrusts 10 times, barely leaning against her, without eliciting a response. The two then feed and rest near each other without further courting for several hours (Schaller, 1973a).

Anderson and Henderson (1961) wrote that in New Zealand tahr "the typical family group consists of a bull, a nanny, its kid, and either the offspring of the previous year or a 2-year-old, and together they move to a well-chosen piece of territory to remain for some 6 to 8 weeks." I saw no evidence in Nepal and New Zealand for such "monogamous grouping" as these authors call it. Rather, the largest male in the group claimed any female in heat, a pattern similar to that of the other Caprini.

Shackleton (1973) observed harem-like herding of ewes by rams in one population of bighorn sheep, behavior which had not been previously reported in caprids. A ram, not necessarily the largest one, followed any female that moved away from the group and cut off her retreat, until in most instances she returned. His movements were slow and his posture similar to the low-stretch except that his shoulders were slightly hunched and his muzzle was pointed at the ground. Territorial male chamois herd females by holding their bodies low, muzzle thrust forward, and by hitting the ground occasionally with one or both forelegs (Krämer, 1969). Saiga males herd females while trotting and grunting (Pohle, 1974).

Once Grubb (1974b) observed a Soay ewe "trotting in tight circles pursued by a ram." This behavior may have been fortuitous, but it does resemble the courtship circling characteristic of many antelopes (Walther, 1974). In this context it is interesting that saiga, which probably belong to the Antilopinae rather than the Caprinae, may court while standing head to tail and circling rapidly (Pohle, 1974).

Prior to mounting, a male usually places his chin on the female's rump, and if she does not avoid him he stands upright with raised head, resting his chest on her as he clasps her lightly with his forelegs, a posture typical of all Caprini. A male chamois leans forward over the female (Krämer, 1969) and a male mountain goat buries his nose in the long hair on the female's back (Geist, 1965). Males of all ages may mount. Class I urial rams were seen to mount once, class IV rams twice, class II rams 12 times, and the others not at all. Wild goat males were observed to mount 24 times, there being 3 yearlings and 16 class V males among the participants, to mention just the smallest and largest classes. Bharal mounted 87 times, and the ratio of observed to expected, based on the number of animal-hours of observation, was 1.0 in yearlings and class I males, 2.0 in class II males, 1.0 in class III males, 0.4 in class IV males, and 1.25 in class V males, indicating that class II and class V males mounted most actively. Bharal seemed to mount more readily than did urial and wild goat. In this respect it may be significant that bharal mount often to express dominance. Mountings, especially during the pre-rut, may have been means of asserting rank rather than just expressions of sexual interest. Geist (1971a) concluded with respect to mountain sheep that young rams are often inept swains, crude and abrupt in their approach. The same is superficially true among bharal. For example, mountings during the pre-rut involved no males older than class II and none of the attempts were preceded by displays. However, dominance may have been a factor in their behavior. Sambraus (1973) noted that domestic rams seldom mounted unless the ewe was ready to tolerate them, whereas domestic goats averaged 6.5 mounts before a copulation was successful. This distinction may not be applicable to free-living sheep and goats. Copulation lasts only a few seconds, and indeed one would expect selection to favor such cursory contacts in societies with excess males whose presence is likely to disrupt the courtship.

Subadult males mount as readily, and in some instances more readily, than adults, but it should not be concluded that they necessarily impregnate many females. Sexual overtures per se are of little consequence. Most yearlings probably lack viable sperm. Yearling Himalayan tahr fail to produce sperm during the rut (Caughley, 1971), but a captive male impregnated several females at the age of 2½ years (K. Tustin, pers. comm.). A free-ranging but tame Kashmir markhor bred 3 domestic goats at the age of 2½ years. To be successful, a copulation must be precisely timed with respect to the female's estrous cycle. Mountings during the pre-rut do little more than perhaps help synchronize the main rut. Estrus in domestic sheep and goats lasts only some one to four days with a peak of about 20 to 36 hours. During this time domestic sheep permit copulation an average of 4 to 7 times, but the optimum time is during the first 12 hours of heat (Jewell and Grubb, 1974; Sambraus, 1973). I have no precise information on the duration of estrus in wild species. Krämer (1969) noted that male chamois show interest in a female for only a few days, and Geist (1971a) reported an estrous period of from 2 to 3 days in mountain sheep. One Kashmir markhor male followed a female for 3 days. In any event, a male must be present during a brief critical period. Normally, an adult male detects the imminence of estrus and then guards the female from young males by forming a tending bond as described for most Caprini. After mating he seeks other females. If more than one female is in estrus, a male may react in one of two ways, both illustrated by wild goat. On one occasion a class V male dashed around frantically trying to guard 3 females from 6 inquisitive competitors, a futile endeavor, whereas on another occasion the three largest males in a herd each courted a female without strife. The latter solution seems to be typical of Caprini. Domestic rams copulate on an average of at least 20 times per day, a rate they can maintain for two weeks or more. (Sambraus, 1973). A ram might inseminate 30 ewes in 4 days (Jewell and Grubb, 1974).

All this does not imply that young males are wholly unsuccessful during the rut. Grubb (1974b) showed that in a herd with 38 Soay rams the 3 highest ranking individuals tended only 29% of the ewes on the important first day of estrus. In Alpine ibex and markhor herds, which often contain just one large male, subordinate individuals probably have an occasional opportunity to mate if more than one female is in heat. In urial and other species with fluid herds the young males may at times find an unattended ewe. If man has altered the normal age and sex ratios of a population, the natural mating order may be seriously disrupted. To note an extreme case: so many adult markhor were shot at Tushi that only class I and II males were with the females during the 1970 rut. The fittest males are most likely to produce the fittest offspring, and one would consequently expect females to be selective in accepting males. A fit male would be one who has shown by his size, vigor, and age that he has the ability to survive well. Geist (1971a) wrote that "the female [mountain sheep] appears to prefer large-horned to small-horned rams and is quite skilled at thwarting the mounting attempts of small rams." Except for yearling and some class I males who were treated as subordinates, the females in the species I studied showed no consistent preference for large males.

DISCUSSION

My broad comparisons and a few additional notes in the literature on Kuban ibex (Steinhauf, 1958), Dagestan tur (Petzsch and Witstruk, 1958), mouflon (Pfeffer, 1967), Marco Polo sheep and Russian markhor (Walther, 1961) indicate that courtship behavior in the various Caprini is remarkably similar, that it has been conservative in the evolution of this tribe. Lacking the constraints of a territory and forming no long-lasting bonds, males have no need to court elaborately: they can simply pursue a female with a few token gestures until she accepts them. There are of course differences in the complexity and frequency of displaying, some of them generic. For instance, goats use odor more than do sheep, the former spraying themselves with urine and wafting scents from their anal glands. The Himalayan tahr deviates most from the others in that males tend to orient displays to the front of the female rather than to the rear and they have incorporated several new patterns into their repertoire. Their exaggerated head-nodding and shaking has its only faint counterpart in the head-bobbing of chamois. The low-stretch, twist, and kick are widespread too, only the aoudad seeming to lack the kick. There are, however, differences in the frequency with which males use these displays. In urial and wild goat the twist is most prominent during the rut, whereas in bharal and mountain sheep it is the low-stretch. The former two may display in long sequences in contrast to bharal, which tend to give an abrupt gesture or two. Geist (1971a) found that bighorn sheep display less than thinhorn ones. "Bighorn rams were more rough and unpolished when courting estrous ewes than Stone's rams. The latter preceded each copulation with a series of front kicks, while bighorn rams tend to mount more spontaneously; they twisted and kicked, but not frequently." The low-stretch seems to be most prevalent in those species which court rather crudely, those whose gestures lack finesse. Such species probably need a gentle display during their initial approach more than do the others. But whatever the differences, they are rather small, probably too small to serve as reproductive isolating mechanisms between species. In this context it is significant that contiguous and closely related species differ greatly in appearance. Wild goat and markhor, ibex and markhor, and Kuban ibex and Dagestan tur are examples emphasizing this point. In areas where two species overlap the appearance of a male might be important in determining whether a female will accept his overtures, and her selection would in such situations affect the incidence of hybridization.

Courtship in Rupicaprini has been little studied except in mountain goat and chamois. The former lacks the twist and the latter both the twist and kick. Reading the accounts of Geist (1965) and Krämer (1969), one receives the impression that these animals display less and behave with greater restraint than do sheep and goats, something not surprising in view of their dangerous horns. In fact mountain goat males are subordinate to females during courtship and they vocalize like young; both practices are apparently means of reducing the chances of being attacked. Courting goral also seem to forgo elaborate displays. According to Dang (1968b) "there was little preamble to the mounting of the doe by a magnificent buck."

11 MOTHER-YOUNG RELATIONS

The social bond between mother and young is close and closed, the only exclusive relationship between any two individuals in caprid herds. Although young are precocial, they are wholly dependent on their mothers until they can subsist on forage. At birth the female licks her offspring, cleaning and drying it, and later she feeds it, assists it in traversing difficult terrain (Couturier, 1962), and defends it against certain predators, either directly or, by such means as luring them away, indirectly. "What the ungulate mother appears to provide is the optimum environment for rapid development and learning" (Lent, 1974). While contact between mother and offspring is close for at least a year, the two seldom interact except when the young tries to suckle. And after weaning the association becomes mostly passive, with the young merely following and resting by its mother. Previous chapters have summarized some of the aggressive and other interactions which involved young. The whole subject of mother-young relations has been well treated by Harper (1970), Autenrieth and Fichter (1975), and Lent (1974), and I shall here add only a few observations, mainly on Punjab urial and wild goat.

A female tends to separate from the herd and isolate herself one or more days before giving birth. Such behavior has been described for mouflon (Pfeffer, 1967), bighorn sheep (Welles and Welles, 1961), Alpine ibex (Couturier, 1962), and feral domestic goats (Rudge, 1970), among others. Urial typically withdrew into the upper reaches of ravines and eroded gulleys and in these shady retreats gave birth. Wild goat sought canyons and deeply slashed slopes. Heavily gravid urial and wild goat tended to concentrate in favored locations but each gave birth in isolation. Mountain sheep ewes dispersed into rough terrain two to three weeks before lambing (Geist, 1971a), but captive Marco Polo ewes avoided rocks and gave birth in the open after scratching shallow beds (Walther, 1961).

Births have been described in various Caprinae. In general, the female lies on her side but occasionally also stands; after the young has appeared she licks it and may prod it as if inducing it to move, a muskox using her head and a Dall sheep her foreleg (see Lent, 1974). Young struggle to their feet soon after birth. Altmann (1970) noted a mouflon first on its feet at the age of 7 minutes, Walther (1961) an Afghan urial at 28 minutes, Grubb (1974a) a Soay lamb at 18 minutes, and Pohle (1974) a saiga at 17 minutes. The social attachment of a mother to her young is rapid: a domestic ewe seems to be able to recognize her lamb within half an hour (Smith et al., 1966) and a domestic goat may need no more than 5 minutes (Klopfer and Klopfer, 1968). Olfaction is the most important sense in forming this social bond, but visual and auditory recognition also plays a role (Haas, 1958; Tschanz, 1962). In contrast to the rapid attachment of a mother to her young is the slow bonding of a young to its mother. Small urial and wild goat young may follow females other than their mothers, or even males if they are near, but horn-threats deter them from associating for long. Such behavior has also been observed in feral goats (Rudge, 1970) and mountain sheep (Geist, 1971a). However, on one occasion I observed a small wild goat young spend over an hour with another female before returning to its own mother. Female urial exuded much material from their preorbital glands during the birth season, the only time of the year they did so. The odor possibly assists in individual recognition.

Young can walk well within a few hours after birth, but, as Walther (1968) has discussed in detail, in some species the newborn then follows its mother whereas in others it remains hidden. At one extreme is the wildebeest infant which follows its mother within minutes after birth (Estes, 1966) and at the other is the kob infant which hides for two to four months (Leuthold, 1967). Sheep have long been considered the quintessential followers. "New-born lambs do not leave their mothers by more than a few feet and follow them constantly" (Hafez and Scott, 1962). However Pfeffer (1967) noted that mouflon lambs remain hidden for about 3 days before they accompany their mothers, and I observed similar behavior in urial. For the first few days the newborn lies quietly in some retreat, its mother rarely more than 100 m distant. When disturbed, the lamb does not freeze rigidly like a newborn gazelle, but it does remain prone with its legs tucked beneath its body and neck extended along the ground (plate 36). If picked up during the first day or two of life, it seldom struggles, lying relaxed in one's arms except perhaps to bleat. But older lambs attempt to escape, and after release may hide themselves again. The lamb decides when and where to conceal itself. On one occasion a ewe and lamb fled. After 100 m the youngster suddenly veered alone into a ravine; another time a lamb crouched beneath a bush while its mother continued in her escape; and on a third occasion the lamb simply cowered in the open. Ewes ignore their young at such times and presumably remember where their offspring stopped to conceal themselves. Once a lagging lamb bleated. Both its mother and a yearling ewe ran back 10 m and waited for it. Urial ewes seemed to thump their hooves when fleeing much more during the birth season than at other times. This behavior may warn the hidden young of danger. Pfeffer (1967) observed that the loud noise of guns caused mouflon lambs to crouch up to the age of 3 or 4 days but not later. Bighorn ewes remain alone with their lamb for 5 to 7 days, but Geist (1971a) does not mention an actual hiding phase.

Goats are also considered to be followers, although Hafez and Scott (1962) noted that mothers may leave their young and graze some distance away. Feral domestic goats often leave their young hidden for 8 hours or more during the first 4 days of life (McDougall, 1975). On returning to the hiding place the mother gives a muted bleat to which the young responds by rushing to her. It then nurses while the female sniffs it (Rudge, 1970). Alpine ibex in captivity do not hide (Walther, 1961), but free-living Asiatic ibex have been reported to cache young for 2 to 3 days (Savinov, 1962, quoted in Lent, 1974) and wild goat seem to do so for a similar period. Variations in the duration of the hiding phase may depend on the terrain. It may not be physically possible for a weak and unsteady newborn to follow its mother up or down a cliff. By remaining hidden until it is well coordinated a wild goat young, for example, would seem to increase its chance of surviving in the maze of canyons in which it is born. At the height of the birth season solitary wild goat females often confined themselves to pieces of rugged terrain where obviously their young were hidden. Once a herd with 3 small young fled up a slope and as it passed through a small thicket one young cached itself. Goats may not be as strong followers as sheep (Walther, 1961) but many do have a short hiding phase, a phase which places them together with sheep on a continuum between true followers and true hiders. Chamois, mountain goat, and muskox are followers (Krämer, 1969; Lent, 1974), but goral and serow may well have a short hiding phase, judging by the fact that the newborn of solitary species living in dense habitats tend to conceal themselves. In at least some species a period of seclusion may be necessary so that mother and young can learn to recognize each other. The only member of the subfamily for which a prolonged hiding phase has been reported is the saiga. According to Heptner et al. (1966), a newborn lies quietly for the first 4 to 5 days. If disturbed after that age, it runs a short distance and hides again. Not until the age of 2 weeks do saiga young begin to follow their mothers.

By the end of the first week all Caprini young have become followers, closely attached to their mothers. When a female walks, her young tends to tag behind, and when she reclines it rests beside or in front of her, often in body contact. A Punjab urial young sometimes walks beside its mother, and, when she halts, it may stop just a little ahead, behavior which according to Walther (1968) is typical of hiders. Chamois young also tend to stand ahead of females (Krämer, 1969). Soay ewes may walk slowly with head lowered and extended, a pose which apparently attracts the lamb (Grubb, 1974a). The following response in urial lambs is strong, and lost ones try to attach themselves to other ewes and even in one instance to a chinkara. One out of twelve newborns persisted in following me after I handled it. Lost lambs also bleat loudly, a sound which other sheep ignore except in one instance when a ewe responded similarly. The two joined and left together. Mothers also search for their lost lambs as this excerpt from my field notes illustrates.

> A bleating ewe trots alone. Seeing a nearby lamb, she sniffs its nose then walks on. Another lamb stands by its mother yet the searching ewe sniffs its rump in passing. As she enters a patch of tall grass a hidden lamb stands up. After stopping a moment, she trots on, calling. About 60 m away is a ewe with young. She walks over to them and forages beside them. Spotting another ewe with a lamb nearby, she hurries to them and sniffs the rump of the youngster. It

urinates. A herd approaches. She feeds with them, then trots on, searching, bleating. A lamb responds by running to her. They halt abruptly one meter apart. And the ewe continues her quest, leaving the lamb behind.

Afghan urial and Marco Polo sheep ewes may stand protectively over their lambs (Walther, 1961), and domestic goat (Collias, 1956), wild goat, and markhor females may jab herd members who crowd their newborns. But on the whole, sheep and goat mothers are not very solicitous. The mountain goat female is an exception in that she may rush 100 m to protect her young from others (Geist, 1965).

Nursing is one of the main forms of contact between female and offspring. The frequency and duration of suckling bouts are useful indicators of population quality. Females from well-nourished populations produce more milk and consequently permit young to suckle longer and terminate fewer bouts than do females from poor habitats (see Shackleton, 1973). Urial lambs attempted to nurse 31 times in April. Of these 24% were unsuccessful in that the ewe trotted ahead, stepped aside, or kicked backwards. Four of the attempts were from between the hindlegs and the rest from the side in a reverse parallel position. The average suckling duration was 25 seconds (S. D. 38.37). Only two bouts lasted longer than a minute—one of 64 and another of 184 seconds. Lambs terminated 3 bouts and ewes the rest. Shackleton (1973) found that lambs in one bighorn sheep population suckled for an average of 14 seconds per bout and in a second population for an average of 28 seconds, the latter animals receiving presumably twice as much milk. Soay lambs nursed on the average of 26–28 seconds during the first two weeks of life and 13 seconds during the next two weeks (Grubb, 1974a). Wild goat young attempted to nurse 26 times during March, always from the side. On 6 of these occasions the mother stepped aside before her young could drink. Average suckling duration was 14 seconds (S.D. 9.88). Only one young ceased to nurse of its own accord.

Other behavior, too, is associated with nursing. In markhor, Alpine ibex (Walther, 1961), mountain sheep (Geist, 1971a), and Dagestan tur (Petzsch and Witstruk, 1958) among others, the female may initiate nursing by calling, and in some species, such as aoudad (Haas, 1958) and muskox (Lent, 1974), both mother and offspring may vocalize. Chamois females may induce young to suckle by presenting the udder, and the reverse—with the young prodding its mother—also occurs (Krämer, 1969). Young sometimes stop in front of their mothers before nursing. Geist (1971a), for example, found that 42% of all bighorn suckling bouts were preceded by this gesture. Krämer (1969) observed the behavior in chamois too and noted that it caused moving mothers to halt. I saw young detouring in front of their mothers once in wild goat, the young brushing slowly past her forelegs, and twice in urial with the lamb circling rapidly past the ewe's head. The females were not moving at the time, and it seemed to me as if the young were identifying themselves before nursing. Later, when suckling, the rapidly beating tail of the young may help to dissipate odors, again promoting individual recognition. Strange young are rebuffed. For instance, when one urial lamb attempted to nurse, the ewe sniffed its rump and then jerked her horns at it vigorously. A second lamb ran up, detoured past her chest and suckled. Grubb (1974a) observed that "the ewe turned round to sniff the rump of her lamb each time it sucked." Walther (1961) suggested that female goats are more prone than ewes to nuzzle or lick the anal area of a suckling young. I observed the behavior 4 times in urial and twice in wild goat. Chamois seldom lick their young (Krämer, 1969).

Young nurse often during the first few days of life. Pitzman (1970, quoted in Lent, 1974) noted that Dall sheep lambs suckle every 90 minutes on the average, and one lamb observed by Grubb (1974a) nursed 45 times in 6 hours on the day of birth. But as they grow older the young nurse less often and for shorter periods. For example, muskox young have 6 to 8 bouts per day each averaging 180 seconds when under 6 weeks old and 2 to 4 bouts per day averaging 30 seconds when older (Lent, 1974). Grass supplements milk at an early age. Mouflon lambs may nibble at grass when 5 days old (Altmann, 1970), and I have seen urial and wild goat young do so within the first week of life. Soay lambs graze at 3 weeks of age (Grubb, 1974a), and bighorns at between 2 and 5 weeks depending on the milk supply of their mothers (Shackleton, 1973). Mountain sheep are essentially weaned by the age of 4 to 5 months (Geist, 1971a), and the same was true of the species I observed. Urial lambs were seen to make only two nursing attempts in September and October, one unsuccessful and the other lasting 5 seconds. At the same time of year, wild goat young made 4 attempts, of which 2 were successful. Markhor young attempted 3 nursing bouts in November and December. In one of these, twins nursed together for 8 seconds, one from between the hindlegs and the other from the side. I witnessed 10 nursing attempts by bharal at Shey, of which 6 were from between the hindlegs. The mothers permitted suckling on three of these instances for periods lasting 3, 9, and 10 seconds. Two Himalayan tahr attempted to nurse at the age of 9 months. Although their mothers did not permit the nursing, they licked their offspring after rebuffing them. Large young cannot reach the udder without lowering their forequarters, and one urial and two wild goat young kneeled on their carpals to suckle.

The social bond between mother and offspring continues long after the latter has been weaned, although contact in most species is no longer as persistent. Young mountain sheep (Geist, 1971a), Alpine ibex (Nievergelt, 1967), bharal, and Nilgiri tahr, among others, may form temporary peer groups. Indeed bighorn sheep lambs often act independently long before weaning. Lambs prefer to stay in familiar terrain, and ewes may leave them in nursery groups while trekking to a salt lick as far as 3 km away (Geist, 1971a). Saiga young are weaned by the age of two months and many separate from their mothers at that time (Heptner et al., 1966). The bond becomes seriously disrupted shortly before a caprid female is due to give birth again. Stone sheep lambs are on their own 2 to 5 weeks before their first birthday (Geist, 1971a). Among urial it was for the first time common to see lambs away from their mothers just at the start of the birth season, and the same applied to wild goat. However, some yearlings attached themselves to their mothers again soon after the birth, in spite of occasional aggression on the part of the adult, and a trio consisting of a female in the lead, followed by a yearling, and then by a newborn, was a typical sight. Mouflon (Pfeffer, 1967) and feral domestic goats (Rudge, 1970) behave similarly. It was my impression that yearling males rejoined their mothers much less often than did yearling females. Some yearling females were still with their presumed mothers at the age of nearly 2 years, especially if the new young had been lost.

Play as an activity of young has not been specifically treated so far. Most authors (Haas, 1958; Grubb, 1974a; Geist, 1971a) include all aggressive and seemingly sexual contacts between young in the category of play. Although young sometimes fight or mount each other with the exaggerated gestures typical of play, the actions may well serve to establish rank. For this reason I have included such behavior in the chapter on aggression. Unambiguous forms of play consisted of gamboling—of running with huge bounds sometimes on a zigzag course, and of bucking with hindlegs thrown up in the air and head waving from side to side. Such exuberance usually stimulated others, even adults, to behave similarly. On one occasion five bharal young, one female, and several class III and IV males dashed around and bucked. Four urial lambs and a yearling ewe behaved like that too, as did four markhor young and two class I males. Four wild goat young once chased each other, and another time several females and young did so. One wild goat young leaped up on the rump of a walking female and balanced there for several seconds, and a yearling male bucked by himself. Twice a young jumped and turned 180° in midair, behavior also seen once each in urial and bharal young. These examples include all my observations of playful running and chasing: young obviously played little. The enervating heat at Karchat and Kalabagh may be partly responsible for this. But nutrition may be a factor too. Shackleton (1973) has shown that bighorn lambs on good range play twice as much as those on poor range. Whether in the heat of the lowlands or the cold of high altitudes, young may not have had much excess energy to dissipate in play.

12 THE RELATION OF SOCIAL BEHAVIOR TO ECOLOGY IN THE CAPRINAE

Body size, sexual dimorphism, habitat choice, group structure, population dynamics, and other social and ecological factors are intricately related; they represent a functional whole. Previous chapters have presented many facts concerning the biology of various Caprinae, based both on my own research and on the literature, and this chapter tries to organize some of the material into a coherent evolutionary framework, drawing particularly on the stimulating work of Geist (1971a, 1974a), Jarman (1974), and Estes (1974) for ideas. Since no Asian or African members of the subfamily have been thoroughly studied, and detailed behavioral research on the European and American species has only begun, any attempt to generalize becomes a speculative venture. But as Charles Darwin noted in his *On the Origin of Species,* when we view creatures as evolutionary products "how far more interesting . . . the study of natural history becomes."

ECOLOGY AND SOCIAL EVOLUTION

Although the subfamily Caprinae contains only about 26 species, its members are physically diverse; among them are the small, plain goral, the markhor with its flowing ruff and elegantly spiralling horns, and the shaggy and bulky muskox. In their choice of habitat the species range from the humid security of the rain forest to the harshness of the Arctic tundra. Some species are essentially solitary whereas others may aggregate by the hundreds and even thousands. The search for an understanding of such diversity has long occupied biologists, but only in recent years has knowledge of the ecology and behavior of ungulates, particularly of the African antelopes, reached a level where the many isolated facts can be combined into a unified theme (see particularly Jarman, 1974).

Small-bodied ruminants have a higher metabolic rate and require a greater daily caloric intake per unit of body weight than large-bodied ones. The great energy requirements of small-bodied species can be satisfied only by a constant supply of highly digestible and nutritious forage such as is found in sprouting leaves and tender green grass. As plant growth slows and stops, the food value falls, the fiber content increases, and the accumulation of aromatic substances makes some species unpalatable. The growing time in most habitats is seasonal, there being an abundant supply of high-energy food available for only a short period of the year. Forage with a high nutritive value may be scarce and scattered even at the best of times and thus difficult for an animal to find. This is especially true for those species that favor browse. Leaves from shrubs and trees fail to grow after being nipped off, whereas grass blades continue to sprout. Although grasslands produce much forage, the nutrient levels of the plants fluctuate seasonally. In contrast rain forests provide little forage near ground level but the supply is relatively constant and of high energy. Given these nutritional parameters, Bell (1971) and Jarman (1974) noted that only a small-bodied species can maintain itself on a constant supply of high-energy forage—a large-bodied one could not find enough to eat. With its high metabolic rate a small animal would lose weight rapidly on a poor diet. Conversely, the energy requirements of a large-bodied species are lower and it loses weight slowly, both adaptations for feeding on forage of low nutrient content.

A species is quite selective in what it eats. Aside from choosing the habitat which best satisfies its requirements, it not only selects as forage certain plants but also specific parts of those plants. As Bell (1971) has shown, the large African ungulates differ greatly in their foraging styles, ranging from highly selective feeders to rather unselective ones. In general, small antelopes weighing less than about 50 kg tend to be highly selective, for they require food of good quality, as noted above. Such food consists often of discrete and scattered plant parts which must be plucked individually. A dispersed resource automatically reduces the number of individuals—the biomass—which an area can support. Large antelopes often crop whole mouthfuls unselectively, especially if their food consists of grass, with the result that an area may maintain a high biomass.

One problem of relating body size to food habits in the Caprinae is the great sexual dimorphism of some species (tables 6–11). Disregarding takin and muskox, no females average more than 75 to 100 kg whereas some males may reach 150 to 200 kg. The subject of sexual differences in the food habits of highly dimorphic species needs study, for it seems probable that males may select coarser forage than females (Geist, 1974b). The smallest Caprinae males weigh 50 kg or less and include saiga, chiru, chamois, goral, and several subspecies of urial. Saiga and chamois eat mainly grasses and forbs and the chiru probably does too, the urial is mainly a grazer, and the goral predominantly a browser (see Chapter 7). The remaining Caprini males as well as mountain goat and serow belong to an intermediate size class. The serow favors browse, though it does eat grass in season; *Capra, Hemitragus, Ammotragus,* and *Oreamnos* are both grazers and browsers, selecting a wide variety of plants, with their choice depending on availability; and *Pseudois* and *Ovis* are grazers by preference although they readily take forbs and browse too. The largest Caprinae, the takin and muskox, eat grasses and the leaves of willow, birch, and others. The main point here is that all species regardless of size tend to be generalists but that some show distinct preferences for either grass or browse. Forbs and browse predominate in the diet of Rupicaprini and takin, at least seasonally, grass is preferred by *Ovis* and *Pseudois,* and the remaining species seem to be intermediate in choice of food. Jarman (1974) divided African antelopes into five feeding types, but the Caprinae fill only three of these:

1. Very selective browsers; use one vegetation type, small home range	none
2. Very selective grazer or browser; use one or few vegetation types; seasonal diet change	Goral? Serow
3. Selective grazer and browser; uses several vegetation types; seasonal diet change; large home area	All other Caprinae
4. Grazes selectively for plant parts; may migrate; poorly defined home area	Saiga Chiru?
5. Feeds unselectively in several vegetation types within a large home range	none

Most Caprinae are flexible feeders, shifting vegetation zones or moving seasonally to adjust diets to whatever plants are most nutritious. Type 3 feeders range in size from 20 kg chamois females to 300 kg muskox males, a variability similar to that found in antelopes. One may wonder why most Caprinae have rather generalized food habits, why they have given up some quality for quantity, in contrast to the 69 species of African antelopes which represent a wide spectrum of feeding types. These antelopes radiated outward from forests into diverse habitats in which they speciated and in the presence of many competitors specialized to such an extent that 20 species of ungulates, each cropping certain plants or parts of plants, may inhabit the same area. In contrast, the Caprinae on leaving the forests penetrated habitats with a simple structure—Alpine and Arctic and semi-desert areas—where primary productivity is low and consequently secondary productivity in the form of vertebrates is also. There is a close correlation between species diversity and habitat diversity (MacArthur, 1965). With the food supply limited, most Caprinae need to be opportunistic feeders, even in the absence of competitors. Tahr, bharal, American sheep, *Capra* goats, and others living on and near cliffs are often the only large ruminants in their respective habitats. Muskox and caribou associate in the Arctic but, according to Wilkinson et al. (1975), the two are not competing for food. Urial, wild goat, gazelle, and wild asses share the same habitat in some areas of Iran; and chiru, argali, gazelle, kiang, and yak associate in Tibet. The coexistence of these species indicates that they are not serious competitors—or some would have become locally extinct—but the details of their habitat division have not been studied (see Chapter 3). With few exceptions the Caprinae are rather adaptable animals occupying large niches in simple habitats.

Ungulate societies are molded most strongly by two selection pressures: the needs to find enough to eat and to escape predators. An animal's food habits influence its group size. Species which browse on small, high-quality items, items which grow scattered and are wholly removed when eaten, generally do not congregate into herds (Jarman, 1974). The dispersed food supply alone would tend to space animals out and it would also be difficult for herd members to maintain contact in the dense habitat this style of feeder usually occupies. At the other extreme are grazers, whose abundant food resources in a typically open environment enable them to forage in herds of hundreds and even thousands. From their feeding style and habitat preference one would predict goral and serow to live singly and in small groups, and this is indeed the case (Chapter 8). Saiga and chiru may at times collect into huge herds in a pattern expected in migratory grazers. The remaining Caprinae occupy almost closed to open habitats and subsist on a diet of intermediate nutritional value, a situation which on the basis of Jarman's (1974) analysis of antelope social structure correlates with small herds of from 6 to 60 with some up to 200. Most Caprinae fit this pattern too. Herd averages range from 5 to 30, except in populations that have been disrupted by man, and maximum herd size seldom exceeds 100. Aggregations of as many as 200 Asiatic ibex, 300 chamois, or 400 bharal have been reported but these are temporary and often the result of a locally abundant food source.

Herd formation per se is, of course, not a consequence of forage quality and quantity: the animals in an area could remain solitary feeders. While herd life confers several advantages, among them a greater efficiency in feeding and mating, the most important probably is protection against predators. The antipredator strategy of a species depends both on its habitat and its size. The best defense for small animals in dense cover is to behave cryptically—to crouch, stand motionless, or sneak away (Eisenberg and Lockhart, 1972). But animals that depend on being inconspicuous cannot form herds. Species in open terrain tend to flee as soon as a predator is deemed dangerous. Only if the prey animal outweighs the predator by a ratio of at least 3:1 may it feel secure enough to attack (Schaller, 1972).

I would expect goral and serow to have some hiding strategy, but I have no observations on this point except to note that animals tend to remain quiet until approached closely by a person; then they bound away, vocalizing loudly. Herd-living animals in the open usually bunch up in response to danger, the cliff-dwellers then fleeing to the nearest precipices and the others across open terrain. One or more individuals may at that time snort, whistle, or bark, depending on the species, and also thump the feet, both being signals warning others of danger (Chapter 7). The bunching of herd members is an important antipredator strategy in the open, just as a dispersed social system is in dense cover, for in both situations the predator must invest time and energy in locating a vulnerable individual. Life in cliffs discourages pursuit by coursing predators. The muskox is well known for its communal protection against wolves (Gray, 1974), the only member of the subfamily which is known to defend itself regularly against large predators. Harassed Japanese serow may however turn on dogs: "when the dogs attacked, it put its back against the cliff, lowered its head to show its horns, and pawed the ground. One dog jumped on the serow, and was thrown eight metres. It is said that dogs were often stabbed and killed when the serow was hunted regularly" (Akasaka, 1974). Serow may even turn on hunters (Dang, 1962). Unlike the antelopes, the Caprinae have been rather conservative with respect to evolving conspicuous visual signals such as striking gaits and flashing colors for use in flight. Most species have a rather subdued pelage, one difficult to spot at a distance. Among the exceptions are the mountain goat and Dall sheep whose white coats are noticeable on cliffs from far away, at least in summer, and the muskox whose blackish color also stands out from the surroundings. But the habitat of the former and the large size and communal defense of the latter reduce the exposure of these species to predation. Several caprid males also molt into a highly conspicuous pelage during the rut, a good indication that success at procreation has priority over escaping predation. The young of some ungulate species avoid predation by remaining hidden for several weeks after birth. With the exception of the saiga, no Caprinae young is known to have a hiding phase lasting for more than about four days, but as yet little is known about the behavior of chiru, serow, and goral (Chapter 11).

While food habits and antipredator strategies do not determine types of social systems, they do limit available options. Rather solitary forest dwellers restrict themselves to small home ranges with a constant food supply and often escape predators by being cryptic. Such species need a land tenure system which spaces individuals out yet permits contact for mating. Muntiac of both sexes maintain territories, the territory of the male overlapping those of several females (Dubost, 1970), and some small antelopes maintain pair territories (see Estes, 1974). Defecation at specific dung sites is a notable marking behavior of these species. On the basis of food habits, body size, habitat choice, and group structure, one would expect territorial behavior in goral and serow. These two species, as well as the ecologically similar muskdeer, also deposit their dung at certain locations, and serow of both sexes mark objects with secretions from their preorbital glands. A territorial system would seem to be the most efficient way for sedentary and essentially solitary species to organize their societies. Males would not only waste little valuable energy in search of females and in competition over them but they would also run little risk of being injured in fights with strangers, an important consideration in such species as goral and serow, which carry lethal weapons. Territorial boundaries can be marked with scent and dung heaps, a cheap method of delineating an area, bioenergetically speaking.

Every herd-living ungulate creates for itself an orderly society during the mating period in one of two ways: one is a territorial system in which the male evicts other males from, or at least claims prerogatives to, females within a specific area, and the other is a dominance system in which all males in a herd establish and maintain a hierarchy based in part on a display of status symbols (see Estes, 1974). A few species possess systems intermediate between these two types. Subordinate white rhinoceros males are tolerated by territorial males (Owen-Smith, 1974), and lesser kudu males neither defend their overlapping home ranges nor maintain a dominance system (Leuthold, 1974). A territorial male's plot of land is his expression of status. But a male in a hierarchical system must carry his status symbols with him, and to be effective these must show a gradation in size, the males with the largest horns, ruffs, and bodies being dominant. A gradation in physical characters among adults indicates that these characters continue to grow for much of an animal's life. Males in dominance-oriented systems thus differ in a basic physiological way from those that are territorial, in that the growth of the former is prolonged (Geist, 1966, 1971a). When there is keen competition for the females in a herd, those males with the most prominent rank symbols would obviously be favored to produce the most progeny. In a society containing many adult males of low rank who have little access to females there must be long-term survival value in mating only rarely until competition is likely to be successful. It is possible that small males are selected for delayed maturation and against being too competitive since a waste of energy in futile pursuits of females might cause them to die during a subsequent harsh winter or drought.

That body size, horn length, and hirsute appendages can respond rapidly to directed natural selection is well shown by the sexual dimorphism in the Caprini. Of course there is also selection against too large a size and elaborate adornments, the final result being an evolutionary compromise. While large size and prominent secondary sexual characters may lead caprid males to reproductive success, these characters may have impacts in other ways too. With respect to predation, the male selects against himself by becoming conspicuous, and he selects for himself by growing larger, making it more difficult for a predator to subdue him. Yet males appear to be more vulnerable to predation than females, judging by the kill records of wolves and snow leopards (Chapter 6). The large males have some advantages over females, among them a lower metabolic rate and perhaps the ability to subsist on coarser forage and tolerate deeper snow (see Nasimovich, 1955). In a cold environment an animal must balance its heat production and loss. The long ruffs of some caprid males might reduce heat loss. But there is no correlation between climate and the presence of hirsute adornments, as shown by aoudad, Himalayan tahr, and markhor. A male perhaps selects against himself by growing heavy horns. Not only do horns lose heat, they also are heavy—3 to 4% of the total body weight of markhor and wild goat, 6.5% of urial, 6 to 8% of Asiatic ibex and bharal, and 10 to 11% of mountain sheep (Chapter 4). To carry such a weight must be a drain on a male's stored energy, and one might surmise that this contributes to his death during lean seasons. Yet it is significant that in many populations the adult sex ratio is about equal (table 13).

Caprinae societies show a close correlation between sexual dimorphism and land tenure systems. The sexes of goral and serow are of similar size, and these species possibly have some form of territorial behavior (see Akasaka and Maruyama, in press). While the social chamois and mountain goat have retained the short pointed horns typical of rupicaprids, sexual dimorphism in body size is greater in these species than in the relatively solitary goral and serow (table 7). Adult males in some chamois populations may become territorial after the snow melts in the spring and maintain their plots until after the rut (Knaus and Schröder, 1975), in contrast to mountain goat males among which territoriality has as yet not been reported. However, female mountain goats may chase males off their home ranges, thereby ridding themselves of food competitors (Geist, 1965), behavior which resembles territorialism. Takin look like overgrown rupicaprids with quite marked sexual dimorphism, but nothing is known of their land tenure system. Muskox form fairly cohesive nonterritorial social units (Gray, 1973), a sensible system for a species living at low densities in a severe environment where feeding sites are dispersed and where during winter the animals need to crowd together for protection from wolves. The Caprinae become progressively more dimorphic starting with the goral and serow and continuing through the chamois and mountain goat to the two Ovibovini and finally the Caprini (see tables 7, 9, and 11, and also the skull measurements in Schaffer and Reed, 1972), and this cline can be correlated with emphasis on dominance, the caprids having marked hierarchies. Different Caprini genera have stressed different status symbols, with for example Himalayan tahr placing their emphasis on coat length and wild goat on horn size (fig. 35), but whatever the final product it reveals a society which places a premium on direct competition between males.

All antelopes except the Tragelaphini are territorial at least during part of their annual cycle (Estes, 1974); the Bovini favor a hierarchical system; and the Caprinae are variable, some Rupicaprini and probably the Saigini being territorial and most if not all others opting for a dominance-oriented society. The male of a herd-living species needs certain ecological requisites before he can maintain a territory. Predictability of food in a limited area throughout the year promotes territoriality, as has already been noted with regard to solitary forest dwellers. A gregarious male needs enough food in his territory to support himself and whatever females he may attract, and this food must be concentrated enough to prevent him from having to waste energy in maintaining an unnecessarily large area.

Geist (1974a) noted that "we cannot expect northern ungulates to be territorial if they rut in fall and early winter" because of the unpredictable food supply. Yet muskdeer, Japanese serow, saiga, chiru, and Tibetan gazelle live in cold climates, rut during winter, and are probably at least seasonally territorial. Some species have devised systems which reduce dependence on food within the territory. Geist (1974a) himself noted the importance of fat reserves. "Since the period of idleness in males coincides with the time of greatest productivity in the temperate north, the males are free to draw from the seasonal superabundance in order to grow large stores of fat and huge horn-like organs. . . . The fat in turn permits the males to spend their time rutting rather than feeding." Horn size cannot be correlated readily with climate, as tables 6 to 12 show, but fat storage is unquestionably important to males. As noted by Caughley (1970d), Himalayan tahr males reach peak fat reserves earlier each year until they become fully adult. "The earlier rise in reserves with age at the end of each successive summer suggests that the early peak anticipates a physiologically demanding period to come. Endocrine rather than environmental influence appears to be responsible." If the rutting season is brief, males may refrain from eating, such abstinence enabling them to maintain small temporary territories. The saiga may have such a system but it is also possible that they establish transient territories in the manner of migratory wildebeest (see Estes, 1974).

Nevertheless the food supply is a major consideration in a territorial system, and it is not surprising that most Caprinae have used the alternate strategy. Food in semideserts is sparse, patchy, and unpredictable. Herds must remain flexible in size and movements to take advantage of locally abundant resources. The fragmented nature of some habitats, especially those on cliffs, may also make territories uneconomical. Similar problems face animals in alpine environments where vertical migrations are often necessary. Since both females and young need a short birth season during a time when food and weather are at their best, the rut tends to be relegated to the winter or dry season, when food is scarce. If both food and females are likely to vanish, a male's most sensible strategy is to stay with the herd. Furthermore it pays an animal to waste as little energy as possible, and in harsh environments the rut tends to be brief and sometimes variable in its onset, depending on the vagaries of weather (table 14, fig. 23).

In arid environments a nonterritorial Caprini often lives in the same habitat as a territorial Antilopini, the urial and chinkara in Pakistan being one example of such an association. Desert caprids such as wild goat and Afghan urial also have a discrete rutting season yet their unpredictable environment seldom offers them a superabundance of food. The energy budgets of such species need to be studied.

While the social Caprinae, as well as most gregarious ruminants, use one or the other of the two basic land tenure systems, the herd structure of most species is similar, it being characteristically flexible, with only a mother and her young and sometimes a yearling, as well, forming a close bond. Such a loose structure is particularly adaptive in unpredictable environments for it enables herds to adjust their size to the available resources. Three kinds of herds exist in most societies: male herds, as well as some solitary males; female herds consisting of females, yearlings, and young; and mixed herds containing adults of both sexes. Most adult males are with or near the females during the rut, but at other times of the year their association with females varies considerably and depends on the species. Geist (1974b) emphasized that it would be adaptive for males to segregate as much as possible, thereby reducing competition for food with pregnant and lactating females. Indeed many mountain sheep, argali, Himalayan tahr, and other males may separate from females so completely that they occupy different ranges. However, among Punjab urial, wild goat, bharal, and Asiatic ibex, to name just four species, many males associate with females outside of the rut (Chapter 8), and the male herds may occupy the same areas as the female ones. Urial and wild goat inhabit rather barren, low-lying terrain which give the sexes little opportunity to seek separate ranges, especially since herds tend to concentrate at favorable feeding sites, but I have no ready explanation to account for the differences between, let us say, mountain sheep and bharal.

Capra, Hemitragus, and *Pseudois* often seemed to be divided into fairly cohesive population units each containing a number of females and subadults which, though split into several herds, tended to frequent a particular slope or cliff. I do not know how long such units maintained themselves and how closed they were, but those on isolated cliffs possibly retained their identity for years. Feral Soay sheep form permanent ewe home-range groups, as Grubb and Jewell (1974) called them, and these recognized and avoided each other. Sociality in marmots increases as communities become more and more isolated (Barash, 1973), a situation which seems to have its ecological equivalent among caprids.

Much of what I have so far written in this chapter is based on the assumption that under similar environmental conditions social animals tend to develop similar social organizations. This hypothesis, first promulgated in detail by Crook (1965), provides a useful framework for correlating social and ecological parameters of species. For example, it is instructive to compare the social systems of rain-forest and open-country Caprinae and to speculate what ecological factors make it advantageous for Caprini to have a dominance-oriented rather than a territorial society. But it is not wholly clear why, for instance, most antelopes living in arid terrain maintain territories whereas caprids living in a similarly unfavorable environment do not. Similar circumstances may produce quite dissimilar social systems for reasons that still remain obscure. An interesting example can be drawn from the deer family. The Père David's deer, Indian barasingha, and South American marsh deer are all large deer adapted to swampy or tall-grass habitat. The first has of course been long extinct in the wild but its widely splayed hooves and propensity to stand in water suggest a preference for marshy terrain. Schaller and Hamer (in press) observed a Père David's deer herd with 7 rutting adult stags at the New York Zoological Park between June and August 1976. Three of the stags were young, slightly over 2 and 3 years old, and the others were fully adult. The stags had a linear dominance hierarchy. The fully adult ones roared, often belabored tree trunks and hooked the soil with their antlers, whipped their penis from side to side thereby splattering themselves with urine, and reclined in wallows while flicking mud over their backs with their antler tines. Their main display was a stiff-legged rolling gait with head slightly lowered. The dominant stag herded hinds persistently, chasing and driving back to the herd those that had strayed, and he often threatened subordinate males, keeping them around the perimeter of the herd. He maintained a harem in the manner of red deer and wapiti (Struhsaker, 1967). Barasingha stags roar and wallow too, but they do not urinate on themselves. Their main display consists of a head-up. Though they resemble rutting Père David's deer in their behavior patterns, the social systems of the two species differ markedly. Barasingha stags and hinds congregate into breeding herds each of which may contain some 10 adult stags. These stags form a linear hierarchy with the highest-ranking animal claiming any hind in estrus. However, competitors are not excluded from the herd, being merely chased from the immediate vicinity of the estrous hind. The marsh deer has not been studied, but my preliminary observations suggest that these animals tend to be unsociable. Among 171 deer tallied in Mato Grosso state of Brazil during July 1976, a time when 91% of the adult stags were in hard antler and the rest still in velvet, there were 97 lone individuals, 29 pairs, 4 groups with 3 individuals, and one group with 4. Seventeen marsh deer seen in October 1975 in Argentina showed similar grouping patterns. These data and scattered notes in the literature suggest that hinds are dispersed singly and in small groups while rutting stags drift through the population in search of estrous ones, a pattern similar to that of *Odocoileus* deer in North America. The factors favoring three different social systems in these large deer occupying similar habitats need explanation.

POPULATION QUALITY

That populations are variable and may differ in the size of individuals, in birth and death rates, and in other characteristics has been well-documented for many years. Indeed rather minor morphological differences between populations have often been assumed to be genotypic and of sufficient magnitude to justify taxonomic distinctiveness. But further research led to the hypothesis that the environment has an influence on many population parameters. For example, among rodents "indefinite increase in population density is prevented through a deterioration in the quality of the populations" (Chitty, 1960). And quality depends largely on habitat condition, at least in ruminants. As Klein (1965, 1968) has shown in his studies of blacktailed deer and reindeer, nutrition may have a drastic influence on individual growth and on the number of young born and raised. Nievergelt (1966a), after observing several Alpine ibex populations, found that large horns and bodies, high fecundity, early maturity, and short life-expectancy were correlated with expanding populations and that the reverse attributes were characteristic of stable and declining ones. Caughley (1970c) studied the demographic statistics of five Himalayan tahr populations in New Zealand, some increasing, one stable, others decreasing, and noted that the condition of the animals, as measured by fat reserves, was best in the increasing populations and worst in the declining ones. Physical condition was dependent on the food supply, populations having increased until forage was depleted and then declined as a result, not of lowered birth rates, but of increased death rates. Combining these ideas, Geist (1971a) then noted that high-quality mountain sheep are typical of expanding populations living on good habitat and that the high level of energy available to them is reflected in various physical, developmental, and behavioral characteristics. "Sheep from a high-quality population are large, early maturing, vigorous individuals which grow large skulls and horns . . . and interact socially frequently and intensely. The ewes of such a population bear a greater number of large vigorous lambs and suckle them adequately; the rams fight more frequently than those in a low-quality population and die early."

While the concept of quality is useful, care must be taken not to apply all the above parameters uncritically to all populations. A case in point is the frequency of fighting between rams. Geist (1971a) equated differences in the frequency and intensity of interactions between rams in Dall, Stone, and bighorn populations with quality, but several other factors can also account for these differences, among them the structure and size of a population and the degree of acquaintance among rams. Genetic differences in the threshhold levels of various activities can also be expected between species, subspecies, and even populations. Thinhorn sheep seldom have broken horns, in contrast to bighorns, and this difference also exists between Marco Polo sheep and Tibetan argali (Chapter 4). Variations in the frequency of horn breakage between species and subspecies may depend not only on horn shape (Shackleton and Hutton, 1971) but also on the amount of clashing, and the latter may have a genetic basis rather than be solely related to quality.

Shackleton (1973) was able to apply many of Geist's (1971a) criteria of quality to two bighorn populations. One population at Radium Hot Springs was expanding and the other in Banff National Park was stable. The sheep at Radium Hot Springs were larger in skull and in horn lengths, more young survived (67 lambs per 100 ewes as compared to 33 lambs), and the lambs suckled longer and played more than those at Banff. The Radium Hot Springs rams also matured earlier, leaving the ewes to join male herds at the age of 16 months as compared to Banff rams who were still with the ewes at 30 months of age. The Lapche and

Shey bharal also fitted the model of low- and high-quality populations well (Chapter 5). The former animals inhabited lush meadows at low densities, reproduction and survival were excellent (88 young per 100 adult females), and few males were old. The range of the latter was ravaged by livestock, females raised relatively few young (40 young per 100 adult females), and over half the males were fully adult. I can only make a few inferences about the quality of some of the markhor and ibex populations in Pakistan. The markhor in the Chitral Gol were expanding after having been decimated by hunters. The twinning rate was high. Although the species' winter range was in poor condition, the animals found ample forage in the form of evergreen oak leaves, and in spring they followed the appearance of the nutritious new plant growth upward into the crags where livestock provided less competition. The Asiatic ibex along the Baltoro Glacier were on small meadows near the limit of vegetation. Caprids in such a situation tend to fill their habitat to capacity—they are K strategists in the ecological lexicon—and one would expect them to be stable and of low quality unless periodically reduced by catastrophic events. The many avalanches in the area may partially serve this function. The life expectancy of animals is shorter in expanding than in stable populations (see Chapter 5). In Chitral, along the Baltoro Glacier, and at Shey I collected the horns of males that were killed by predators, in avalanches, or in other natural mishaps (table 16) and the mean ages at death were: markhor 4.8 years, ibex 8.0 years, and bharal 8.3 years. These figures follow expectations.

The species discussed so far live at high altitudes or latitudes where the environment is fairly predictable and where nutritional levels tend to be relatively constant though seasonally variable. How does quality affect urial and wild goat, species living in semidesert conditions where the availability of good forage is highly erratic and competition between livestock and wildlife for food is keen? The Kalabagh urial population was stable. Rams were small, weighing about 40 kg, as compared to the 60–80 kg reached by Kopet Dagh rams on the good upland pastures in Iran (table 10). Although different subspecies are involved, it is worth noting that the Armenian urial, which occupy a habitat much like that of the Kalabagh urial, also weigh about 40 kg. About a third of the lambs died soon after birth, a figure which was fairly consistent over several years. Some lambs were obviously weak and lethargic. If ewes have ample nutritious forage the fetus deposits fat and this affects birth weight and subsequent chances of survival (see Sadleir, 1969). I weighed 37 young ranging in age from newborn to several days. Twelve males averaged 2.2± .53 kg (.9–2.9 kg) in weight and 25 females averaged 2.3± .39 kg (1.6–3.2 kg). One male weighed 1.8 kg at birth and 2.3 kg four days later. Newborn lambs in the Kopet Dagh weigh 3 to 4 kg (Decker and Kowalski, 1972) but the ewes there are also about 50% heavier than those at Kalabagh. Thus there seems to be no striking difference in birth weights between these two populations. However, the available data lack precision. Average suckling duration at Kalabagh was 25 seconds per bout as compared to 28 seconds reported for bighorn lambs on good range (Shackleton, 1973). But a distinction must be made between a lamb suckling during its hiding phase after its mother has been absent for several hours and a lamb suckling while it is following its mother. If I delete my two observations of nursing that lasted over a minute, the average drops to 14 seconds, the figure given by Shackleton (1973) for a bighorn population of poor quality. Milk production in ewes depends on nutrition, and the growth rate of lambs is in turn influenced by the amount of milk they obtain. Those which receive little milk and which must depend on grass early in life are likely to be stunted (Sadleir, 1969). Judging by suckling durations, the early age at which young ate grass, the low frequency with which lambs played (Chapter 11), and the small ultimate size of adults, the Kalabagh population seemed to be of poor quality. A high twinning rate is perhaps the best indicator of quality in those species that normally may have multiple births. Fewer than 10% of the ewes had twins at Kalabagh as compared to a nearly 50% multiple birth rate in the Kopet Dagh (Decker and Kowalski, 1972). Since ewes with twins do not produce twice as much milk as ewes with single lambs (Alexander and Davis, 1959), the selective advantages of single births under poor conditions are obvious. Species with long life-spans may readily sacrifice some yearly production if by so doing they increase their own chances of survival (Elliott, 1975). A high fecundity in yearling females is also a sign of good condition (Caughley, 1970c), but I lack precise data on this point in urial.

The habitat of the Karchat wild goat is more arid than that of the Punjab urial, there sometimes being little or no rain for years. The goat population was aged, with few subadult and many adult males. Survival of young was always poorer than at Kalabagh, one drought year producing only 19 young per 100 adult females (table 13). Many newborns were weak, they rarely played, average suckling time was 14 seconds, and females terminated almost all nursing bouts (Chapter 11). I saw no twins in 1973 and 1974 and yearlings seldom conceived. The population was obviously of poor quality. However, in a letter dated February 21, 1976, T. Roberts wrote me: "I hardly recognized the area because of the changed landscape following very good rains last summer and considerable rains this winter. There was green grass and vegetation everywhere and the place looked like savannah rather than desert . . . . We had nice views of 2 wild goats going to where they had concealed their babies and were pleased to note twins in both cases."

Up to now I have not referred to horns as indicators of population quality. Horns grow in an annual cycle beginning usually in spring when nutritious forage becomes available and ceasing in autumn. There is little growth in winter, even in the desert where conditions are often favorable at that season, possibly because increased testosterone and other hormone levels during the rut are known to inhibit horn growth (see Shackleton, 1973). Annual horn increments are at first longer in expanding than in stable populations as the animal gains weight rapidly and reaches maturity early (Nievergelt, 1966a). The effects of a constant supply of good food on maturation were evident in two urial rams at Kalabagh which had been caught as lambs and raised in an enclosure situated in the same habitat as that of the free-living animals. At the age of 2½ years these captive rams had the horn, ruff, scrotum, and body size of 3½ to 4-year-old rams in the natural population. Since horn growth is affected by nutrition, the average annual increments in the horns of ibex, bharal, and other species living in predictable environments should be less variable than for those living in unpredictable environments. My data follow this prediction (table 45). High-altitude species or species which make their food supply predictable by migrating have smaller coefficients of variation than, for instance, urials, wild goat, and Sulaiman markhor whose nutritional level is dependent on sporadic desert rains. Horn growth would thus be a poor general indicator of quality among the latter species.

Markhor reveal interesting horn growth patterns (fig. 21). Growth rates are similar in all populations until the animals are about 4 years old. After that the average annual increments of the straight-horned markhor become smaller than those of the flare-horned ones. Since relative horn length in the two types is similar, the data suggest that the straight-horned markhor grow little in body or in horn length after they become fully adult and that the general size difference between the two subspecies can be related not to an initial growth spurt in subadults but to differential growth rates among adults. The fact that flare-horned markhor usually live in a more favorable environment than do straight-horned ones is no doubt significant.

Caughley's (1970c) studies of Himalayan tahr in New Zealand have shown that population quality can fluctuate quite rapidly. After being introduced to the country in 1904, some populations erupted, declined, and were again on the upswing during the 1960s. These animals lived in a predictable environment. But species inhabiting an unpredictable environment may change in population quality from year to year, as intimated above when I discussed variability in horn growth and changes in twinning rates of wild goat. Fecundity, nursing duration, frequency of playing, and other such parameters of quality are closely related and may be typical of only one particular season. Similarly, the quality of a male as measured by horn growth may change several times during his life. This variation produces populations which in effect contain males of high, low, and intermediate quality. An urial ram born, let us say, at the onset of a four-year good spell would mature early and grow well, producing horns with long annual increments. During the following poor spell the fecundity and other measures would indicate a population of low quality, except that many males would obviously be of high quality and some old ones of intermediate quality. In a ram assemblage of mixed quality an age hierarchy based on horn size would be less well defined than in bighorns, and this might have an influence on the outcome of competitive situations such as priority rights in mating. It might also be difficult to equate population quality with intensity of social activity because the presence of rams of different quality further obscures such attenuating factors as weather and the composition of herds.

In species with a long life the average age at death may be a good indicator of quality, as noted earlier. Urial however, regardless of quality, have a short life (Chapter 5). Mean age at death of the Kalabagh urial was 5.4 years. Marco Polo rams resemble urials in longevity, the mean age of death of picked-up horns being 4.6 years (table 16). The Karchat wild goat lived an average of 5.9 years, a low figure 'for a population of presumed low quality. Genotypic differences in longevity probably exist, a fact which may affect comparisons between species. A short life does not of course mean that the male leaves few progeny. Urial on good habitat often have twins, and a ram of this species might in four years produce what a bighorn does in eight.

Urials and argalis have short lives when compared to mountain sheep and probably snow sheep. Other species with a long lifespan include ibex, Himalayan tahr, and bharal. Markhor and wild goat seem to be intermediate (table 16). When one looks for correlations, two environmental variables obtrude: (1) Species live longer in a habitat with a predictable than with an unpredictable food supply, and (2) species dwelling on and around cliffs live longer than those on flat to rolling terrain. The first point is invalid, at least with the scant and coarse data available at present, but the second one seems to be true. The short life of Eurasian sheep indicates that the rams need to grow faster and age physiologically and behaviorally more rapidly than do mountain sheep. Some urial rams rut vigorously at the age of 3½ years and all do so at 4½ years, unlike mountain sheep which do not enter the rut fully until the age of about 6½ years (Geist, 1971a). Thus urials and probably argalis live hard and die young, superficially behaving like high-quality animals, if mountain sheep are uncritically taken as a basis for comparison.

The concept of population quality can be applied to species in unpredictable environments only with the realization that the parameters may change drastically from one year to the next. A trend may be apparent but a sudden reversal is possible if the weather so dictates. Furthermore, individuals in a population differ in their genotypes, and any parameter, whether it be longevity or aggressiveness, will be influenced to some extent by this. For example body weight has been mentioned here repeatedly as a variable of population quality. Falconer (1953) selected for large and small body size in captive mice and in 11 generations achieved a mean adult weight in one population almost twice that of another population even though food was not a variable. Size can obviously respond rapidly to natural selection, and discretion must be used when ascribing differences in size to quality. Quality depends on a syndrome of characters, and in each population the characters must be evaluated anew.

If the environment predictably affects a population, the genotype of certain characters may be suppressed or even selected against, with the result that variability is reduced. But in urial and wild goat populations which contain animals of differing quality—with all this implies regarding maturation rates, life expectancy, ability to compete for females, and so forth—genotypes can find expression and maintain their variability. It is generally recognized that genetic variation is desirable in a population for it in effect preadapts some individuals to new conditions and enables them to colonize unoccupied terrain successfully. Sheep and goats living in the harsh and unpredictable environment of the desert ranges may have retained a greater physical and physiological adaptability than those confined to high mountain pastures. Some of the data in this report support this idea.

BEHAVIOR IN CAPRID SOCIETIES

A minor physical structure may have a profound influence on social structure, as Geist (1965) has shown for mountain goat. While adapting to a social existence, this species has retained the short, pointed horns typical of rather solitary rupicaprids. Possessing dangerous weapons, members of a herd interact in a characteristically restrained fashion. Individuals tend to be discretely spaced and to use displays rather than physical contact to assert themselves. To this end both sexes have evolved a conspicuous pelage. Geist (1974a) explained the similarity in the appearance of the sexes in rupicaprids as a case of the female mimicking the male, on the supposition that by looking like a male the female reduces her chances of being attacked. The fact that the sexes of all rupicaprids resemble each other is a good indication that aggressive and other social interactions between males and females are equally important, but a lack of dimorphism is not in itself mimicry. There may however be sexual convergence in certain characters such as ruffs. Mountain goat ears seem to mimic the horns in color and shape, apparently on the principle that redundancy increases the strength of a signal (Guthrie and Petocz, 1970).

Mountain goat females are so highly protective of their young that males may retreat from the females, even being subordinate during part of the rut in spite of their somewhat larger size. Dangerous weapons obviously impose constraints on some interactions but they may also promote contact: mountain goat males may court and breed by the age of 2½ years, or as soon as they can do so physiologically, the older animals being unable to exclude them from the rut because of the younger ones' prickly defenses. Males risk death if they fight with sharp horns. For example, Wilkinson et al. (1976) found 7 fresh muskox carcasses on Banks Island of which at least 6 had been killed in rutting fights.

The situation is different among the Caprini, none of which carry deadly weapons. As Schaffer (1968) and Reed and Schaffer (1972) have shown, the Caprini have horns and frontal bones that are relatively larger than those of the Rupicaprini. Sexual dimorphism is great, it being least so in tahr and aoudad and most so in sheep and bharal. Females have in general retained rupicaprid-like horns except that these tend to be relatively blunt. Since horns are used for aggression, I would expect females to become more aggressive as their horns increase in size relative to those of males. Furthermore, the larger and more permanent a society the more often are members likely to find themselves in competitive situations. The data generally follow expectations. Himalayan and Nilgiri tahr females are quite aggressive (tables 32 and 33). Among captive aoudad "fighting was almost as common among the ewes as among the rams" except during the rut (Katz, 1949). It is interesting that aoudad females, and sometimes Himalayan tahr females, have small ruffs which give them the apparent status of young males. *Capra* females have substantial horns although these are much shorter and more spindly than those of males. Markhor and wild goat females are quite aggressive (tables 34 and 35). Nievergelt (1974) observed little fighting among Walia ibex, possibly because population density was low. In a year of observation he "saw 6 fights between females.... Although females do not fight much, there are many fights or

play-fights within the female-young groups, mainly by young."
Urial ewes with their stubby horns seldom interact (table 36).
Mountain sheep seem to fight somewhat more often than do urial
(Geist, 1971a; Shackleton, 1973). Collias (1956) noted that domestic goats fight more than sheep. An exception to the trend is the
bharal, a species in which females clash often and hard but have
insignificant horns. One can only speculate about the selective
advantages which females derive from increased aggressiveness.
Female goats generally have more contact with males than do
sheep and they live in such restricted and difficult terrain that
social contact is often forced upon them. In such situations
females may need to be able to assert themselves not only in claiming small feeding sites, scarce resting places, and rights of way
along ledges but also in protecting their young from the attentions
of others. Bharal females look like sheep but fight as often as
goats, an anomalous situation for which I have no satisfactory
explanation.

Horn length in relation to shoulder height is highly variable in
Caprini males. Tahr have proportionately by far the smallest
horns, followed by aoudad, Kuban ibex, Dagestan tur, and bharal
(table 8). Himalayan tahr and aoudad have conspicuous ruffs,
age-graded rank symbols similar to the horns. But ruffs are not
weapons per se, and I would expect male Himalayan tahr and, to
a lesser extent, aoudad to threaten more and fight less than other
species. Indeed, male tahr fight rarely. Aoudad have not been well
studied. If, however, a species has both a ruff and large horns
then direct aggression may remain common, as shown by markhor and urial.

Schaffer and Reed (1972) tried to correlate horn size in males
with environmental harshness, harshness being defined by the
shape of the survivorship curve of the population. According to
their definition a steep curve represents harshness and a gentle
one a benign environment. They concluded that "increased environmental harshness is associated with smaller horns in the
males." Species do show differences in survivorship curves but
these may have little to do with environmental harshness. For
example, the short-lived urials and argalis have horns of the same
relative size as the long-lived mountain sheep. Tibetan argali and
bharal inhabit the same environment but differ in horn length

and also in their survivorship curves. Thus my data do not support Schaffer and Reed's (1972) generalization.

Whatever their size, all male Caprini except tahr have prominent horns that vary conspicuously according to age and represent not only rank symbols but also weapons with which an animal can threaten or attack. Since the horns are designed for ramming rather than puncturing, fights seldom cause injury. All species spar and exchange blows, two combatants either clashing in unison or one delivering the blow and the other catching it with his horns. Such fights, which are so precisely timed and require the active participation of both animals, were compared by Schaffer and Reed (1972) to the jousting of medieval knights. Fighting techniques vary, but the most violent horn-to-horn contacts are so powerful that the animals obviously need special adaptations to absorb the impact. Schaffer (1968), Geist (1966, 1971a), and Schaffer and Reed (1972) discuss these adaptations in detail. Only a few are mentioned here.

Male Caprini have two basic cranial shapes, according to Schaffer and Reed (1972). In one the braincase projects behind the bases of the horn cores, a pattern found in *Capra*, Nilgiri tahr, and mouflon males, among others. In the other the foramen magnum is beneath the horn bases; this condition occurs in Himalayan tahr, bharal, aoudad, and the argalis. Bighorn sheep are intermediate. These shapes are thought to be related to fighting styles. If the impact occurs somewhat away from the base of the horns, as it sometimes does in goats, or if only one horn strikes a blow, as may occur in bighorn sheep, then the great lateral rotation of the head could cause injury unless counteracted. *Capra* have large neck muscles which when contracted just prior to impact will resist such rotation. Such muscles need a large area for attachment and this the long braincase provides (Schaffer, 1968). Species which specialize in clashing hard seem to have particularly wide occipital condyles and a broad atlas, both apparently designed for absorbing torque by locking the skull against the atlas (Geist, 1971a). Animals such as sheep, aoudad, and bharal which bash horns frontally—though usually not just horn-to-horn as Schaffer and Reed (1972) assumed—have generally broader horns (fig. 8) and skulls than do *Capra* and tahr. With the foramen magnum brought beneath the horns the whole neck helps to absorb the force of the blow (Schaffer and Reed, 1972). The impact tends to rotate the head downward, but a mass of fibrous tissue connecting the occiput to the vertebrae counteracts this. In urial this tissue forms a conspicuous bump on the back of the head.

The skulls of Caprini males are also adapted to withstand severe blows. Two layers of bone separated by sinuses protect the brain. Each sinus is confined to one frontal bone—of which the horn cores are outgrowths—and each has several supportive bony struts (Reed and Schaffer, 1972). The protective layer of bone and airspace may reach an average thickness of 6 cm in adult bighorn rams (Geist, 1971a) and 5.5 cm in old male bharal, providing indeed much shock-absorbing capacity. "Our conclusion is that the more frontal sinuses overlie the brain and the more complex the pattern of struts in the sinuses—within both the horn core and the frontal bone—the more forceful the male's butting and ramming behavior is" (Reed and Schaffer, 1972), and, as a corollary, the greater will be the recipient's ability to withstand the clash. Furthermore, the heavier the horns and skull the more forceful will the blow be (Geist, 1971a). A basic assumption here is that skull morphology is correlated with behavior. While this is unquestionably true in a broad context, much too little is known about the frequency, force, and the techniques of clashing to compare the behavior of various caprids. Rupicaprids which clash little or not at all have thin frontal bones and tiny, round horn cores, *Capra* which deal out moderately heavy blows have frontals of intermediate thickness and horn cores of intermediate circumference, and sheep and bharal which bash horns vigorously have thick frontals and massive cores (Chapter 4). Wild goats and Asiatic ibex have horns of about the same relative length but those of the former are much lighter. Do ibex clash more often or harder than do wild goat? No data are available. In some sheep the horns weigh proportionately much more than in others. One cline, listed in the order of increasing weight, would consist of Punjab urial, Marco Polo sheep, and Altai argali; another would be snow sheep, Stone sheep, and bighorn sheep. I would expect the heavy-horned forms to be hard fighters, given more to direct than indirect aggression. Geist (1971a) noted that bighorn sheep clash more than thinhorns and that the latter seldom break their horn tips, but no comparable information is available for Eurasian sheep except that the animals with heavy horns seem to break their tips more often than do those with light ones (Chapter 4). However, horn shape also has an influence on the frequency of breakage (Shackleton and Hutton, 1971). The tahr presents some puzzling features. According to Schaffer and Reed (1972) the Nilgiri and Himalayan tahr have different skull shapes; from this one would infer that their fighting techniques differ significantly, but I saw no evidence of such difference. Himalayan tahr have quite thick skulls with much pneumation—the heads of active fighters—yet adult males rarely clash.

Although caprids differ in horn shape and size, their methods of clashing remain generally similar. As they clash, one combatant has his head cocked sideways, or both may have their heads twisted in opposite directions, so that the horns do not meet frontally but interdigitate, the blow landing along the inner edges of the horn bases and sometimes touching the forehead as well (fig. 33A, C). Among mountain sheep "the force of the blow is focused on that one narrow horn keel ... then, after making contact with one horn, he rotates his head and brings the other horn into contact with the opponent. The ram performs in effect a double blow" (Geist, 1971a). Whether other species perform a double blow remains to be investigated. Noting differences in horn keels, angles, and shapes among Eurasian sheep, Geist (1971a) then speculated that methods of horn contact would differ. This too remains to be investigated. The round-horned aoudad and bharal clash much like Punjab urial, whose clashes in turn resemble those of mountain sheep; wild goat, markhor, and tahr clash in a similar manner in spite of differing horn shapes. While methods of clashing are influenced by horn shape—as for example the battering rams of *Ovis* compared to the thin wands of *Capra*—many idiosyncrasies in the horn are not related to specific fighting styles any more than are the plethora of horn shapes in African antelopes. Horns are malleable in evolution, becoming readily modified, sometimes randomly in isolation and at other times probably under the selection pressure of making an animal look different from a neighboring and closely related form.

Aside from such a widespread form of bovid horn contact as sparring, the Caprini clash in two specialized ways: they may rear bolt upright, standing balanced on their hindlegs, and lunge down to hit the horns of a waiting opponent (fig. 33D), or they may run at an opponent from a distance and ram him head-on.

Occasionally both combatants rear and clash in unison. The former method is practiced by *Capra* and rarely by *Hemitragus,* the latter by *Ovis* and *Ammotragus. Pseudois* usually clash like goats but occasionally they also run at each other like sheep. Domestic sheep and urial charge on all fours whereas argalis and mountain sheep partially rear up and, running unbalanced on their hindlegs (fig. 33B), fling themselves into a clash. These distinct fighting styles are designed to increase the force of the clash, and like all evolutionary innovations they must have appeared in response to some selection pressures. Goats and tahr differ from the other species in their predilection for precipitous terrain. To run rapidly along a cliff and clash horns violently and with precision is not only difficult but also potentially hazardous. It would seem that stationary combatants dealing out moderately strong blows would have the most suitable fighting style on cliffs. An upright stance is also of great advantage when browsing on trees, and goats are particularly partial to this form of foraging (Chapter 7). Living as they do in rolling terrain or on slopes near cliffs, sheep are not compelled to restrain themselves when fighting. The evolutionary history of rearing to clash remains unknown, but two hypotheses seem plausible: (1) The common ancestor of both sheep and goats rose up to clash. The fact that on rare occasions chamois may rear and clash is significant in this context. Goats retained and emphasized the behavior but sheep deemphasized it as they ventured into less precipitous areas. Some sheep still rear to clash, others like the urial rear only as a form of threat, and domestic sheep have essentially discarded the behavior. (2) Rearing evolved independently and convergently in goats and sheep. The argalis and mountain sheep retained the pattern but the urials almost lost it. That bharal fight essentially like goats suggests that they evolved as cliff animals which then penetrated a more moderate habitat. Krumbiegel (1954, quoted in Geist, 1974b) noted that behavior patterns may exist long after the habitat has changed or morphological characters have been modified. Aoudad do not fit my line of reasoning in that they seem to prefer "incredibly rough" terrain (Rodd, 1926) yet fight like open-country sheep, but nothing more can be said until someone studies the behavior of this species in the wild. Muskox may run at each other to clash, and they select flat terrain for such combat (P. Lent, pers. comm.).

I have discussed horns and their uses at some length because they feature so prominently in caprid society, and in fact it can be said that they symbolize it. Females are relatively passive, concerned with saving energy for reproduction. But males are socially active, asserting themselves and testing their rank and for a brief time of year displaying themselves to females. What are the rules of conduct in caprid society? Geist (1971a) presented a stimulating analysis of mountain sheep society, a valuable study against which information from other species can be tested. So many variables—population quality, snow cover, age and sex structure of herds—have an influence on the frequency of interactions that direct comparisons between subspecies and species are possible only in broad terms. Chapters 9 and 10 describe the details of aggression and courtship, and I limit myself here to raising several points.

There is a premium on large horns in caprid society. By physiologically prolonging their body and horn growth, males can add considerably to their size and hence status. Many ruminant males have an adolescence of a year or so, a period when they are mature but unable to compete for females with much success. In sexually dimorphic species this period has been extended to at least 2 to 3 years. Urial reach a fully competitive age at 4½ years, bharal and wild goat at 5½ years, and bighorn sheep, according to Geist (1971a), not until at least 6½ years. A two-year difference in length of adolescence between urial and bighorns should have an effect on behavior because physical and behavioral maturation occur together. Geist (1971a) has shown that bighorn rams use fewer overtly aggressive acts and more displays toward rams of similar horn size as they grow older—rams seem to become more subtle in their contacts as they age. A similar trend was not apparent in the species I studied. Wild goat males of all ages show direct aggression (jerk, lunge, jump, butt) toward other males but almost totally exclude displays as a means of asserting rank. Markhor behave similarly. The reverse is true among Punjab urial rams which seldom assert themselves with direct aggression, except to clash (table 36), preferring instead to use such indirect forms as kicking. Bharal reveal yet another pattern in that males of classes IV and V exhibit 69% of all direct aggressions and 95% of all indirect ones (kick, twist, low-stretch, mount, mouth penis). Yearling and class I bharal males seldom interact (table 37).

One of the key concepts in Geist's (1971a) interpretation of mountain sheep society is included in the following passage.

> The male groups are homosexual societies in which the dominant acts the role of the courting male and the subordinant the role of the estrous female. The dominant male treats all sheep smaller than he is, irrespective of sex and age, like females: it is his prerogative to act sexually, but it is the subordinate's prerogative to act aggressively. Most overt aggression is directed by subordinates at dominants, not vice versa.

To fully understand the meaning of this quote one should remember that rams display the low-stretch, twist, kick, and mount not only as a form of threat toward other rams but also toward ewes during courtship. A male can only be said to treat males and females alike if he displays to them with the same frequencies. As noted above, *Capra* males seldom display to other males, so Geist's finding in mountain sheep does not extend to this genus. Urial and bharal males kick and otherwise treat both sexes similarly but they do so at quite different frequencies (fig. 42). For example, urial rams kick other rams significantly more often than they kick ewes, and bharal males twist to females more often than they twist to other males. Geist further noted that subordinate mountain sheep rams show direct aggression mainly toward dominant rams during dominance interactions. This was not the case among wild goat, urial, bharal, and the other species I observed. Jerks, lunges, chases, and butts were directed mainly by dominant males at subordinates, and clashes predominated between individuals of the same horn-size class (fig. 32). Thus rules of conduct in mountain sheep society differ in several basic aspects from those in other caprids.

Based on social selection for sexually dimorphic males, Caprini society consists of a cline of individuals differing in size and appearance, grading from a young, yearling, and ewe to the largest males. Each age and sex class differs in body size and horn length and males also emphasize their sex with such characters as pelage color, ruff length, and scrotum size. The last is a particularly conspicuous status symbol when a Punjab urial is seen from the rear. In fact a ram seems to have duplicated its effect in front: his white bib appears to mimic a scrotum in size, color, and shape. The scrotum of wild goat is greyish and not obvious. However, adult males have small and very white rump patches which re-

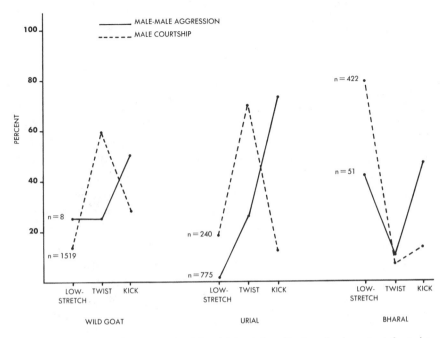

Fig. 42. Frequency (in percent) with which males display the low-stretch, twist, and kick toward males when aggressive and toward females during courtship.

semble scrota, whereas young males and females have diffuse rump patches. The possible relationship between scrota and rump patches as status symbols needs to be investigated. The cline in physical characters also represents a dominance hierarchy except that females and yearling males are often similar in appearance and unsettled in rank. Males prefer to associate mainly with males of about equal rank, and females and subadults tend to favor each other's company, with the result that the sexes tend to separate into different herds. One reason for such sexual segregation may be that females give unsatisfactory social responses to males for much of the year (Geist, 1971a). Females tend to withdraw when males display to them or invite them to spar. Only during estrus do females tolerate this behavior, and at that time they may even retaliate by butting. Such reversals in behavior prompted Geist (1971a) to conclude:

> The female sheep has two behavioral phases. Normally she behaves like a juvenile, but during estrus she acts like a subordinate, *young*

male. Conversely, one can claim that the young male acts like an estrous female, mimicking her behavior and appearance, which allows him to live side by side with larger males. The female is paedogenic, representing throughout life the immature form of the male. Except for old males all sheep are juveniles, some sexually mature, some not, at various stages of development.

While it is true that both estrous females and subordinate males may tolerate the attentions of a dominant male, the behavior of the latter is highly variable depending on the situation and on the species. Subordinate *Capra* males usually withdraw when threatened overtly, as they almost always are. Subordinate urial rams may accept the displays of dominant ones but they may also retreat or retaliate with kicks or twists. Since it is not "the subordinate's prerogative to act aggressively" in most species, as noted earlier, an estrous female cannot be said to act like a subordinate young male, or vice versa, except perhaps in mountain sheep. Geist's notion that the female represents an immature form of the male does little, to my mind, to help explain caprid society. It might be more accurate to say that the female represents the conservative physical type, reflecting the rupicaprid ancestry of caprids, and that males constitute physical and behavioral elaborations on this basic theme.

Adult males tend to be the most socially active members of caprid society, especially if some resource is in short supply—such as an estrous female. Males may initiate aggressive contact with any herd member, in contrast to females, yearling, and young which usually interact amongst themselves. However, males prefer to interact with males of their own horn-size class. Direct forms of aggression are evident in animals of both sexes since birth, whereas indirect forms are most evident in males, the patterns maturing slowly and usually not appearing until the yearling stage (Chapter 9). Adult males also court females more actively, or at least more successfully, than do the young males, as Grubb's (1974d), Geist's (1971a), and other studies have shown. This is of course to be expected, for a dominance-oriented society is based on the premise that the largest males have priority of access to females. Young males also court, often doing so vigorously, but they tend to be most active during the pre-rut when females are not yet in estrus (see fig. 41).

Aggression pervades caprid society like no other type of behavior, yet only a few species have friendly gestures which low-ranking males may use to appease and make contact with those of high status. *Ovis* males rub each other's faces and bharal rub each other's rumps with their muzzles. Geist (1971a) found that rubbing comprised 61% of all behavior patterns directed by subordinate Stone sheep rams at dominant ones. Since Geist and I quantified our data somewhat differently, direct comparisons of results are not possible. But taking all forms of direct aggression (table 36) as well as the indirect forms of mounting, low-stretching, twisting, and kicking, urial rams of a smaller horn-size class initiated 93 of these patterns toward males of a larger class, and they also initiated 50 rubs. Thus rubbing comprised 35% of these interactions. In bharal the percentage was 77%. Bharal use friendly gestures proportionately over twice as often as do Punjab urial. As noted earlier, bharal use mainly direct forms of aggression when interacting, in contrast to urial, which exhibit predominantly indirect ones. Frequent friendly overtures by subordinate bharal males may help to promote peace. But *Capra* males, which show proportionately even more direct aggression than do bharal, lack a friendly gesture. In societies in which males display mainly overt aggression one would expect subordinates to avoid dominants, for contact may be risky. Furthermore, such contact may be irrelevant. It may be that goats, having a more stable herd structure than sheep, know each other fairly well as individuals. Subservient greetings are of particular value when dealing with strangers whose responses are unpredictable. *Capra* society has almost discarded indirect aggression possibly because there is no selective advantage in using such behavior toward someone whose fighting potential is known. The situation is different in the fluid urial society. These sheep interact mainly with indirect forms of aggression, if clashing is excluded from consideration. Such a system allows subordinates to express themselves with relative impunity, for retaliation, if any, is likely to be mild. A friendly gesture is valuable in such a society, for not only is there much contact between strangers but also subordinate males need some behavior pattern which enables them to reaffirm their subordinate positions after having been aggressive. Bharal lie in the middle of an aggressive cline between *Capra* and *Ovis*. The relative proportions of indirect and direct aggression are more evenly balanced in bharal than in the other two genera with the result that subordinate males are presented with rather ambivalent situations in which the intensity of response is unpredictable. Males have a choice of either discarding friendly overtures or emphasizing them. They have chosen the latter.

Selection has decided on a broad theme for the functioning of caprid society: to court successfully is the prerogative of the largest and hence most dominant males. But the details of organizing each society have been left to each species. Striking differences exist in such matters as the relative frequencies of direct and indirect aggression, the responses of subordinate to dominant males, and the frequencies and kinds of displays directed by males toward females. Social and ecological selection have molded each species so intricately that the data collected so far can do little more than point to problems that need to be solved before the rules of conduct in caprid society can be understood. Just as poor-quality forage stunts a population, so does the lack of high-quality data hamper the further development of solid concepts and the adequate testing of current ideas.

EVOLUTION OF AGGRESSIVE DISPLAYS IN THE CAPRINAE

The intense period of mountain building in the Himalaya and elsewhere during the mid-Miocene and then during the late Pliocene and Pleistocene created new habitats and extended old ones. The Caprinae took advantage of the situation and radiated into many genera and species, some now extinct, as they occupied the newly available niches. The survivors range from small species without prominent display paraphernalia such as the goral to large impressive beasts with sweeping horns such as the ibex. Although each species represents the end of a lineage adapted to a certain environment, it is valid to infer that morphologically conservative forms may resemble the ancestral ones in appearance and behavior. This is especially true of species living in rain forests and other conservative habitats. A phylogenetic reconstruction would thus use the goral and serow as a basis for comparison with the more extravagantly fashioned Caprinae. That this approach is legitimate has been shown by the fossil record. The earliest Caprinae were goral-like creatures of the Miocene; the tahr-like *Tossunnoria* appeared during the Pliocene; and *Capra* and *Ovis* in their present form are no older than the late Pliocene and Pleistocene. When one attempts to construct evolutionary hypotheses by comparing the behavior of living species, when in effect one reconstructs the behavior of animals that died 8 million or more years ago, one must make three assumptions: (1) contemporaneous patterns were present in the common ancestors of the species; (2) a widespread pattern is more likely to be ancestral than one found in only a few species; (3) the least complex of several patterns is the most primitive.

The earliest Caprinae were probably small, plain, and essentially solitary forest dwellers much like the goral in appearance. Like the goral they may well have preferred slopes and rough terrain, thereby becoming preadapted to exploiting the new ecological opportunities in the mountains. But a small body restricts movement and a large surface-to-mass ratio limits tolerance to cold. Pioneer Caprinae that left the forest either for the open mountain slopes or the plains must have been under strong selective pressures to grow larger. Deprived of continuous protective cover and presented with abundant food resources, evolution favored a social existence for these pioneers. Herd life also increases genetic variability which then enables animals to adapt more rapidly to new conditions. The change from a solitary to a herd life may not have been drastic. If a male and one or more females occupy overlapping ranges and tolerate each other, as is the case among goral and serow, then lasting associations can be formed. At this stage some species remained conservative, and these were the ancestors of chamois and mountain goat. Others began to emphasize frontal combat and a display of status symbols, an evolutionary trend seen in many ruminant families, with the result that most social signals became concentrated and elaborated on the head (Geist, 1966). Morphologically conservative species fight quite differently than the others, as Walther (1961, 1974) has shown, and the main purpose here is to briefly compare some aggressive patterns of the Caprinae.

Present-day ungulates without horns or with small horns may provide an idea of the spectrum of fighting possibilities open to an ancestral caprid. Primitive deer and antelope usually fight by standing head-to-tail while attacking with the canines (Ralls, 1975). Fighting zebra circle in a reverse parallel position as they bite and push each other with the shoulders (Klingel, 1967). Mountain goats also assume this fighting position but they use their horns, not their teeth, to slash each other (Geist, 1965).

Young chamois may stand head-to-tail during combat, whereas adult ones usually confine themselves to chasing and hooking with the horns. I would expect goral and serow to fight much like mountain goats. Another form of direct aggression among ungulates with small horns is the neck-fight during which one individual places its neck over the neck or shoulder of another and pushes downward. I have observed this behavior in such diverse forms as nilgai and domestic camel. Krämer (1969) reported neck-fighting in young chamois, but the behavior has not been observed in other rupicaprids so far; however, P. Lent (pers. comm.) has seen behavoir reminiscent of neck-fighting in muskox young.

Lateral displays are widespread among ruminants, suggesting that they are phylogenetically old forms of intimidating opponents (see Walther, 1974). Mountain goats have a conspicuous head-down posture during which the back is humped as the animal presents its impressive profile (fig. 34). Chamois display in somewhat similar fashion, as do probably goral and serow. Another lateral display is the head-up during which the animal raises its muzzle, a gesture which makes the animal look tall and which in many species, including various antelope and deer (see Walther, 1974), exposes a white throat patch. The chamois displays a head-up, and since both goral and serow have white throats I would expect them to do so too.

Most horned animals thrash bushes and other objects during aggressive encounters, the behavior being particularly prominent in mountain goat and chamois because their post-cornual glands are located at the base of the horns. Aggressive pawing and stamping with a foreleg is another widespread ruminant trait found in both mountain goat and chamois.

Frontal fighting is not limited to ungulates with elaborate horns. Although horns act as defensive shields and hold combatants together as they twist and push (Geist, 1966), females and young with insignificant horns also bash heads, as do such small-horned species as Maxwell's duiker (Ralls, 1975). However, clashing with pointed spikes is dangerous, and rupicaprids make frontal contact seldom or not at all. The immediate precursors of the Caprini may well have had blunted horns, not pointed ones, to encourage the evolution of clashing. A miniature takin would serve as an admirable ancestor. It is significant that evolution in the Caprinae progressed from animals with deadly weapons to ones with rather ineffective weapons.

This brief overview of rupicaprid fighting shows that most combat and display techniques used by tribe members can be considered phylogenetically old ones which are widespread among ungulates. Some species have evolved some new methods of displaying, with, for example, chamois shaking themselves and mountain goats digging rutting pits, but these represent rather minor modifications of the basic repertoire.

Before the Caprini are discussed, two other tribes need comment. The Saigini became an independent branch so early that it remains unclear whether the animals belong to the Caprinae or Antilopinae. In any event, aggressive behavior of saiga and chiru has not yet been studied. The Ovibovini are essentially bulky rupicaprids which have evolved rather unsophisticated frontal methods of combat that in the case of muskox include vigorous head-on clashes. The lateral displays of both takin and muskox (plates 23, 24) resemble those of mountain goat.

The tahr represent morphological links between the rupicaprids and the advanced caprids, and in this respect their behavior has particular interest. The head-to-tail is a prominent form of combat in Nilgiri tahr and less so in the Himalayan tahr, the animals either just pushing each other or also hooking with their horns. Neck-fighting is rare in Himalayan tahr and it has as yet not been observed in Nilgiri tahr. Both species exhibit a prominent lateral display, a head-down posture which closely resembles the one of rupicaprids. Although tahr have retained these old combat methods, they have also adopted typical caprid forms of fighting and displaying. Being essentially goats, they may rear up and fall into a clash and they may also mouth their penises and spray urine on themselves (table 39). Thus tahr seem to be split personalities with respect to their aggressive behavior in that they fight and display like both rupicaprids and caprids.

Several phylogenetically old combat methods have been relegated to minor positions in most advanced Caprini. The head-to-tail is rare among *Ovis* and *Capra* (table 39), but it has been retained as a common pattern by *Ammotragus*. All Caprini except the tahr have almost replaced the head-to-tail with the shoulder-push, possibly because this fighting form enables combatants to clash horns and otherwise interact head-to-head as they shove. Neck-fighting is rare or absent in most Caprini, the mouflon and aoudad being notable exceptions. Lateral displays retain their importance among sheep, goats, and bharal but their emphasis has changed, having been transformed from patterns containing strong components of direct aggression to displays of indirect aggression. The head-down, so prominent a display in rupicaprids, takin, and tahr, seems slated for slow oblivion among the advanced caprids, for it remains important only in bharal. But the head-up and a few other lateral displays are still found in several species.

The low-stretch is a common posture among ungulates and the kick has a wide distribution among the Bovidae (see Walther, 1974), suggesting that these two gestures have been in the repertoire of hoofed stock for a long time. However, in most species the males exhibit the behavior only during courtship, not during interactions with other males. Mountain goat, muskox, and *Capra,* for example, kick while courting but not while fighting, except on extremely rare occasions. Sheep and bharal males differ from the others in that they often low-stretch and kick when threatening other males. The twist is commonly associated with the other two displays among the Caprini and it also occurs in muskox during courtship, but I have no information about its occurrence outside of the subfamily.

Some caprid genera have evolved distinctive modes of threatening. *Hemitragus* and *Capra* spray themselves with urine. Chamois and takin do too, but their methods of doing so differ from those of goats. The behavior probably evolved independently in the various tribes. All caprid genera except *Ovis* mouth their penises, and bharal also use the extended penis as a threat symbol.

The Caprinae have obviously been conservative in the evolution of their aggressive displays. A few species have discarded certain patterns or added new ones, but most have retained those of their rupicaprid past, sometimes only as vestiges and at other times in unchanged or slightly modified form. Frontal combat has become the evolutionary fashion in all except the rupicaprids, but the change has been one of emphasis rather than one of repertoire.

The taxonomic affinities of aoudad and bharal have been a matter of dispute, yet it may be possible to clarify relationships by comparing their behavior with that of sheep and goats. The bharal will be discussed in the next section but I would like to

comment briefly here on the aoudad. The aoudad is morphologically for the most part a goat but serologically a sheep (Chapter 2).
Males favor such primitive forms of combat as neck-fighting and
the head-to-tail. Goat-like, they insert 'their penises into their
mouths but do not urinate on themselves. They do not use the
jump as a form of threat, as do sheep and goats. Two combatants
may clash by running sheep-like at each other on all fours.
Aoudad are not as specialized as true goats, prominently displaying some primitive fighting styles, foregoing urine marking, and
clashing without first rearing up. On the basis of the behavioral
evidence I surmise that aoudad split from the goat stock shortly
after the sheep and goats diverged onto their separate evolutionary paths but before the ancestors of *Hemitragus* and *Capra* ac-
quired their specialized urine spraying and distinct method of
clashing.

Differences in aggressive behavior might also help clarify rela-
'tionships between species of the same genus. Among *Capra,* wild
goat and markhor were much alike not only in their aggressive
patterns but also in the frequencies with which these were displayed. The head-down was not definitely observed in markhor
and the head-to-tail seemed to be lacking in wild goat, to mention
two of several differences (table 39), but more work is certain to
extend the known repertoire of both species. Wild goat and markhor seem closely related and of recent evolutionary divergence
in spite of their different appearance. At least 8500 years of
domestication have affected the aggressive behavior of domestic
goats in only minor respects, the animals still resembling their
probable progenitor, the wild goat, in most aggressive patterns
(see Schaller and Laurie, 1974). The Asiatic and Alpine ibexes
lack the head-up, and in general their lateral displays seem vestigial and are rarely exhibited. Neck-fighting and the head-to-tail
have not been observed. Ibex have dispensed with most displays
and emphasized direct aggression. This species differs from the
wild goat and markhor more than the two differ from each other,
suggesting that the ibex have had a longer period of independent
evolution. It would be interesting to find out if the Spanish goat,
Nubian ibex, and Dagestan tur resemble the wild goat or the
Alpine and Asiatic ibexes most closely in behavior.

Among *Ovis,* neck-fighting is common in mouflon, vestigial in
mountain sheep, and seemingly absent in Punjab urial although

perhaps expressed only rarely. The head-to-tail has so far been reported only in mountain sheep and Soay sheep, and the head-down in Punjab urial. The mouflon is thought to be the ancestor of Soay sheep. The neck-fight is a prominent pattern in mouflon and at best vestigial in Soay sheep whereas the reverse is true of the head-to-tail. Mouflon readily jump as a form of threat but Soay sheep do not. Huddling has been reported in mouflon but Grubb (1974b) is uncertain whether this behavior occurs in Soay sheep. The reasons for such changes in the fighting repertoire of a subspecies or species remain unknown. Commenting on differences among sheep, Schaller and Mirza (1974) wrote: "Marco Polo sheep, and most likely all argalis, rear up before clashing, contrasting in this respect with other Eurasian forms but resembling the American sheep. Mountain sheep lack a submissive posture whereas Eurasian sheep have one. On the basis of available data, no clear behavioral demarcation between American and Eurasian members of the genus *Ovis* is evident."

Is the Bharal a Sheep or a Goat?

Bharal have the physical attributes of both sheep and goats, but taxonomists have generally held that they are aberrant goats with sheep-like affinities. However, as Geist (1971b) has pointed out, taxonomists use for classification mainly such structures as horns, which are under strong pressures of social evolution. Species living in similar habitats may evolve similar societies and somewhat similar structures without being closely related. When comparing bharal with sheep and goats, one needs to decide first which aspects of their behavior correlate with ecological conditions, and second which traits the bharal have in common with sheep or goats.

Capra prefer cliffs, a habitat choice which is reflected in the stocky build of the animals as well as in certain aspects of their behavior. Confined as they often are to isolated cliffs, herds tend to be more cohesive than those of sheep. Males spend much time with females even outside of the mating season, and during it they wander little. Tending to be acquainted with each other, males have little need to test an opponent's strength by indirect means and instead threaten overtly when necessary. With food, rest sites, and other resources on cliffs localized, females may not tolerate intrusions into their individual space, and they defend their young readily against conspecifics. Since a goat population tends to remain localized, a strong leader-follower relationship is not necessary. Besides, it may not be adaptive for all herd members to crowd onto the same ledges. Whatever the reason, Hafez and Scott (1962) noted that domestic goats are not as persistent followers as sheep. In contrast, Eurasian *Ovis* are lithely built animals, adapted to flat or rolling terrain. With their habitat extensive and food resources often patchy, herds tend to be fluid. The animals are good followers. Rams spend little time with ewes outside of the rut and they rove much during it. Ewes are rather passive creatures. Combating rams tend to be strangers, making indirect forms of aggression adaptive even though differential horn size relegates individuals to a certain rank. While observing domestic sheep and goats, Collias (1956) noted that sheep were less aggressive than goats but that differences were mostly quantitative. Habitat may thus have a great influence on the behavior of Eurasian sheep and goats, if my basic assumptions are valid. A simple ecological parameter cannot of course be used to categorize complex societies but it can serve as an obvious variable against which certain traits can be compared.

Snow sheep, American sheep, and bharal occupy a habitat which is neither as precipitous as that used by goats nor as flat as that preferred by Eurasian sheep. Although they escape to cliffs in times of danger, American-type sheep and bharal spend most of their time on nearby slopes. One would expect these species, which are ecologically similar, not only to behave much alike but also to be intermediate in some traits between *Capra* and the Eurasian *Ovis*. Adapted to a mountain life, bharal and American-type sheep are stockily built, like goats. Mountain sheep rams spend little time with ewes, even less time than do urial, and during the rut they roam widely. Indirect aggression is important in their society. Judging by Geist's (1971a) and Shackleton's (1973) accounts, mountain sheep ewes may be more aggressive toward each other and in defense of their young than are urial ewes. In general, mountain sheep have adapted physically to steep terrain, but in some basic aspects of their behavior they resemble their Eurasian relatives.

Some bharal populations appear to be quite cohesive, but this point needs verification. Males readily associate with females throughout the year, and in this respect bharal resemble goats.

Only a few males roam during the rut, behavior which in its intensity and degree is intermediate between sheep and goats. Bharal males threaten each other directly as well as indirectly. They threaten indirectly more commonly than do goats, and in this aspect of behavior bharal are also intermediate between *Ovis* and *Capra*. Female bharal are aggressive like goats. Bharal society obviously is goat-like but with some sheep-like traits. I would surmise that the ancestors of bharal evolved on cliffs but that they later occupied less precipitous terrain where new selection pressures modified their society somewhat in the direction characterized by sheep. By contrast, the immediate ancestors of mountain sheep probably evolved on relatively flat terrain and their society continues to reflect this. Having moved onto steep slopes, a change not as drastic as the ancestral bharal's venture away from cliffs, mountain sheep found that the new selection pressures required only minor adjustments by their society. If this was so, then selection operated much more drastically on mountain sheep morphology than on social behavior. Be that as it may, it seems that bharal and mountain sheep arrived in similar habitats with different ecological pasts and that their societies reflect both this past and the selective pressures of the present.

So far I have only compared the general behavior of sheep, goats, and bharal without discussing specific displays and other patterns. In regard to aggression, the three genera show only minor differences in their forms of fighting (table 39). Those differences that do exist usually reveal no definite pattern. For example, neck-fighting is rare in *Capra*, common in mouflon but not other sheep, and seemingly absent in bharal. There is a major difference in the way goats and sheep clash, as noted in the previous section. Goats rear bolt upright whereas sheep run at each other either on all fours or unbalanced on their hindlegs. Bharal fight like goats, although they may also run at each other like sheep. *Ovis*, *Capra*, and *Pseudois* share most indirect forms of aggression too. Shank (1972) stated that lateral displays are not present in sheep, but table 39 shows that they occur in *Pseudois* and *Capra* as well as in *Ovis*. Mounting, low-stretching, twisting, kicking, and several other displays are also part of the combat repertoire of goats, sheep, and bharal though used with widely varying frequencies. Goats may spray themselves with urine and they may mouth their penises, behavior not typical of sheep. Bharal also mouth their penises but they do not urinate on themselves. They also use their extended crimson penises as display gestures, an elaboration seemingly not practiced by goats. A behavioral cline exists, running from sheep, which merely unsheath their penises, to bharal, which unsheath and mouth them, to goats, which unsheath and mouth them and also urinate on themselves. If one function of a goat's beard is to serve as a sachet, then its absence in bharal is explainable. Only bharal (as well as Himalayan tahr) commonly use the humped approach, although bighorn sheep also show the posture in vestigial form.

The huddle, during which males cluster and interact rather randomly, occurs in sheep, except possibly in domestic sheep, and in bharal but not in goats. Such behavior would not seem well adapted to precipitous terrain. However, on one occaison I observed a free-for-all in Asiatic ibex during which males in a seemingly playful manner clashed horns often but not hard and without regard for the opponent's size. The animals were on an alpine meadow, not a cliff. This may represent the behavioral equivalent of a huddle.

Capra males dispense with indirect forms of aggression when dealing with subordinates, and these retreat or behave innocuously without assuming a specific submissive posture. Mountain sheep seem to lack a submissive posture too, but Eurasian sheep, Himalayan tahr, and bharal assume a characteristic neck-low posture. Subordinate sheep and bharal males, but not goats, rub each other as a friendly gesture, and such behavior can be correlated with frequency of indirect aggression and social fluidity, as noted in the previous section. Bharal are more like sheep than goats in their submissive behavior.

Sheep, goats, and bharal test females for estrus and court them in much the same way. There are variations in the frequencies of some displays, but since single-gene polymorphisms and maturational factors might be sufficient to produce such changes they are not useful indicators of taxonomic affinities. By virtue of their anal glands, *Capra* use odor more than do *Ovis* during courtship. Rutting male goats typically walk around with their tails folded up over their rumps, whereas sheep seldom raise their tails above the horizontal. Bharal carry their tails vertically. I do not know if bharal have anal glands, but the fact that subordinates rub the rumps of dominants suggests that glands may be there. One possible function of face rubbing in sheep is for a low-ranking individual to acquire the scent of a high-ranking one from the suborbital glands. Lacking suborbital glands, bharal may have oriented their rubbing toward some glandular area near the anus instead.

When signaling the presence of danger, *Capra* may hit their forelegs in unison on the ground, *Ovis* may thump all four feet, and *Pseudois* was once observed to hit all feet too, behaving in this respect like a sheep. Other antipredator responses show no generic distinctiveness (Chapter 7).

Capra and *Hemitragus* lick themselves more often than they scratch with the hindlegs, and in *Ovis* the reverse is true. Bharal resemble goats in their grooming frequencies (table 25).

What then can be deduced about the taxonomic affinity of bharal on the basis of their behavior patterns? It depends partly on how much emphasis is placed on the various differences. Two behavior patterns—the raised tail and the bounding gait—are contingent on the presence or absence of certain glands. The Caprinae are strikingly variable in their choice of glands, and I am uncertain how useful glands are as indicators of affinity. The raised tail of bharal resembles that of goats and the thumping gait is similar to that of sheep. Both bharal and sheep rub as a friendly gesture. The part of the body they rub is divergent but the evolution of the behavior seems convergent. The huddle of bharal is sheep-like, except that animals stand at odd angles rather than in a neat circle, and I feel that this behavior is also convergent, both habitat and society having favored some means by which males can interact informally. Bharal mouth their penises like goats but they do not soak themselves with urine, behaving as if they were in an evolutionary stage prior to the one reached by *Capra*. Bharal clash like goats but selection for a specific fighting style has been less rigid away from the cliffs, with the result that some sheep-like patterns have appeared.

In general, the behavioral evidence confirms the morphological evidence that bharal are basically goats. Many of the sheep-like traits of bharal can be ascribed to convergent evolution, the result of the species having settled in a habitat which is usually occupied by sheep. But bharal do more than just show that a presumed change in habitat may modify behavior. Although bharal have clearly adapted to a mountain life, they remain rather generalized. Their horns are short, unembellished bashing instruments not quite like those of sheep or of goats. Their glands are either tenuously present or entirely absent. In their use of the penis the bharal also show an evolutionary hesitation to specialize beyond displaying it as a status symbol. The species has straddled an evolutionary fence, and if it had to make a choice of whether to become an *Ovis* or a *Capra* it could become either one with only minor alterations. Like the aoudad, the bharal probably split early from the ancestral goat stock. If I had to design a hypothetical precursor from which the sheep and goat line diverged, it would in many ways resemble a bharal in appearance and behavior. That such an animal still survives in the probable evolutionary center of many alpine vertebrates is suggestive.

EVOLUTION OF THE CAPRINAE

The Miocene in Asia was a period of geological unrest when mountains rose and the Tethys Sea retreated. After this turbulent epoch came the serene Pliocene. With a climate that was warmer than today, vast steppes covered Asia in what is now either desert or tundra, and on these grassy plains thrived a rich fauna including *Hipparion*, many antelopes, and several bovids related to muskoxen (Kurtén, 1972). Then, late in the Pliocene and in the Pleistocene, the earth once again buckled and climates cooled, turning the huge tract into wastes. As the grasslands in Central Asia became increasingly arid during the late Pliocene, as the climax vegetation shifted, some species became extinct. But others adapted to the changes, some like the muskox moving north to become tundra animals and others such as the saiga, Przewalski's horse, and yak retreating southward to the upland steppes (see Guthrie, 1968). The Caprinae seem to have had two major spurts in evolution, both during periods of mountain building. One was during the late Miocene and early Pliocene when members of the sub-family spread over Eurasia and probably parts of Africa, venturing from the forests and in the process becoming markedly larger; the other was in the late Pliocene and Pleistocene, a time which witnessed a remarkable radiation of mountain forms. Some of the new Pleistocene species differed from their ancestors particularly in the length and massiveness of their horns. The evolutionary paths of today's species are difficult to trace not only because the fossil record in mountains is poor but also because environmental change places such a premium on adaptability that speciation may have proceded so rapidly that no recognizable intermediate forms were left behind.

Geist (1974b), in what he termed his dispersal theory, postulated how such rapid evolution could occur.

> After glacial withdrawal ungulates began to colonize the vacant habitat of the once glaciated zones. . . . Until the population reached carrying capacity, the individuals would be confronted by a super-abundance of forage to which they would respond in a predictable fashion. Birthrates, birth weights and milk production would increase. Consequently neonatal survival and the growth rates of young would go up. The young would mature early and reach an adult size close to their genetic maximum. . . . Under such conditions large-bodied individuals as well as those with larger horns or antlers, which improved combat techniques . . . would have the advantage . . . . Large body and horn size can be achieved by increasing the intensity and duration of growth. This can be done by enlarging the physiological mechanisms characteristic of juveniles. If this happened then ungulates should become not only larger, but also more juvenile-like during early post-glacial evolution . . . . It could be shown that in all parameters tested the behaviour of the advanced bighorn sheep . . . was more juvenile-like than that of the more primitive Stone's sheep.

Geist (1974a) also noted that:

> the further the species disperse, the greater the difference between individuals of the parent population and the population of the fringe. Since selection for neotony comes to an end when the habitat is filled to carrying capacity, the gene flow is broken by a fragmentation of the species into isolated populations (see Geist, 1971), and individuals from the populations between parent population and the colonizing fringe will form a cline.

Geist's theory needs rigorous testing before it can be accepted in its present form, and with my limited data I can do little more than make some queries and point to some possible weaknesses. The response of individuals to good food is well known, and, given a species' inherent plasticity with respect to size, populations penetrating wholly new terrain might produce larger animals—if the food supply was unlimited. But how much new habitat is available to a species once it has adapted to certain conditions? Animals in the mountains would tend to follow retreating glaciers at the limit of plant growth, as alpine species do in Asia today,

filling the habitat to capacity before it has the opportunity to grow
a superabundance of forage. Animals may disperse into and fill a
habitat so rapidly that genetic and phenotypic changes may not
have time to appear or to assert themselves. For example, chamois
after being introduced into New Zealand dispersed at the rate of
8.6 km per year and Himalayan tahr at 1.8 km (Caughley, 1963),
and aoudad in Texas dispersed at the rate of 3.2 km per year
(Evans, 1967). Given good forage, animals would be larger, but
their horns would become only relatively, not absolutely, larger.
As tables 8 and 10 show, the horns of sheep species living in
formerly glaciated terrain are relatively and absolutely no greater
in length in most instances than those of their close relatives inhabiting other areas, and the same applies to goats. Genetic
changes favoring absolutely larger horns could occur in populations which have reached phenotypically their largest size on good
habitats, and this obviously has occurred in the evolution of the
caprids. However, it remains unclear whether this occurred rarely
or repeatedly in the history of the Caprini. Geist's theory
may lay too much stress on social selection at the expense
of natural selection. For example, body size is not just influenced
by the food supply but also by the ruggedness of the terrain, snow
depth, predator pressure, the optimum metabolism for a particular climate, and so forth. Once individuals found it selectively
advantageous to become dominance-oriented, they had a choice
of emphasizing horns or other features of the body such as prominent humps or hairy mantles. The Caprini favored one alternative and the Bovini the other, and it seems likely that environmental pressures had an influence on those choices.

On the basis of chromosone studies, Nadler et al. (1974)
suggested a brief scenario of caprid evolution.

> The cytogenetic evidence suggests there are two main lineages
> among surviving caprines. Starting from a hypothetical
> rupicaprine-type ancestor with a primitive 2n-60, FN=60
> karyotype, one lineage evolved through an intermediate aoudad-
> like form to the true sheep, with reductions in diploid number. In
> the true goat (*Capra*) lineage, morphological differentiation pro-
> ceeded while the karyotype remained morphologically close to the
> hypothetical rupicaprine ancestor while its chromosome number
> has been reduced (2n=48). The 5th member of the tribe Caprini,
> the bharal (*Pseudoid nayaur*) has a reduced chromosome number

(2n=54) but morphologically exhibits convergence toward true sheep [see fig. 43 in this book].

Tahr, aoudad, and bharal have been discussed in previous sections but *Capra* and *Ovis* need additional comment, especially with reference to Geist's (1971a) dispersal theory. Both sheep and goats evolved into their present forms during the turmoil of the ice age and both probably had the center of their evolution in the mountains of Asia where alpine ungulates reached and maintained their greatest diversity. But after that the history of *Capra* becomes obscure. Three evolutionary lines survived—the ibexes, the Dagestan tur, and the wild goat-markhor, with the Spanish goat probably belonging to the ibex line. Geist (1971a), noting that the Asiatic ibex has the largest body and longest horns of all ibexes, suggested that it is the most advanced race, it having "penetrated deepest along once glaciated mountain ranges into Central Asia." It does indeed seem likely that the ibexes are more advanced than the other *Capra* lineages, but relative and absolute horn lengths of the ibexes show little correlation with penetration into formerly glaciated terrain in Europe, the Caucasus, and in central Asia (table 8).

The precursors of *Ovis* and *Capra* presumably evolved together, the former adapting to rolling terrain and the latter to steep slopes and cliffs. Because mountain systems were continuous toward Europe, both sheep and goats could expand westward from Asia. But a lack of suitable habitat halted the advance of goats toward the northeast, toward Siberia. However, sheep could penetrate plains, and a probable ancestor of the argalis and urials crossed the Central Beringian and Mongolian plateaus to the ranges beyond. Finding an empty habitat, one usually occupied by goats in other Eurasian mountains, these pioneering sheep moved onto the steep slopes and with time became isolated in their mountain refugia as a change in climate late in the Pliocene eliminated the good grazing on the steppes and either exterminated plains-living sheep or forced them to retreat southward. The Siberian sheep evolved distinctive features, becoming the snow sheep, and probably during the Illinoian glacial period spread into North America together with the mountain goat. Blocked from penetrating southward by glaciers, they remained in their Alaska-Yukon refugium until during the Sangamon interglacial they could perhaps expand their range again. The Wisconsin glaciation divided the population which then evolved into thinhorn and bighorn sheep (Cowan, 1940).

The evolution of urials and argalis is more complex than that of mountain sheep. Nasonov (in Sushkin, 1925) arranged sheep into two evolutionary lines of which the Armenian and Cyprian urials were considered to be the most primitive. One line supposedly led through Kopet Dagh urial to Marco Polo sheep, and the other through Ladak urial and Altai argali to mountain sheep. Geist (1971a) arranged the Eurasian sheep into an "*ammon* cline" starting with aoudad, continuing through western and eastern urials, and ending with *polii, karelini,* and *ammon* argalis, in that order. Armenian urial were thought to be the most primitive and Altai argali the most advanced sheep. The ruffs, small rump patches, and light horns of urials were felt to be primitive characters, and the lack of ruffs, large rump patches, and heavy horns advanced characters. He further noted:

> All the mountains inhabited by the argalis today were then probably under ice (Frenzel, 1968). Since the glaciations in central Asia were restricted due to insufficient moisture, it appears that large cold deserts surrounded glaciated mountains, making it unlikely that sheep survived there. Asiatic sheep for all their plains adaptations are restricted to mountainous terrain and grasslands, not plains and deserts. This hypothesis implies that the sheep we find today in central Asia are of rather recent evolutionary origin, which surged from unglaciated terrain into the mountain ranges after glacial withdrawal.

Geist realized that his supposed cline presented some problems when he noted that "for Asiatic sheep this view cannot be verified or rejected at present." *Ammotragus* is morphologically too much of a goat to belong in a sheep cline, although an unspecialized ancestor of aoudad or bharal might properly be placed there. The western urials have smaller ruffs and larger rump patches than the eastern ones, the reverse of the expected trend. The ruffs of urials and argalis have quite different growth patterns, suggesting convergent evolution rather than a cline. Furthermore, the Altai, the uplands of Tibet, and most other ranges in central Asia were not wholly "under ice." While glaciers were more extensive during the Pleistocene than at present, the uplands were so arid that vast glaciers, such as those found in Europe and North America, could not develop. Even at the height of the glaciations, huge upland tracts remained free of ice (Frenzel, 1968; Kukla, 1976) and these probably provided refugia for various vertebrates, many of which survive in the area today. The ancestors of the argalis probably inhabited the region throughout the Pleistocene, shifting only as needed with the changing climate. More precipitation may actually have made the Pleistocene pastures more equitable than those of today.

Nadler et al. (1973) suggested yet another pattern of Eurasian sheep dispersal. The western urials have 2n=54 chromosones, the eastern ones 2n=58, and the argalis 2n=56. Since Robertsonian fusion commonly reduces chromosome numbers (see Chapter 2), the species with the largest number is thought to possess the primitive complement. On this basis the eastern urials would be most primitive. "We think that evolution of the sheep in the regions west of Iran and Afghanistan was in the direction of decreasing size, becoming mouflon-like and developing into a 2n=54 chromosome form, whereas in the east, in the high mountains of Central Asia and southern Siberia, sheep evolved larger body and horn size of the arkhar/argali with 2n=56 chromosomes" (Nadler et al., 1973). These authors then suggested two possible patterns of chromosome dispersal: (1) The eastern urials developed a polymorphism which included diploid numbers of 54 to 58, the characteristic forms spreading east and west; and (2) the ancestral 2n=58 population developed 2n=54 and 56 populations in geographic isolation, and each then dispersed. However, chromosome numbers by themselves may have limited use in plotting dispersal patterns. Some snow sheep are 2n=52 whereas their descendants in America are 2n=54 (Korobitsyna et al., 1974), the reverse of the expected trend.

There is little evidence to support the idea that urials are more primitive than argalis and that the latter are merely neotonous urials, as Geist (1971a) phrased it. A case can be made for argalis having evolved first. Upper Pliocene fossils of argali-type sheep have been found in China, and mid-Pleistocene ones in the Caucasus (Chapter 3), indicating an early and wide dispersal of large sheep. Herre and Kesper (1953) suggested that these sheep gave rise to mouflon. Urials are morphologically more variable

than argalis, they have not stabilized phenotypically and genotypi-
cally, and this could indicate that they have evolved recently.

The probable relationships among the Caprinae are outlined in
figure 43, which lists only living species, not the many extinct
ones. Published sources do not agree on when the various
evolutionary lines diverged. For example, Pilgrim (1947) postulated that *Ovis* separated from the goats during the mid-
Miocene, whereas Payne (1968) felt that the radiation may have

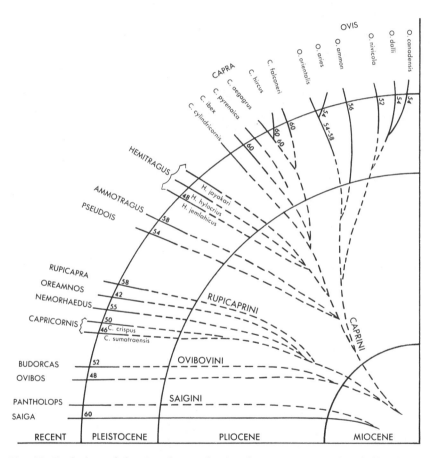

Fig. 43. Evolution of the Caprinae. The 2n chromosome number is listed with
each species. Chromosome data are based mainly on Wurster and Benirschke (1968), Benirschke et al. (1972), Nadler et al. (1971, 1973), and
Bogart and Benirschke (1975).

occurred as late as the Pleistocene. Thenius and Hofer (1960) present the best general outline of Caprinae evolution, and figure 43 is partly based on their work.

CONSERVATION

Biological evolution created complexity, but one of its products was the king of ashes, as Harry Martinsson called man, who has reduced the earth's diversity. In the vast Himalayan region the signs of human existence seem tiny in relation to the bleak sea of ridges and crags, yet man has had a devastating effect, as I have tried to document in Chapters 1 and 3 and other parts of this report. Forests have turned into timber and firewood, steep slopes have become terraced fields, and the grass has vanished into live-stock. Wind and rain then worry the soil until it erodes away, leaving the earth's bones bare. With livestock occupying its ranges and man questing after its meat, wildlife has retreated into remote and inhospitable terrain where in many areas the remnants are making a last stand. Man now realizes that in an effort to master the environment he has tampered with it too much, that he has been using the capital of his finite resources rather than the in-terest. In recent years he has defined the moral imperatives of a land ethic and in general established a design for coexistence with the natural world. But to translate philosophical propositions into acceptable ecological guidelines is a monumental task. Aside from the fact that pleas for moral enlightenment have seldom deterred greed, the needs of rapidly growing populations who are wholly dependent on a few sparse resources are difficult to reconcile with conservation practices. Although it may seem simple to preserve wildlife in some mountain valley, the traditional grazing rights of the local population must also be considered, as must the fact that the market value of an ibex or urial is such that an inhabitant can hardly afford not to hunt (see Roberts, 1973). Yet a tenuous harmony between the needs of man and the other species must be achieved. The knowledge needed for this is for the most part available; it must now be applied in its biological and social contexts. No single blueprint can help to rehabilitate an area as politically and ecologically complex as the Himalaya, but I would like to make two points.

The earth is remarkably resilient, and habitats can ultimately recover unless their constituent species have been extirpated.

Plate 1. Typical Himalayan tahr habitat along the upper Bhota Kosi River in Nepal. Serow and goral also occur along the canyon (March, 1972).

Plate 2. The High Range in south India as viewed from Anamudi Peak. Nilgiri tahr inhabit the cliffs in the foreground (October, 1969).

Plate 3. The summer range of the hangul in the Dachigam Sanctuary of Kashmir at 3400 m consists of alpine meadows and patches of conifers and birch. Ibex inhabit the cliffs beyond (October, 1968).

Plate 4. In the Chitral Gol, stands of evergreen oak and conifers grow on the precipitous slopes in the winter range of the Kashmir markhor (December, 1972).

Plate 5. Astor markhor habitat on the lower slopes of Nanga Parbat mountain near Chilas, Gilgit Agency. A village clings to the slope, and in the foreground is a pile of markhor horns (June, 1973).

Plate 6. Downstream from Skardu the Indus River cuts between the Karakorâm and Great Himalayan ranges. The lower treeless slopes represent Astor markhor habitat, the upper ones Asiatic ibex habitat (June, 1973).

Plate 7. Asiatic ibex habitat at 4700 m near Dorah Pass in the Hindu Kush, northwestern Chitral (July, 1972).

Plate 8. The porters of a mountaineering expedition ascend the Braldo Valley at 3200 m near the Baltoro Glacier in Baltistan, Gilgit Agency. Ladak urial inhabit the lower slopes and Asiatic ibex the upper ones (May, 1975).

Plate 9. Extensive glaciers cover the Hindu Raj Range in Chitral at 4500 m. My yak caravan approaches Darkot Pass (July, 1973).

Plate 10. The valley of the Yarkhun River in the Hindu Kush Range of north-eastern Chitral is barren except where irrigated. The photo shows an oasis of crops and fruit trees. Tirich Mir, 7695 m high, towers in the distance (July, 1973).

Plate 11. The southern part of the Chiltan Range near Quetta, showing the principal habitat of the Chiltan wild goat. Forest covered the slopes until this century (November, 1970).

Plate 12. The Gadabar Ghar is typical of the desolate mountain ranges in Baluchistan. Both wild goat and straight-horned markhor are said to inhabit this range. One of the graves in the foreground is marked with a markhor horn (April, 1973).

Plate 13. Andrew Laurie scans the cliffs of the Karchat Hills in Sind for wild goat. This area is now a part of Kirthar National Park (September, 1972).

Plate 14. The Punjab urial reserve at Kalabagh in the Salt Range consists of sparsely wooded hills deeply dissected by ravines (November, 1972).

Plate 15. The rolling uplands at 5000 m near Khunjerab Pass on the Hunza-Sinkiang border are a favored habitat of Marco Polo sheep. A Marco Polo sheep skull lies in the foreground. This area is now part of Khunjerab National Park (November, 1974).

Plate 16. The rolling hills in the upper Namdo Valley of the Dolpo District in Nepal is typical Tibetan argali habitat. *Lonicera* shrubs cover the slopes. My camp is at 4600 m (December, 1973).

Plate 17. The Kanjiroba Range near Phoksumdo Lake in Dolpo District, Nepal. Musk deer live in the birch and conifer stands and bharal occupy the slopes above timberline (October, 1973).

Plate 18. The Shey monastery lying at 4500 m in Dolpo District, Nepal, was our home base during the bharal study. Bharal at that season concentrate on snow-free slopes, such as the one in the foreground (November, 1973).

Plate 19. A bharal herd forages near Shey in Dolpo District, Nepal. *Lonicera* and *Caragana* shrubs grow on the slopes, as do dark clumps of procumbent junipers (November, 1973).

Plate 20. Bharal habitat at 4200 m near Lapche Monastery in northeastern Nepal. Tingri Himal in Tibet rises along the skyline and a domestic yak forages in the foreground (March, 1972).

Plate 21. An adult female mountain goat at the Alberta Game Farm shows the stiletto-like horns typical of rupicaprids. Note the dark post-cornual gland at the base of the horn.

Plate 22. A female Burmese gray goral. (Photograph courtesy of the San Diego Zoological Society.)

Plate 23. A male muskox in the Calgary Zoo stands in a lateral display, showing off his impressive profile.

Plate 24. A female takin in the New York Zoological Park presents the head-down display with her head slightly averted.

Plate 25. A young adult male Himalayan tahr stands on a grass-covered ledge along the upper Bhota Kosi River of Nepal in March.

Plate 26. Adult female and yearling Nilgiri tahr in the High Range of India line up to watch their photographer.

Plate 27. An adult male Nilgiri tahr. His gray saddle and the white spots above the knees are clearly evident.

Plate 28. A hangul stag in the Dachigam Sanctuary during the October rut.

Plate 29. An adult male Kashmir markhor in the Chitral Gol. Ahead of him is a female and tagging behind are her twins (December, 1970).

Plate 30. A tame but free-ranging Kashmir markhor female in Chitral resting with one foreleg extended.

Plate 31. A male straight-horned markhor in the Toba-Kakar Ranges of Baluchistan. (Photograph by S. A. Khan.)

Plate 32. A male Alpine ibex. Note the short beard and small rump patch. (Photograph by Frederick Stoever.)

Plate 33. A mixed herd of wild goat in Kirthar National Park during March. The males are in their inconspicuous summer coats.

Plate 34. Domestic goats may browse on low-hanging foliage while balancing on their hindlegs.

Plate 35. A 2½-year-old Stone sheep ram lip-curls in the Calgary Zoo. His stocky build is typical of all American sheep. Note the large white rump patch.

Plate 36. A newborn Punjab urial young lies motionless after being ear-tagged.

Plate 37. A 5½-year-old Punjab urial ram. Note his white bib and pointed goat-like ears.

Plate 38. A herd of Punjab urial rams in April after they have shed their ruffs and are in their summer coats.

Plate 39. Several Punjab urial rams and one ewe in December. All rams sport conspicuous throat ruffs.

Plate 40. Two Marco Polo sheep ewes and a yearling ram, illustrating the typical body shape of argalis. (Photograph courtesy of New York Zoological Society.)

Plate 41. A male aoudad. Note long tail, lack of rump patch, long ruff, and hairy chaps. (Photograph courtesy of the New York Zoological Society.)

Plate 42. A yearling male chinkara in the Kalabagh Reserve.

Plate 43. A mixed bharal herd during the pre-rut at Shey.

Plate 44. A bharal male approaches a female in the low-stretch display.

Plate 45. Four bharal males stand in line, those behind rubbing the rumps of the ones in front.

Plate 46. A bharal male extends and nibbles his penis while standing broadside to another male, who averts his head. A young male rubs the rump of the latter.

Plate 47. Two bharal males rear and fall into a clash. Note how the males have their heads turned to catch the blow between the horns.

Plate 48. A young bharal male rears up in threat before an adult male who stands with humped back and unsheathed penis while flicking his tongue. On the right, one male rubs the rump of another.

Plate 49. A threatening male Alpine ibex rears before another. (Photograph by Frederick Stoever.)

Plate 50. A female snow leopard in the Chitral Gol Reserve.

Plate 51. A partially concealed snow leopard stalks toward its kill, apprehensive because a human observer is nearby.

Plate 52. Brown bear cubs are captured and trained to dance by itinerant entertainers, a practice which has seriously decimated the species in Pakistan.

With many uplands already so seriously modified that the future of some plants and animals is in jeopardy, it is essential that a network of relatively unmodified areas be preserved as national parks and reserves. Some day man may want to rebuild what he has squandered, and from such samples of original habitat he can then not only draw genetic stock but also learn how the ecological pieces have adjusted to create a harmonious system. "The day is rapidly approaching when the remnants of the natural environment will be contained in a patchwork of parks and reserves. Much of the world's biological endowment will then be locked into insular refugia that are surrounded by an inhospitable landscape, through which dispersal to the next refuge is slow or non-existent" (Terborgh, 1974). The emphasis of most conservation efforts has been on individual species rather than on whole communities. Recent work has shown that this approach may be inadequate. MacArthur and Wilson (1963) developed an important idea which in brief states that small islands and islands far from others support relatively fewer species and that extinction rates on such islands are remarkably high. This concept has been expanded to include isolated habitats such as those on mountains. Large species are particularly prone to extinction, when limited to small areas, because of their intrinsically low population densities. "These should be sobering thoughts for conservationists who would like to save most or all the species of the world in a representative set of sanctuaries. Noah's Ark can offer only very short-term relief—species will not be saved two by two. After forty days and forty nights a sizable proportion would become extinct by chance" (Brockelman, 1976). Large reserves are obviously needed to preserve mountain ungulates, especially those which require vertical migrations between their summer and winter ranges. The same applies to large predators. Sullivan and Shaffer (1975) estimated that in the western United States a minimum of 600 to 760 sq km are needed to maintain just one pack of wolves or eight puma. Few mountain reserves in South Asia are this large, the Kirthar and Khunjerab National parks in Pakistan, the Nanda Devi Sanctuary in India, and the Mount Everest National Park in Nepal being notable exceptions.

Reserves can only preserve fragments of habitat. In addition, countries need to examine their whole land-use pattern and to develop integrated programs of forest, range, and wildlife management. In some areas it would be possible to establish management units under the control of local village councils. With time it might be possible to limit the amount of livestock to the carrying capacity of the land, and such animals as ibex, urial, and wild goat might be allowed to increase in certain parts of each management unit until they were abundant enough to be cropped for meat and trophies (see Schaller, 1976a). Programs of restoration require surveys of the remnants. Simple criteria must be found by which untrained persons can evaluate local conditions. For example, among urial and wild goat the number of multiple births are useful indicators of environmental quality. I hope that some of the facts in this report may be of value in future restoration programs, but above all I hope to convey the realization that the mountain habitats have been rapidly deteriorating and that an effort is needed to halt this trend and to restore the quality of life for man and beast. The prognosis for the future is not all bleak, for governments are now concerned about the denudations of watersheds and the decline in the productivity of land. Some reserves have been created, and ordinances prohibit the indiscriminate killing of wildlife and cutting of trees in most areas, although the enforcement of regulations has not been notably vigorous.

In writing about this vanishing mountain fauna I feel a special urgency. All too often in history the last of a species has disappeared into the belly of a hungry hunter, its epitaph a belch. Perhaps my effort to accumulate knowledge about certain species will merely enable someone to compose learned obituaries about them at a future date. I am reminded of Hudson's (1892) words:

> It is hardly to be supposed or hoped that posterity will feel satisfied with our monographs of extinct species, and the few crumbling bones and faded feathers, which may possibly survive half a dozen centuries in some happily-placed museum. On the contrary, such dreary mementoes will only serve to remind them of their loss; and if they remember us at all, it will only be to hate our memory—this enlightened, scientific, humanitarian age, which should have for a motto "Let us slay all noble and beautiful things for tomorrow we die."

The high altitudes are a special world. Born of the Pleistocene, at home among pulsating glaciers and wind-flayed rocks, the animals have survived and thrived, the harshness of the environment breeding a strength and resilience which the lowland animals often lack. At these heights, in this remote universe of stone and sky, the fauna and flora of the Pleistocene have endured while many species of lower realms have vanished in the uproar of the elements. Just as we become aware of this hidden splendor of the past, we are in danger of denying it to the future. As we reach for the stars we neglect the flowers at our feet. But the great age of mammals in the Himalaya need not be over unless we permit it to be. For epochs to come the peaks will still pierce the lonely vistas, but when the last snow leopard has stalked among the crags and the last markhor has stood on a promontory, his ruff waving in the breeze, a spark of life will have gone, turning the mountains into stones of silence.

ACKNOWLEDGMENTS

Several institutions and many individuals made this study possible, and I would like to express my gratitude to them all. The project was financed by the New York Zoological Society and National Geographic Society and it was conducted under the auspices of the Center for Field Biology and Conservation of the New York Zoological Society. Without the enthusiastic backing and help of William G. Conway, the general director of the New York Zoological Society, this project would not have been possible. To convey my gratitude for his support, as well as to express my admiration for his efforts on behalf of conservation, I have dedicated this book to him. My research in Pakistan was done under the local auspices of World Wildlife Fund Pakistan, and the generous support of Syed Babar Ali, the secretary of that organization, is particularly acknowledged. For permission to work in Nepal I am indebted to His Majesty's government.

Several individuals accompanied me on trips and collaborated in collecting data, and the measure of help of some of these is evident from the list of joint publications in the reference section of this report. A few words of thanks cannot do justice to the assistance given by Major S. Amanullah Khan, the first administrator of World Wildlife Fund Pakistan. Intermittently over a period of four years we travelled together from one end of Pakistan to the other, on foot, by horse, and by car, enduring hardships and sharing joys. His help was essential while making wildlife surveys, for by his congenial nature and fluency in local languages we were able to overcome the natural distrust of remote inhabitants toward outsiders. Without his constant assistance and cheerful interest my work in Pakistan would have been far more difficult. Because he is one of the most knowledgeable persons about wildlife in Pakistan, his observations became useful additions to my notes. Pervez A. Khan accompanied me on three long

339

trips through the Hindu Kush and Karakorams. A dedicated and experienced outdoorsman and mountaineer, he smoothed our way with recalcitrant porters and suspicious officials, and he was a fine friend and companion during our arduous treks. Zahid Beg Mirza was a valuable collaborator for nearly three months in 1970, and as the current administrator of World Wildlife Fund Pakistan he is continuing his wildlife work. Andrew Laurie enthusiastically joined me for a month to observe wild goats. Peter Matthiessen was a stimulating companion during our journey to northern Nepal to study bharal. And Melvin Sunquist was a most capable collaborator for nearly two months while we sought the elusive snow leopard of Chitral.

Everywhere in Pakistan, government officials and private individuals made extra efforts to help me, and their hospitality remains among my most pleasant memories of the country. In Chitral the royal family assisted me in many ways, and I am most indebted to H. H. Saif-ul-Mulk and to Shazadas Asad-ur-Rehman and Burhan-ud-Din. I was the guest of Malik Muzaffar Khan, the Nawab of Kalabagh, during my urial studies, and a few words are insufficient to express by gratitude to him and his staff. W. A. Kermani, Secretary of Forests of the government of Sind, generously supported my research on the wild goat, and it was in part through his farsighted efforts that Kirthar National Park became a reality. I owe a special debt to S. M. H. Rizvi who first encouraged me to study in Sind and whose splendid hospitality and help made my work in the area most enjoyable. No private citizen in Sind works more actively to conserve wildlife than does S. M. H. Rizvi. T. J. Roberts knows the fauna of Pakistan better than anyone. It was he who first aroused my interest in that country and subsequently provided constant encouragement, hospitality, and assistance. For help in various ways I am also indebted to Mumtaz Bhutto, K. M. Khan, Hamid Ali, Ghaus M. Khattak, Mirza M. Hassan, M. Ashraf, Ikramullah Khan, Syed Asad Ali, Sher Panah, Raja Hussain Ali Khan, Raja Feroz Khan Soofi, C. D. Savage, M. Afzal, Imdad Ali Shah, J. Eckert, Ghulam M. Beg, Husain Mahmud, Husain Mehdi Kazmi, I. Grimwood, and many others.

Rashid Wani generously assisted me in Kashmir, and one hangul census was done in collaboration with C. Holloway. In South India the hospitality and dedicated help of E. R. C. Davidar and J. C. Gouldsbury enabled me to observe Nilgiri tahr.

For generous assistance in Nepal I am particularly indebted to John Blower, who suggested several study areas to me. I would also like to thank the sherpas, especially Phu-Tsering and Jangbo, who ably and cheerfully accompanied me on my travels. M. Cheney and R. Fleming, among others, provided local help and hospitality.

For making my trips into New Zealand tahr and chamois country possible I would like to thank K. Tustin and C. Clarke of the New Zealand Forest Service. R. Valdez generously accompanied me into the field in Iran.

Joanne Hess of the National Geographic Society provided constant encouragement in my efforts to film wildlife, and R. Altemus ably edited some of my footage into a short film entitled *The Blue Sheep of Shey Gompa*.

Several individuals kindly identified specimens. G. Paine and D. Bankov, both formerly at the Veterinary Research Institute in Lahore, identified internal parasites, and M. Kaiser and H. Hoogstraal some ticks. T. Shresta, Herbarium of Nepal, and M. S. Zahur, University of the Punjab, worked on the plants. G. Musser and M. Lawrence, American Museum of Natural History, identified rodents. M. Hornocker, University of Idaho, had the proximate analysis of oak leaves done.

For assistance with literature references and other information I am indebted to K. Benirschke, C. Reed, K. Kawata, J. McNeely, R. Barrett, G. Rowell, K. Tustin, L. Simmons, A. Laurie, E. Dolensek, and others.

P. Lent, R. Valdez, A. Aeschbacher, C. Nadler, and T. Akasaka generously permitted me to read their manuscripts before publication.

C. Hill of the San Diego Zoological Society, F. Stoever, and S. A. Khan kindly provided me with some photographs.

The excellent sketches (plates 6 and 7) were made by R. Keane.

B. Bruning prepared the figures and A. Hamer assisted with typing the manuscript.

I owe a great debt to J. Eisenberg and P. Lent for reading the whole manuscript critically and to V. Geist for checking Chapters 9 to 12. Their many comments and criticisms improved the work considerably.

My last and greatest expression of gratitude is reserved for my wife, Kay. This was by far the most difficult of the projects in which she has participated, as it was necessary for her to make a home for our sons in Lahore while I roamed the mountains in search of wildlife. She participated in the work when possible and she edited, criticized, and typed drafts of this report.

APPENDIXES

Common and scientific names of wild bird and mammal species mentioned in the text.

Mammals
Order Marsupialia
    Red kangaroo   Macropus rufus
Order Insectivora
    Musk shrew   Suncus murinus
    Stripe-backed shrew   Sorex cylindricauda
Order Primata
    Common langur   Presbytis entellus
    Orang-utan   Pongo pygmaeus
Order Lagomorpha
    Black-naped hare   Lepus nigricollis
    Wooly hare   Lepus oiostolus
    Pika or mouse-hare   Ochotona roylei
                                    (and other species?)

Order Rodentia
    Indian porcupine   Hystrix indica
    Himalayan marmot   Marmota bobac
    Long-tailed marmot   Marmota caudata
    Red squirrel   Tamiasciurus hudsonicus
    Kashmir flying squirrel   Hylopetes fimbriatus
    Little Himalayan rat   Rattus eha
    —————   Rattus turkestanicus
    House mouse   Mus musculus
    Yellow-necked field mouse Apodemus flavicollis
    Sikkim vole   Pitymys sikimensis
                                      Pitymys irene
    —————
    Stoliczka's high mountain vole Alticola stoliczkanus
Order Carnivora
    Wolf   Canis lupus
    Asiatic jackal   Canis aureus
    Indian wild dog or dhole   Cuon alpinus
    Red fox   Vulpes vulpes
    Asiatic black bear   Selenarctos thibetanus

Brown bear	Ursus arctos
Striped hyena	Hyaena hyaena
Spotted hyena	Crocuta crocuta
Lynx	Felis lynx
Caracal	Felis caracal
Temminck's or golden cat	Felis temmincki
Leopard cat	Felis bengalensis
Puma or mountain lion	Felis concolor
Lion	Panthera leo
Tiger	Panthera tigris
Leopard	Panthera pardus
Snow leopard	Panthera uncia
Cheetah	Acinonyx jubatus
Stone or beech marten	Martes foina
Order Proboscidea	
Indian elephant	Elephas maximus
Order Perissodactyla	
White rhinoceros	Ceratotherium simum
Kiang or Tibetan wild ass	Equus kiang
Asiatic wild ass	Equus hemionus
Order Artiodactyla	
Hippopotamus	Hippopotamus amphibius
Giraffe	Giraffa camelopardalis
Muskdeer	Moschus chrysogaster and M. sifanicus
Muntjac or barking deer	Muntiacus reevesi
Chital or Axis deer	Axis axis
Caribou	Rangifer tarandus
Reindeer	Rangifer tarandus
White-tailed deer	Odocoileus virginianus
Black-tailed deer	Odocoileus hemionus
Sambar	Cervus unicolor
Barasingha or swamp deer	Cervus duvauceli
Red deer	Cervus elaphus
Kashmir stag or hangul	Cervus elaphus hanglu
Manchurian wapiti	Cervus elaphus xanthopygus
Sikkim stag or shou	Cervus elaphus wallichi
Wapiti or elk	Cervus canadensis
Marsh deer	Blastoceros dichotomus
Père David's deer	Elaphurus davidianus
Nilgai	Boselaphus tragocamelus
Yak	Bos grunniens
Gaur	Bos gaurus
African buffalo	Syncerus caffer
Lesser Kudu	Tragelaphus strepsiceros
Eland	Taurotragus oryx
Maxwell's duiker	Cephalophus maxwellii
Uganda kob	Adenota kob

Wildebeest	Gorgon taurinus
Dik dik	Madoqua kirki
Blackbuck or Indian antelope	Antilope cervicapra
Chinkara or Indian gazelle	Gazella gazella
Goa or Tibetan gazelle	Procapra picticaudata
Persian or goitered gazelle	Gazella subguttorosa
Chiru or Tibetan antelope	Pantholops hodgsoni
Saiga	Saiga tatarica
Goral	Nemorhaedus goral
Serow	Capricornis sumatraensis
Japanese serow	Capricornis crispus
Rocky Mountain goat	Oreamnos americanus
Chamois	Rupicapra rupicapra
Muskox	Ovibos moschatus
Takin	Budorcas taxicolor
Aoudad or Barbary sheep	Ammotragus lervia
Bharal or Blue sheep	Pseudois nayaur
Himalayan tahr	Hemitragus jemlahicus
Nilgiri tahr	Hemitragus hylocrius
Arabian tahr	Hemitragus jayakeri
Ibex	Capra ibex
Dagestan or East Caucasian tur	Capra cylindricornis
Spanish goat or ibex	Capra pyrenaica
Wild goat	Capra aegagrus
Markhor	Capra falconeri
Urial or mouflon	Ovis orientalis
Argali	Ovis ammon
Snow sheep	Ovis nivicola
Thinhorn sheep	Ovis dalli
Bighorn sheep	Ovis canadensis

Birds
Order Falconiformes
    Himalayan griffon vulture   Gyps himalayensis
    Lämmergeier or bearded vulture   Gypaëtus barbatus
    Golden eagle   Aquila chrysaëtas
Order Galliformes
    Chukar partridge   Alectoris chukar
    Snowcock   Tetraogallus tibetanus
                                      and T. himalayensis
    Monal pheasant   Lophophorus impejanus
Order Columbiformes
    Blue rock pigeon   Columba livia
Order Passeriformes
    Jungle crow   Corvus macrohynchos
    Magpie   Pica pica
    Treepie   Dendrocitta vagabunda
    Yellow-billed chough   Pyrrhocorax pyrrhocorax

TABLE 1
Observation times and weather at various Caprini study sites

Species	Place	Altitude (m)	Date	Temperature °C Mean max.	Mean min.	No. days with precipitation	Total no. hrs. obs.
Asiatic ibex	Shah Salim, Chitral	3350	July 22–31, 1972	24 (20–29)	7 (4–13)	2	20
	Besti, Chitral	3450	Aug 11–19, 1972	25 (22–27)	5 (4–6)	1	
	Golen Gol, Chitral	2075	Feb 5–17, 1974	5 (2–7)	−6 (−10--2)	2	
	Besti, Chitral	3175	Feb 21–25, 1974			4	
	Kilik, Hunza	4270	Nov 17–20, 1974	−9	−18		
	Baltoro, Baltistan	3385	May 7–15, 1975				
Wild goat	Kirthar, Sind	200	Sept 6–Oct. 3, 1972	40 (36–43)	25 (23–32)	1	130
			Mar 1–19, 1973	32 (25–37)	17 (9–24)		
			Sept 4–9, 1973				
			Mar 20–31, 1974	39 (33–42)	24 (22–29)	1	
			Oct 9–13, 1974	34 (33–36)	24 (21–26)		
	Chiltan, Baluchistan	1800	Nov 15–18, 1972				
Markhor	Chitral Gol, Chitral	2375	Nov 26–Dec 22, 1970	5 (2–8)[a]	−1 (−3–+3)	5	90
		2000	Aug 2–6, 1972	30 (29–30)	20 (18–22)		
			Nov 27–Dec 14, 1972	10 (6–13)	2 (0–4)	1	
			Jan 9–17, 1973	3 (2–5)	−4 (−5--2)	2	
			Jan 22–Feb 3, 1974	3 (1–5)	−4 (−8--2)	10	
	Tushi, Chitral	1750	Dec 15–17, 1972;			5	
			Jan 18–29, 1973;				
			Feb 19–20, 1974				
Bharal	Lapche, Nepal	3800	Feb 29–Mar 1, 9–19, 1972	7 (5–10)	−4 (−8--2)	3	181
	Shey, Nepal	4475	Oct 28–Dec 6, 1973	2 (−4–+7)[b]	−11 (−15--9)		

Himalayan tahr	Lamnang, Nepal	2590	Feb 16–28, 1972	14 (6–20)	−2 (−7–+1)	1	92
	Lamobagar, Nepal	2440	Mar 20–Apr 5, 1972	24 (22–26)	8 (7–9)	8	
	New Zealand	975	May 16–20, 1974				
Nilgiri tahr	Nilgiris, S. India	2000	Sept 29–Oct 7, 9–16, 1969			9	53
	High Range, S. India	2000	Oct 22–Nov 13, 1969			3	
Punjab urial	Kalabagh, Punjab	450	Oct 6–Nov 9, 1970	31 (28–35)	21 (17–26)		157
			July 11–14, 1972	37 (36–38)	26 (22–29)	1	
			Oct 27–Nov 7, 14–18, 1972	30 (28–33)	17 (16–17)		
			Dec 28, 1972–Jan 1, 1973	18 (17–19)	5 (4–6)		
			Apr 14–28, 1973	38 (32–41)	23 (17–27)		
			July 14–16, 1973			1	
			Apr 21–28, 1974	41 (39–42)	26 (23–28)	1	
			Oct 27–Nov 1, 1974				

[a] Temperatures taken to Dec. 14 only
[b] Temperatures taken from Nov. 8

TABLE 2

Mean monthly rainfall (in mm) at various stations

Location	Altitude (m)	J	F	M	A	M	J	J	A	S	O	N	D	Total
Baluchistan														
Ft. Sandeman	1413	18	27	35	24	23	16	54	48	8	3	3	14	273
Quetta	1668	49	50	44	25	10	4	12	8	1	3	7	25	238
Panjgur	1310	22	24	15	10	3	3	20	9	1	—	1	14	122
N.W. Himalaya														
Drosh	1465	35	39	95	104	49	17	15	17	18	32	9	30	460
Gilgit	1630	6	7	20	24	20	9	10	14	10	6	1	3	130
Ladak														
Leh	3506	10	8	7	6	6	5	12	15	7	3	1	5	85
W. Himalaya														
Srinagar	1586	74	72	92	93	60	36	59	61	39	30	11	34	661
C. Himalaya														
Simla	2196	61	69	61	53	66	175	424	434	160	33	13	28	1577
Jumla	2310	36	38	44	25	29	52	200	167	103	39	2	12	747
E. Himalaya														
Pokhara	760	35	30	49	82	174	654	807	925	515	170	22	14	3477
Jomsom	2750	28	24	30	28	6	14	46	45	40	29	2	3	295
Namche Bazar	3400	34	24	32	27	43	139	212	210	140	51	15	13	940
Darjeeling	2259	15	27	45	97	217	622	807	652	460	112	20	5	3079

SOURCE: Data from Ahmad (1969), Schweinfurth (1957), Stainton (1972), and others.

TABLE 3
Warmest and coldest months at various stations

		Temperature °C			
		Warmest month (July)[a]		Coldest month (Jan)	
Location	Altitude (m)	Mean max.	Mean min.	Mean max.	Mean min.
Ft. Sandeman	1413	37	24	15	−1
Quetta	1668	34	18	10	−2
Panjgur	1310	40	25	17	4
Drosh	1465	36	23	8	0
Gilgit	1630	35	22	8	0
Leh	3506	25	10	−1	−13
Srinagar	1586	31	18	5	−3
Simla	2196	23	16	8	2
Kathmandu	1333	30	16	18	2
Darjeeling	2259	19	14	8	2
Tezpur	78	32	26	23	11

SOURCE: Data from Ahmad (1969) and Anon. (1958).
[a]In Simla the warmest month is in June and in Kathmandu in May.

TABLE 4
Subspecies of Capra falconeri in Pakistan

Horn type	Current scientific name	Common name	Proposed scientific name
Straight-horned	C.f. megaceros Hutton 1842	Kabul	C.f. megaceros
	C.f. jerdoni Hume 1874	Sulaiman	
Flare-horned	C.f. falconeri Hügel and Wagner 1839	Astor	C.f. falconeri
	C.f. cashmiriensis Lydekker 1898	Kashmir	
Chiltan	C.f. chialtanensis Lydekker 1913	Chiltan	C. aegagrus

TABLE 5

Some physical characteristics of the subspecies of urial, Ovis orientalis

Subspecies	Common name	Horn winding	Ruff	Saddle patch
Musimon	European	Homonym	Short-black	White
Ophion	Cyprus	Heteronym	Short-black	White
Gmelini (urmiana, anatolica armeniana)	Armenian	Heteronym	Short-black	White
Laristanica	Laristan	Homonym	Short-black	White
Isphahanica	Isphahan	Side of neck	Long-black	White
Arkal (dolgopolovi)	Kopet Dagh	Homonym	Long-white	Black
Cycloceros (blanfordi)	Afghan	Homonym	Long-black, white bib	Black
Punjabiensis	Punjab	Side of neck	Long-black, white bib	Black and white (variable)
Vignei (bochariensis)	Ladak	Heteronym	Long-black, white bib	Black and white (variable)
Gmelini × arkal (erskinei, orientalis)	Elburz hybrid[a]	All types	Variable	Variable

NOTE: Data for several Iranian sheep were taken from Bunch and Valdez (in press).
[a]Two other hybrid populations are not listed (see text).

TABLE 6

Some weights (in kg) and measures (in cm) of adult Saigini, Ovibovini, and Rupicaprini males

Species	Shoulder height[a]	Average weight[b]	Average[c] max. horn length	Record[d] horn length	Ratios Av. horn length / shoulder height	Ear length / shoulder h.	Tail length / shoulder h.	Sources[e]
Chiru	94	36	69	70	.73	—		Engelmann, 1938
Saiga	73	43 (33–51)	33	37	.45	.15	.14	Bannikov et al., 1967
Goral (India, China)	70	28[f]	20	23	.29	.18	.20	Allen, 1940; Burton, 1915
Goral (Russia)	72.5 (73, 74)	32 (30, 34)	20		.28	.19		Bromlei, 1956
Serow	97	92 (61–140)	26	32	.27	.19	.10	Wallace, 1913; Ward, 1924; Allen, 1940
Mt. goat	98	69 (56–88)	29	30	.28	.13	.10	Brandborg, 1955
Chamois	89	37	29	32	.33	.15	.07	Christie, 1964
Muskox	138	334 (218–460)	72	84	.52	.11	.08	Jonkel et al., 1975; Tener, 1965
Takin	122	249 (198, 229)	58	64	.48	.09	.15	Cooper, 1923; Wallace, 1913

[a]All measurements in tables 6 to 11 are rounded to the nearest centimeter or kilogram.

[b]All weights are listed if the sample includes only two or three animals.

[c]These and other horn measurements are based on the average of the 10 largest horns listed in Dollman and Burlace (1922) and Anon. (1971a).

[d]If horn lengths on a trophy differ the longest one is listed here.

[e]Measurements of body dimensions can be found in Stockley (1928), Harper (1945), Allen (1940), Clark (1964), and others; the sources listed in the tables provide most body weights. The sources for female weights are the same and are not listed again in subsequent tables.

[f]Only one animal weighed.

TABLE 7

Some weights and measures of Saigini, Ovibovini, and Rupicaprini females

Species	Average shoulder height	Average weight	Average horn length	Ratio Female weight / Male weight
Chiru	—	—	None	—
Saiga	68	31 (21–41)	None	.71
Goral (India, China)	62[a]	27[a]	15	.96
Goral (Russia)	72.5 (71, 74)	32.5 (32, 33)	—	1.01
Serow	88	91 (54–131)	23	.98
Mt. goat	87	53 (49, 53, 58)	23	.76
Chamois	80	23	19	.62[b]
Muskox	123[a]	266 (212–302)	—	.79
Takin	115[a]	225–270 est.	37[a]	—

[a]Only one animal weighed or measured.

[b]This figure is based on New Zealand animals. Knaus and Schröder (1975) found that in Bavaria the males are only 10% heavier than the females (males weigh an average of 20 kg), giving a ratio of .90.

TABLE 9

Some weights and measures of Ammotragus, Pseudois, and Capra females

Species	Shoulder height	Av. weight	Av. horn length	Ratio Female weight / Male weight
Aoudad	77	52 (50, 54)	40	.50
Bharal	87	39	20	.65
Himalayan tahr	81	59[a]	20	.58
Persian wild goat	74[a]	30[a]	25	.43
Cretan wild goat	56 (49–60)	20 (18–23)	18	.59
Russian markhor	—	36.5 (32, 41)	25	.44
Alpine ibex	74 (70, 78)	32 (24–63)	24	.38
Kuban ibex	88	50–60 est	22	
Asiatic ibex (Tien Shan)	71 (67–74)	50 (40–56)	29	.71
Asiatic ibex (Him.)	—	55 (47, 62)	24	.71

[a]Only one animal weighed or measured

TABLE 8
Some weights and measures of Ammotragus, Pseudois, and Capra males

Species	Shoulder height	Average weight	Av. max. horn length	Record horn length	Ratios			Sources
					Av. horn length Shoulder height	Ear length Shoulder height	Tail length Shoulder h.	
Aoudad	97	104 (68, 139)	69	84	.71	.09	.26	Rothschild, 1921; Bigourdan and Prunier, 1937.
Bharal	91	60	78	84	.86			Schäfer, 1937
Himalayan tahr	97	101 (95, 99, 108)	37	39	.38	.08	.09	Schäfer, 1950
Nilgiri tahr	100	—	41	44	.41			
Arabian tahr	62	23[a]	29	29	.48			Harrison, 1968a
Spanish goat	72	65–80	75	85	1.04			Couturier, 1962
Persian wild goat	85	77 (90, 95, 46)	126	144	1.48	.17	.19	Couturier, 1962; Wahby, 1931; Valdez, pers. comm.
Cretan wild goat	68 (59–73)	34 (26–42)	58[b]	—	—	.17	.14	Papageorgiou, 1974
Straight-horned markhor	97	—	82[c]	123	—			Ward, 1924
Flare-horned markhor	102	104 (99, 108)	143	165	1.40			Couturier, 1962
Russian markhor	88–100	83 (80, 86)	—	—	—	.14		Heptner et al., 1966
Alpine ibex[d]	86 (63–93)	84 (47–117)	85	113	.99	.13	.16	Couturier, 1962
Kuban ibex	96	65–80 (to 100)	86	102	.90	.13	.14	Couturier, 1962
Nubian ibex	84	52 (52,[e] 56.5,[e] 48.5)	116	127	1.38	.13	.09	Heptner et al., 1966; Powell-Cotton, 1902
Walia ibex	97	120 (115, 125)	108	112	1.11			Powell-Cotton, 1902; Flower, 1932

(80–110)

						Couturier, 1962; Heptner et al., 1966
Asiatic ibex (Him.)	102	78 (75, 80)	128	140	1.25	Ward, 1924
Dagestan tur	80 (70–94)	65–80 (to 100)	103	117	1.29	.15 Heptner et al., 1966; Nasimovitch, 1955

aOnly one animal weighed.
bBased on 10 largest heads in population studied by Papageorgiou (1974).
cHorn sizes of straight-horned and flare-horned markhor cannot be compared because the former were traditionally measured in a straight line and the latter around the curve.
dIncludes only animals 4 years old and older.
eWeight based on captives.

TABLE 10
Some weights and measures of Ovis males

Species	Shoulder height	Av. weight	Av. max. horn length	Record horn length	Ratios Av. horn length / Shoulder height	Ear length / Shoulder h.	Tail length / Shoulder h.	Sources
Urials								
Mouflon	69	25–55	83	98	1.20			Pfeffer, 1967
Cyprus	66	32 est	57	69	1.16			Harper, 1945
Armenian	84	43±	77	103	.92			Nasimovitch, 1955; R. Valdez, pers. comm.
Kopet Dagh	95(92–98)	57–83	97	115	1.02			Nasimovitch, 1955; Decker and Kowalski, 1972
Afghan	81	56–59 est	96	105	1.19			Stockley, 1928
Punjab		41 (39, 41, 42)	84	98	1.04	.13	.13	This study
Ladak	91	56 est	95	99	1.04			Ward, 1924
Argalis								
Marco Polo	112	118 (99, 118, 138)	179	191	1.60			Ward, 1924; Clark, 1964
Tien Shan	112	101–130 (to 180)	143	180	1.28	.11		Heptner et al., 1966
Altai	127	180 est	154	169	1.21			Clark, 1964
Mongolian	110	101 est	113	130	1.03	.10	.10	Dollman and Burlace, 1922
Tibetan	118	105 (92, 95, 108, 126)	127	145	1.08	.12	.08	Ward, 1924; Stockley, 1928; Phil Acad. Sci., pers. comm.

	99	86–128	91	100	.92	.09	
Snow sheep (Kamchatka)	99	86–128	91	100	.92	.09	Heptner et al., 1966; Nasimovitch, 1955
Thinhorn, Dall	99	81 (78–83)	117	128	1.18		Ulmer, 1941
Thinhorn, Stone	99	77[b] (to est. 122)	117	128	1.18	.10	Geist, 1971a
Bighorn, Desert[a]	93	70 (57–86)	105	112	1.13	.08	Blood et al., 1970
Bighorn, Rocky Mt.	101	93 (72–119)	113	126	1.12		Blood et al., 1970

[a]Mostly *nelsoni*.
[b]Only one animal weighed.

TABLE 11

Some weights and measures of Ovis females

Species	Shoulder height	Weight	Av. horn length	Ratio Female weight / Male weight
Armenian urial	67 (65–68)	27±		.62
Kopet Dagh urial	81 (77–85)	34–41		
Tien Shan argali		60 est.	36	
Tibetan argali	108 (104, 112)	68 est.	37	
Snow sheep (Kamchatka)	85.5 (76–94)	47–54	30	
Bighorn, Desert		44 (33–51)		.63
Bighorn, Rocky Mt.		72 (54–90)		.77

TABLE 12

Measurements (in cm and kg) of 3 Punjab urial rams and a male chinkara from Kalabagh.

Species	Date	Age (yrs)	Total length	Tail length	Ear length	Hind foot length	Shoulder height	Longest horn	Horn circum.	Weight
Urial	May 3, 1973	7	120	11.1	11.1	34.3	79.7	79.5	24.3	39.1
Urial	Oct 29, 1974	5½	—	10.9	11.7	—	82.3	57.4	20.3	40.9
Urial	Oct 30, 1974	4½	128	11.9	10.6	33.5	81.3	62.5	20.8	41.6
Chinkara	Oct 30, 1974	ad.	124.5	14.5	14.2	—	67.6	28.1	10.9	23.4

TABLE 13
Adult and yearling sex ratios and subadult/female ratios in several caprid populations

Species	Locality	Date	Sample size	Ad.♂♂/100 ad.♀♀	Yearl.♂♂/100 yearl.♀♀	Young/100 ad.♀♀	Yearl./100 ad.♀♀
Himalayan tahr	Bhota Kosi, Nepal	Mar–Apr 1972	246	48	155	67	49
	Macauley, N.Z.	May 1974	141	111	42	76	48
Nilgiri tahr	Nilgiris	Oct 1969	164	63	55	75	55
	High Range	Oct–Nov 1969	260	58	—	88	51
Bharal	Lapche	Mar 1972	216	116	109	88	82
	Shey	Nov–Dec 1973	1952	134	61	40	29
	Phoksumdo	Oct–Nov 1973	47	128	—	50	57
	Dhaulagiri	Oct 1973	46	69	—	88	31
Wild goat	Chiltan[a]	Aug 1971	164	34	67	54	34
	Karchat	Sept 1972	828	114	72	45	63
		Mar 1973	439	56	60	31+	43
		Sept 1973	259	117	50	19	42
		Mar 1974	463	49	54	49	21
		Oct 1974	236	103	90	38	21
Kashmir markhor	Chitral Gol	Dec 1970	328	42	93	128	53
		Dec 1972	276	63	52	130	57
		Jan 1974	241	80	84	137	54
	Tushi, Chitral	Jan 1973	184	10	79	104	49
		Feb 1974	78	14	33	107	57
Asiatic ibex	Besti	Aug 1972	82	292	—	85	77
		Feb 1974	40	127	75	64	64
	Kilik	Nov 1974	59	33	50	67	50
	Baltoro	May 1975	65	110	133	80	35

	Oct–Nov 1970	1367	102		73	55
	July 1972	736	86	69	63	35
	Oct–Nov 1972	725	98	97	65	31
	Apr 1973	1010	99	50	—	60
	July 1973	126	122	228	57	62
	Apr 1974	426	131	100	—	43
	Oct 1974	534	86	93	63	30
Ladak urial Baltoro	May 1975	37	100	300	54	50

aData collected by Z. B. Mirza.

TABLE 14
Mating seasons, gestation periods, and usual number of young in a sample of Caprinae

Species	Locality	Approx. main mating season	Gestation period (in days)	Usual no. of young	Source
Saiga	Russia	Nov–late Dec	145	1–2	Bannikov et al., 1967
Chiru	Tibet	Late Nov–Dec	—	1	Rawling, 1905
Muskox	Canada, Alaska	Aug–Sept	246	1	Tener, 1965; P. Lent, pers. comm.
Takin	China	Late July–Sept	200–220	1	Schäfer, 1933; Wallace, 1913
Goral	Russia	Late Sept–Nov	180	1	Heptner et al., 1966
Japanese serow	Japan	Oct–Nov	210–220	1	Komori, n.d.
Mountain goat	Idaho	Nov–early Dec	178	1	Brandborg, 1955
Chamois	Switzerland	Dec–mid Jan	165	1	Couturier, 1960
Himalayan tahr	Nepal	Mid Oct–mid Jan	180	1	Caughley, 1970c; Schaller, 1973a
Nilgiri tahr	India	June–Aug	180(?)	1	Schaller, 1971
Wild goat	Sind, Pak.	Aug–Oct	165[a]	1–2	This study
Alpine ibex	Switzerland	Dec–mid Jan	165–170	1	Couturier, 1960
Asiatic ibex	Pamirs	Dec–Jan	170–180	1–2	Heptner et al., 1966
Walia ibex	Ethiopia	All year, peak March–May	—	1	Nievergelt, 1974
Nubian ibex	Egypt	Sept–Oct	150	1–2	Flower, 1932; Couturier, 1962
Flare-horned Markhor	Chitral	Mid Dec–early Jan	155	1–2	Haltenorth, 1958; this study
Dagestan tur	Caucasus	Late Nov–early Jan	150–160	1	Heptner, et al., 1966
Mouflon	Corsica	Mid Oct–Dec	148–159	1–2	Pfeffer, 1967
Punjab urial	Punjab, Pak.	Mid Oct–Nov	150	1–2	This study
Marco Polo sheep	Pamirs	Dec–early Jan	150	1–2	Lambden, 1966; Heptner et al., 1966
Bighorn (desert)	California	June–Dec; peak July–Oct	170–180	1	Welles & Welles, 1961

Table 14 (Continued)

Species	Locality	Approx. main mating season	Gestation period (in days)	Usual no. of young	Source
Bighorn (Rocky Mountain)	B.C., Canada	Late Nov–Dec	175	1	Geist, 1971a
Aoudad	Air, Niger	July–Aug	160	1–2	Brouin, 1950
Bharal	Nepal	Late Nov–Dec	160	1	Crandall, 1964; this study

[a]T. Roberts told me that his captive wild goat gave birth after gestation periods of 162, 162, and 170 days.

TABLE 15
Progression of births in wild goat and Punjab urial during 1973

| Date | Wild goat | |
	% females pregnant	% females accompanied by young
Mar 1–5	72	1
Mar 6–10	50	4
Mar 11–15	25	20
Mar 16–19	6	35
	Punjab urial	
Apr 14–18	91	1
Apr 19–23	49	13
Apr 24–28	23	29
May 3	8	33

TABLE 16
Age at death of males in several caprid species in Pakistan (expressed as % of sample)

Species	Source	No. in Sample	1–2	2–3	3–4	4–5	5–6	6–7	7–8	8–9	9–10	10–11	11–12	12–13	13–14	14–15
Wild goat	Sind trophy	17				6	18	29	12	12	24					
	Chiltan trophy	19	5	16	42	16	16	5								
	Karchat Hills pick-up	12			8	33	25	17	8					8		
Markhor	Sulaiman trophy	49	2	4	22	22	24	12	4	8						
	Kabul trophy	16		6	6	6	25	19	6	19	13					
	Astor trophy	23		4	4	13	17	22	9	17	13					
	Kashmir trophy	46				2	13	22	20	15	20	2	4			
	Kashmir pick-up	12	17	17	8	17	8	17	8		8	8		2		
Ibex	Trophy	91		4	2	7	13	14	16	14	12	13		2		1
	Baltoro pick-up	12			8			17	8	33	33					
Bharal	Shey pick-up	19			5			26	11	32	5	11	5			5
Punjab urial	Trophy	10			10	20	50	10	10							
	Kalabagh pick-up	20		15	10	10	30	10	20	5						
Ladak urial	Trophy	14	7		14	43	29	7								
Afghan urial	Trophy	56	2	11	20	36	16	14	9	9		2				
Marco Polo sheep	Trophy	23				4	9	35	39	9	4					
	Pick-up	15		7	33	27	13	13	7							

TABLE 17

Percent of males in various age classes

Species	Locality	Date	Class and age in years				
			Yearl. (1–2)	I (2–3)	II (3–4)	III (4–5+)	IV + V (most 5+)
Wild goat	Karchat	Sept 1972	19	14	10	13	44
Markhor	Chitral Gol	Jan 1974	23.5	21.5	21.5	21.5	15
Asiatic ibex	Besti	Jan 1974	18	18	23	36	5
	Baltoro	May 1975	18	6	12	18	47
Bharal	Shey	Nov 1973	8	8	8	17	59
	Lapche	Mar 1972	27	26	16	22	9
Urial	Kalabagh	Oct 1974	15	15	24	24	20

TABLE 18
Frequency of occurrence of parasites in feces of domestic goats and several wild ungulates in Pakistan

	Kashmir markhor Chitral Gol (10 samples)	Domestic goat Chitral Gol (10 samples)	Domestic goat Kalabagh (7 samples)	Punjab urial Kalabagh (14 samples)	Chinkara Kalabagh (7 samples)	Wild goat Karchat (7 samples)
Eimeria parva	5	3				
Eimeria intricata	1	1				
Eimeria arloingi			1			
Eimeria sp.				1		3
Nematodirus sp.	5	9	1		1	
Trichuris sp.	1	2			1	
Trichocephalus ovis			2	7	3	
Haemonchus sp.				1	2	1(?)
Oesophagostomum sp.				4		
Unid. Trichostrongylidae			7	10	5	1
Moniezia sp.		1				
Cystocaulus oeratus				8		

TABLE 19

Frequency of occurrence (in %) of food items in jackal and red fox droppings

| | Jackal | | Red fox | |
| | High Range, India Oct–Nov 1969 | Baltoro May 1975 | Dhaulagiri Oct 1973 | Chitral and Golen Gol Dec–Feb 1973–74 |
Food				
Rodents[a]	94	43	100	42
Hare	1	8		1
Flying squirrel				2
Sambar	3			
Ladak urial		16		
Markhor				4
Ibex		8		
Livestock				10
Unid. hair		3		1
Feathers		22		4
Egg shells		16		
Lizard and snake	29			
Land crab	10			
Snail	1			
Insects	7		16	
Cloth				3
Vegetation[b]		14	16	63
Fruit[c]	6			25
Earth				1
Total no. droppings	119	57	12	113

[a]Includes pika at Baltoro.
[b]Includes buds, leaves, grass.
[c]*Pistacia, Zizyphus, Rosa,* acorns, maize.

TABLE 20
Frequency of occurrence (in %) of food items in wolf droppings

Food	Chitral and Gilgit Agency	Shey area, Nepal
Ibex	37	
Ladak urial	2	
Marco Polo sheep	2	
Bharal		38
Livestock	38	29
Marmot	17	32
Hare	2	3
Grass	2	
Total no. droppings	63	34

TABLE 21
Frequency of occurrence (in %) of food items in bear droppings

	Black bear			Brown bear
Food	Dachigam[a] Oct 1968	Dachigam Nov 1969	Chitral Gol Winter den 1972	Baltoro May 1975
Celtis	40			
Walnut	33			
Oak	12	1		
Grape	9			
Zizyphus	5			
Apple	5			
Maize	4			
Apricot	2			
Rosa	1	100		
Grass			78	69
Unid. leaves		2	6	
Unid. roots				34
Feathers	1			
Ibex				16
Markhor			17	
Livestock		1		
Unid. hair	1		6	
Wasp	1			
Total no. droppings	82	94	18	29

[a]Data are from Schaller (1969b).

TABLE 22

Frequency of occurrence (in %) of food items in leopard and snow leopard droppings

Food	Leopard Karchat	Snow leopard Shey	Lapche	Chitral
Wild goat	70			
Urial (?)	1			
Bharal		50	73	
Markhor				40
Chinkara	1			
Livestock	14	13	9	45
Striped hyena	1			
Hare	1			
Porcupine	9			
Marmot		31	9	5
Unid. hair		6	9	
Grass and leaves	13	38	18	10
Zizyphus	4			
Earth	1			5
Total no. droppings	70	16	11	20

TABLE 23
Some food species of caprids in my study areas

Species	Wild goat	Kashmir markhor	Asiatic ibex	Bharal	Punjab urial
Shrubs and trees					
Acacia modesta					+
Acacia senegal	+				
Artemisia maritima		+	+		
Berberis sp.				+	
Capparis cartiliginae	+				
Capparis aphylla	+				
Capparis spinosa	+				
Caragana brevifolia				+	
Cotoneaster microphyllus				+	
Cotoneaster sp.		+			
Crataegus sp.			+		
Ephedra gerardiana				+	
Grewia sp.	+				
Gymnosporea sp.	+				
Hippophaë rhamnoides					+
Indigofera gerardiana		+			
Inula graveolens	+				
Juniperus sp.				+	
Launea orientalis	+				
Leptadenia pyrotechnica	+				+
Lonicera spinosa				+	
Moringa oleifera	+				
Oeriploca aphylla					+
Pistacia sp.		+			
Quercus ilex		+			
Rhazya stricta	+				
Salvadora oleoides	+				+
Solanum incana					+
Tecoma undulata	+				
Zizyphus nummularia	+				+
Forbs					
Acanthophyllum sp.		+			
Anaphalis contorta				+	
Arenaria kashmirica			+		
Astragulus sp.			+		
Descruannia cruciferum			+		
Erymurus sp.		+			
Indigofera sp.			+		
Lindolphia spectabilis			+		
Napeta sp.			+		
Oxytropis sp.			+		

TABLE 23 (continued)

Some food species of caprids in my study areas

Species	Wild goat	Kashmir markhor	Asiatic ibex	Bharal	Punjab urial
Pleurospermum candollei			+		
Polygonum sp.				+	
Rheum emodi		+			
Rumex hastatus		+			
Rumex nepalensis				+	
Thermopsis barbata				+	
Tricholepis furcata		+			
Scrophularia sp.		+			
Vicatia sp.			+		
Fern				+	
Grasses and sedges					
Agrostis pilosula				+	
Aristida adscendens		+			
Arundinella nepalensis				+	
Bromus sp.			+		
Carex sp.			+		
Cenchrus ciliaris	+				
Cenchrus pennisetiformis					+
Chloris villosa					+
Cymbopogan jawarancusa	+				+
Deschampsia caespitas				+	
Dichanthium annulatum	+				
Digitaria bicornis					+
Danthonia schneideri				+	
Eleusine flagillifera					+
Eriochloa procera	+				
Festuca sp.				+	
Heteropogon contortus					+
Oryzopsis monorii		+	+		
Pennisetum sp.	+				
Poa pagophila				+	
Saccharum spontaneum					+
Saccharum munja		+			
Stipa sp.	+				
Trisetum sp.				+	

Table 24
Percent vegetation cover in two alpine areas at various altitudes in Chitral

Location	No. plots	Altitude (m)	Bare soil and rock	Grass-sedge	*Artemisia*	*Ephedra*	Forbs
Near Dorah Pass	40	3600	53	1	27	3	16
	60	4000	70	2	10	1	17
	20	4330	75	11	0	0	14
	20	4450	85	1	0	0	14
Near Besti	30	4400	91	4	0	0	5
	20	4900	91	2	0	0	7

Table 25
Type of body care (in %) shown by several caprid species

Species	No. observations	Scratch with horn	Lick	Scratch with hindleg	Rub body on object
Wild goat	284	14.5	63	20	2.5
Asiatic ibex	26	23	46	27	4
Kashmir mark-hor	97	16.5	49.5	34	0
Himalayan tahr	108	1	78	21	0
Bharal	486	9.5	61	29	0.5
Punjab urial	110	4.5	34	54.5	7
Stone sheep (Geist, 1971a)	639	24	12	23	41

TABLE 26

Herd sizes (mixed herds and female herds combined) of various caprid species at different seasons

Species	Area	Total no. herds	Rut Mean	SD	Birth season Mean	SD	Others Mean	SD
Wild goat	Karchat	130	23.8	20.60	21.4	23.33		
	Chiltan[a]	60	5.6	2.68			4.3	1.84
Markhor	Chitral Gol	114	9.0	5.83				
	Tushi	18	16.1	21.30				
Asiatic ibex	Pakistan	24					9.4	6.04
Himalayan tahr	Nepal	36					6.5	4.92
	New Zealand	14	8.8	5.04				
Nilgiri tahr	India	23					24.1	22.36
Bharal	Shey	93	18.4	14.93				
	Lapche	18					11.2	5.88
	Sang Khola	8					4.8	3.62
Punjab urial[b]	Kalabagh	733	7.4	5.81	5.8	5.97	9.5	7.10

[a]Some data provided by Z. Mirza.
[b]The differences in the means are statistically significant. (<0.01).

TABLE 27

Sizes and compositions of several population units of various caprid species, based on the largest herds seen

Species	Male					Female				Total
	V	IV	III	II	I	Yearl.	Adult	Yearl.	Young	
Markhor										
(Chitral Gol)		1			1	2	6	6	8	24
		2	2	1	4	1	10	3	12	35
		1		1	2	1	8	1	10	24
Nilgiri tahr										
(High Range)			1	1	2	2	17	3	8	34
			1	2	3	5	12	7	9	39
			1	1	6	1	16	4	9	38
Himalayan tahr										
(Nepal)			2	2	2	2	8	2	5	23
			1	1	2	4	2	2	12
			1		1	1	3	1	2	9
Bharal (Shey)	5	4	6	2	3	3	27	4	7	61
	1	5	1			1	11	2	7	28
	4	2		1	2		15	2	7	33

TABLE 28

Percent of class II and III Nilgiri tahr males seen alone, in male herds, and in mixed herds

Date	No. males in sample	% males alone	% males in male herds	% males in mixed herds
Sept 30–Oct 16	31	19	29	52
Oct 17–31	75	9	72	19

After Schaller, 1971.

TABLE 29
Percent of bharal males of classes I–V seen alone, in male herds, and in mixed herds at Shey

Date	No. males	% alone	% in male herds	No. male herds	Av. size male herds	SD	% males in mixed herds	No mixed herds	Av. size mixed herds	SD
Nov 1–9	233	0.5	27.5	9	7.1	4.14	72	20	20.7	13.90
Nov 10–18	180	0	33	6	10.0	4.43	67	16	20.9	13.54
Nov 19–27	231	1	39	13	7.0	3.56	60	16	26.7	18.42
Nov 28–Dec 5	219	4	16	10	3.5	2.51	80	25	17.5	14.33

NOTE: Some variations in the averages are due to sampling bias. For example, the same large male herd was seen repeatedly between November 10 and 18.

TABLE 30
Mean herd sizes of Punjab urial at various seasons

		Male herd	SD	Female herd	SD	Mixed herd	SD
April	No. herds	49		156		50	
	Mean herd size	7.6	6.15	4.3	4.67	10.3	7.43
	Largest herd	24		51		39	
July	No. herds	22		38		37	
	Mean herd size	4.4	3.43	5.5	3.52	12.7	8.11
	Largest herd	18		14		30	
Oct 23–Nov	No. herds	104		121		270	
	Mean herd size	2.9	1.35	4.1	2.23	8.9	6.23
	Largest herd	8		11		35	
December	No. herds	6		18		41	
	Mean herd size	8.3	2.73	4.9	2.37	12.0	7.64
	Largest herd	11		9		33	

NOTE: The seasonal changes in herd size of male and mixed herds are significant at least at the 5% level, except that changes are insignificant between December and April.

TABLE 31
Horn-size class differences of partners in urial ram pairs

Class difference	0	1	2	3	4
No. observations	41	37	6	1	0

TABLE 32
Type and frequency of direct aggression in Himalayan tahr in Nepal

Age and sex class	Jerk	Lunge and chase	Jump	Butt	Clash	Head-to-tail	Total No.	Total %	Approx. population comp. (%)	No. per animal-hour of observation
Male III							0	0	6	0
Male II	2				(1)		3	3	6	.09
Male I	1			1			2	2	7	.05
Yearl. male	1			1 (2)	6 (9)	(3)	22	25	11	.30
Adult female	4 (6)	3		2 (12)	1 (10)	(3)	41	47	38	.18
Yearl. female				1	3		4	5	7	.08
Young	1		2	5	6	2	16	18	25	.10

NOTE: Interactions during courtship are in parentheses. Data are based on 603 animal-hours of observation. (Adapted from Schaller, 1973a.)

Table 33

Type and frequency of direct aggression in Nilgiri tahr

Age and sex class	Side clash	Head-on clash	Head-to-tail	Total No.	%	Approx. population comp. (%)
Males II and III	3	0	1	4	4	5
Male I	2	13	9	24	23	6
Adult female	15	6	0	21	20	38
Yearl. (male and female)	13	19	16	48	45	20
Young	3	6	0	9	8	31

Adapted from Schaller, 1971.

TABLE 34

Type and frequency of direct aggression in Kashmir markhor in the Chitral Gol

Age and sex class	Jerk	Lunge and chase	Jump	Butt	Clash	Shoulder-push	Head-to-tail	Total no.	Total %	Approx. population comp. (%)	No. per animal-hour of obs.
Male IV	4	6			1			11	4	3.5	.58
Male III	8				4			12	4	5	.44
Male II	4	7			3			14	5	3.5	.47
Male I	5		1		3			9	3	6	.20
Yearl. male	2		1		43			46	17	6	.96
Adult female	21	7	3	4	35			70	27	29	.33
Yearl. female	2	1	1	1	22			27	10	9	.37
Young	6		3	2	61	2	2	76	29	38	.28

NOTE: Data are based on 725 animal-hours of observation.

TABLE 35
Type and frequency of direct aggression in wild goat (Sept–Oct)

Age and sex class	Jerk[a]	Lunge and chase	Jump	Butt	Clash	Neck-fight	Shoulder-push	Total No.	%	Approx. population comp. (%)
Male V	90			10	31			131	19	8
Male IV	22		4	4	98			126	18.5	11
Male III	6		1	2	21			29	4	7
Male II	1				38	4		43	6	5
Male I	1			1	43		4	49	7	6
Yearl. male	3		3	2	41		6	55	8	7
Adult female	28	6	4	14	66			118	17.5	33
Yearl. female	6	1	6		46			59	9	10
Young	1		2		64			67	10	13

NOTE: A. Laurie helped in collecting these data.
[a]Because of difficulties in quantification, jerks and lunges have been combined in adult males.

TABLE 36
Type and frequency of direct aggression in Punjab urial (Oct–Nov)

Age and sex class	Jerk	Lunge and chase	Jump	Butt	Clash	Horn-pull	Shoulder-push	Total No.	Total %	Approx. population comp. (%)
Male IV	12	5	2	8	149	3	2	181	43.5	9
Male III	9	4	2	2	73	1		90	21.5	8
Male II	24	3	1		46			75	18	9
Male I	7			2	41			50	12	8
Yearl. male	2			1	4			7	1.5	6
Female (adult and yearl.)	5		1		2			8	2	37
Young			1		5			6	1.5	23

TABLE 37
Type and frequency of direct aggression in bharal

Age and sex class	Jerk	Lunge and chase	Jump	Butt	Clash	Bite	Shoulder-push	Total No.	Total %	Approx. population comp. (%)	No. per animal-hour of obs.[a] (Shey)	No. per animal-hour of obs.[b] (Lapche) Normal	Saltlick
Male V	14	10	46	10	98		1	179	25	15.5	.32	.03	.50
Male IV	8	3	34	3	49		1	98	14	13	.23	.14	.72
Male III	3	1	24	4	19			51	7	8	.19	0	.33
Male II	2		1	1	3			7	1	3.5	.05	.12	0
Male I	6		2	2	2			12	2	3	.07	.23	.28
Yearl. male	1	1	8	2	20			32	5	4	.22	.07	.08
Adult female	34	22	66	10	131	6		269	38	33	.21	.05	.08
Yearl. female	1		12		25			38	5	6	.12	.01	.05
Young	2		4	2	16			24	3	13	.05		

[a] The data are based on 1788 animal-hours of observation.
[b] Data are based on 634 animal-hours of observation in a normal situation and 89 hours at a saltlick (after Schaller, 1973b).

TABLE 38

Number of direct aggressions per animal-hour of observation in bharal during the pre-rut and rut

Age and sex class	Nov 2–10	Nov 11–19	Nov 20–28	Nov 29–Dec 5
Male V	.26 (.14)	.30 (.14)	.30 (.21)	.63 (.14)
Male IV	.52 (.13)	.19 (.14)	.19 (.10)	.16 (.09)
Male III	.58 (.27)	.05 (0)	.06 (.01)	.39 (.17)
Male II	.03 (0)	.13 (.08)	0 (0)	.13 (.04)
Male I	.21 (.06)	.17 (0)	.03 (0)	0 (0)
Yearl. male	.37 (.26)	.59 (.35)	.19 (.11)	.03 (.03)
Adult female	.46 (.22)	.04 (0)	.20 (.10)	.05 (.03)
Yearl. female	.17 (.08)	.39 (.35)	.07 (.04)	.08 (.07)
Young	.17 (.13)	0 (0)	0 (0)	.02 (0)
Total	2.77	1.86	1.04	1.36

NOTE: The first figure is based on the total number of aggressions and the one in parentheses on clashes only.

TABLE 39

Direct and indirect visual forms of aggression between males in several Caprinae

	Himalayan tahr	Nilgiri tahr	Kashmir markhor	Wild goat	Asiatic ibex	Domestic goat	Punjab urial	Mouflon	Mountain sheep	Soay sheep	Bharal	Aoudad	Mountain goat	Chamois	Muskox
DIRECT															
Jerk	+	+	+	+	+	+	+	+	+	+	+	+	+	+	+
Lunge	+	+	+	+	+	+	+	+	+	+	+	+	+	+	+
Chase	+		+	+	+	+	+	+	+	+	+	+	+	+	+
Jump	(+)	(+)	+	+	+	+	+	+	+	+?	+		+?	(+)	+
Butt	+	+	+	+	+	+	+	+	+	+	+	+	+	+	+
Clash	+	+	+	+	+	+	+	+	+	+	+	+		(+)	+
Head-to-tail	+	+	(+)			+?			(+)	+		+	+	(+)	
Shoulder-push			(+)	(+)	(+)	+	(+)	+	+	+	(+)	+			
Neck-fight	(+)			(+)				+	(+)	+?		+		(+)	
Horn-pull							(+)	+	(+)			+			
Bite				+?							(+)				

TABLE 39 (Continued)
Direct and indirect visual forms of aggression between males in several Caprinae

	Himalayan tahr	Nilgiri tahr	Kashmir markhor	Wild goat	Asiatic ibex	Domestic goat	Punjab urial	Mouflon	Mountain sheep	Soay sheep	Bharal	Aoudad	Mountain goat	Chamois	Muskox
INDIRECT															
Head-up			+	+		+	+	+	+	+	+			+	+
Head-down	+	+	+?	+			+		+?	+?	+		+	+	+
Broadside	+		+	(+)									+		
Parallel walk		(+)	(+)		+	(+)	+		+	+	+			+	+
Block				·	+		+	+			+				
Body shake														+	
Humped approach	+										+				
Spray urine	+		+	+	+	+								+	
Mouth penis	+		+	+	+	+					+	+			
Mount			(+)	(+)		+	+	+	+	+	+	+			
Low-stretch				(+)			+	+	+	+?	+				
Twist		(+)	(+)	(+)			+	+	+	+	(+)				
Kick				(+)			+	+	+	+	+				
Tongue-flick				+			+	+	+	+	+				+
Lip-curl														+	
Poke							(+)		(+)						
Dig rutting pit													+		
Head-shake					+			+	(+)		(+)?				+
Paw or stamp		+		(+)		+	(+)	+			(+)		+	+	+
Huddle							+	+	+	+?	+				
Horn object	+	+	+	+	+	+	+	+	+	+	+	+		+	+
Rub eye gland on leg															+

NOTE: This table includes my own observations and those of Walther (1961), Geist (1965, 1971a), Pfeffer (1967), Grubb (1974b, d), Haas (1958), Krämer (1969), and Gray (1973).
+ = trait observed.
(+) = observed but seems to be rare or found only in young.
+? = possibly present but observation not clear-cut.
Blank space = trait not observed so far or absent.

TABLE 40

Number of occurrences of three bharal displays per male-hour of observation before and during the main rut

Display	Nov 2–10	Nov 11–19	Nov 20–28	Nov 29–Dec 5
Penis mouthing	.03	.03	.06	.12
Mounting	.04	.07	.08	.06
Rump rubbing	.05	.23	.22	.22

TABLE 41

The kind of stimulation used by some Caprini males that induced females to urinate (in %)

Species	No. observations	Sniffing	Displaying[a]	Displaying and sniffing	Following	Mounting or attempting to mount
Wild goat	65	43	17	6	34	0
Bharal	265	59.5	11	25	3	1.5
Punjab urial	138	52	28	11	8	1

[a]Low-stretching, twisting, kicking.

TABLE 42

Courtship display frequencies (in %) by wild goat during various stages of the 1972 rut

	Sept 6–12	Sept 13–19	Sept 20–26	Sept 27–Oct 3	Total No.	%
No. display sequences	7	44	95	100		
No. sequences per hour observation[a]	.4	1.6	4.1	9.7	285	
% low-stretch	86	51	17	5	194	13
% twist	14	39	59	62	895	59
% kick	0	10	24	33	430	28
Average no. displays per sequence[b]	1.0±0	1.8±0.3	6.7±1.1	7.9±0.9		

After Schaller and Laurie (1974J.
[a]Weekly increases are significant (<0.05). Twists and kicks also increase significantly (p<0.001) from week to week.
[b]A t test on standard error of differences shows that all are significantly different (p<0.001) except for the third and fourth weeks.

TABLE 43

Courtship display frequencies (in %) by Punjab urial during various stages of the rut

	Oct 6–22 1970	Oct 23–31 1970	Nov 1–9 1970	Nov 15–18 1972	Total No.	%
No. display sequences	4	27	43	19	87	
% low-stretch	25	39.5	35	3	43	18
% twist	0	37	44	95	168	70
% kick	75	23.5	21	2	29	12
Av. no. displays per sequence	1.0±0	1.7±1.44	1.7±.91	7.8±14.46		

NOTE: Low-stretches and kicks decrease significantly between the third and fourth weeks and twists increase (p<0.01).

TABLE 44

Courtship display frequencies (in %) by bharal during various stages of the rut

	Nov 2–10	Nov 11–19	Nov 20–28	Nov 29–Dec 5	Total No.	%
No. display sequences	3	4	116	196	319	
No. sequences per hour observation[a]	.09	.1	3.2	8.5		
% low-stretch	33	75	77	82	336	80
% twist	33	0	6	7	29	7
% kick	33	25	17	11	57	13
Average no. displays per sequence	1.0±0	1.0±0	1.3±.72	1.3±.78		

[a]Increases from second to fourth week are significant.

TABLE 45

Coefficient of variation (CV) in the lengths of horn annuli in several Caprini populations from Pakistan and Nepal

Age	Wild goat		Chiltan goat		Sulaiman markhor		Kabul markhor		Kashmir markhor		Astor markhor		Asiatic ibex		Bharal		Punjab urial		Afghan urial		Marco Polo sheep	
	No.	CV	No.	CV	No.	CV	No.	CV	No.	CV	No.	CV	No.	CV	No.	CV	No.	CV	No.	CV	No.	CV
1½–2½	37	29.69	19	19.38	45	25.28	14	18.82	54	19.60	23	12.61	92	16.83	21	24.26	30	27.42	50	27.85	35	20.76
2½–3½	36	30.16	12	25.76	33	21.21	14	18.29	54	17.92	21	18.23	85	17.47	21	20.81	24	26.92	37	27.54	34	16.29
3½–4½	32	29.36	6	20.37	22	22.25	13	19.77	52	17.20	20	11.61	80	16.48	20	17.12	20	27.17	28	24.24	26	17.62
4½–5½	24	27.05			12	29.38	11	21.14	44	17.30	15	17.00	73	19.29	20	13.02	13	53.19	16	31.15	22	20.30
5½–6½	16	35.25			8	33.21	6	19.08	33	19.37	11	17.70	62	20.56	18	21.64	7	46.77	10	25.79	12	20.10
6½–7½	9	31.73			4	24.00	5	25.63	22	26.76	7	12.75	40	18.83	14	14.09			7	32.85	4	27.07

REFERENCES

Adams, A. 1858. Remarks on the habits and haunts of some of the Mammalia found in various parts of India and the Western Himalayan Mountains. Proc. Zool. Soc. London. 26:512–31.

Aeschbacher, A. In press. Brunftverhalten des Alpensteinwildes (*Capra ibex ibex* L.).

Ahmad, K. 1969. A geography of Pakistan. Karachi: Oxford Univ. Press.

Akasaka, T. 1974. Japanese serow in the wild. Wildlife. 16 (10): 452–58.

————, and Maruyama, N. In press. Preliminary report on social organization and habitat use of Japanese serow. J. Mammal. Soc. Japan.

Alexander, G., and Davis, H. 1959. Relationship of milk production to number of lambs born or suckled. Aust. J. Agric. Res. 10:720–24.

Ali, S. 1949. An ornithological pilgrimage to Lake Manasarowar and Mount Kailas. J. Bombay Nat. Hist. Soc. 46:285–308.

Allen, G. 1940. The mammals of China and Mongolia. Natural History of Central Asia, vol. 11, no. 2. New York: Am. Mus. Nat. Hist.

Altmann, D. 1970. Ethologische Studie an Mufflons, *Ovis ammon musimon* (Pallas). Zool. Gart. 39:297–303.

Anderson, J., and Henderson, J. 1961. Himalayan thar in New Zealand. N. Zealand Deerstalkers' Ass. Special Publ. No. 2. 36 pp.

Anonymous. 1894, reprinted 1972. The gazetteer of Sikkim. Bibliotheca Himalayica, vol. 8. Delhi: Manjusri Publ. House.

————. 1958. Tables of temperature, relative humidity and precipitation for the world. Part 5; Asia. London: Her Majesty's Stationery Office.

————. 1971a. North American big game. Pittsburgh: Boone and Crockett Club.

————. 1971b. Pakistan wildlife news. The Outdoorman (Karachi). 2:9–10.

————. 1972. The snow leopard in Pakistan. Animals. 14:256–59.

————. 1974. Increase in Walia ibex. Oryx. 12:405.

Aung, H. 1968. A note on the birth of a Mishmi takin at Rangoon zoo. *In* Int. Zoo Yearbook, vol. 8, ed. C. Jarvis, p. 145. London: Zool. Soc. London.

Autenrieth, R., and Fichter, E. 1975. On the behavior and socialization of pronghorn fawns. Wildl. Monogr. No. 42. 111 pp.

Backhaus, D. 1959. Experimentelle Untersuchungen über die Sehschärfe und das Farbsehen einiger Huftiere. Z. Tierpsych. 16:445–67.

Bailey, F. 1915. Notes from southern Tibet. J. Bombay Nat. Hist. Soc. 24:72–78.

———. 1945. China-Tibet-Assam. London: Jonathan Cape.

Baker, M., ed. 1964. Records of North American big game. New York: Holt, Rinehart and Winston.

Baldwin, J. 1876. The large and small game of Bengal and the northwestern provinces of India. London: Henry King.

Bannikov, A., Zhirnov, L., Lebedeva, L., and Fandeev, A. 1967. Biology of the Saiga. Jerusalem: Israel Program Scient. Trans.

Barash, D. 1973. The social biology of the Olympic marmot. Anim. Beh. Monogr. 6:172–244.

Barrett, R. 1966. History and status of introduced ungulates on Rancho Piedra Blanca, California. M.S. thesis, Univ. Michigan.

Bartz, F. 1935. Das Tierleben Tibets und des Himalaya-Gebirges. Wissenschaftliche Veröffentlichungen des Museums für Länderkunde zu Leipzig. 3:115–177.

Bell, R. 1971. A grazing ecosystem in the Serengeti. Scientific Am. 225:86–93.

Benirschke, K., Soma, H., and Ito, T. 1972. The chromosomes of the Japanese serow, *Capricornis crispus* (Temminck). Proc. Japan Academy. 48:608–612.

Berwick, S. 1976. The Gir Forest: an endangered ecosystem. Am. Scientist. 64:28–40.

Bibby, G. 1972. Looking for Dilmun. New York: A. A. Knopf.

Bigourdan, J., and Prunier, R. 1937. Les mammifères sauvages de l'ouest africain et leur milieu. Montrouge: Imprimerie Jean de Rudder.

Blanford, W. 1888–91. The fauna of British India. London: Taylor and Francis.

Blood, D., Flook, D., and Wishart, W. 1970. Weights and growth of Rocky Mountain bighorn sheep in Western Alberta. J. Wildl. Manage. 34:451–55.

Bogart, M., and Benirschke, K. 1975. Chromosomes of a male takin, *Budorcas taxicolor taxicolor.* Mammal. Chrom. Newsletter. 16:18.

Bonatti, E. 1966. North Mediterranean climate during the last Würm glaciation. Nature. 209:984–85.

Brandborg, S. 1955. Life history and management of the mountain goat in Idaho. Dept. Fish and Game, Bull. No. 2, Boise, Idaho. 142pp.

Brehm, A. 1918. Brehms Tierleben, vol. 4. Die Säugetiere. Leipzig: Bibliographisches Institut.

Brockelman, W. 1976. The conservation of diversity. Tigerpaper. 3:2–5.

Bromlei, G. 1956. Goral (*Nemorhaedus caudatus raddeanus* Heude 1894). Zool. J. 35 (9): 1395–1405. (In Russian.)

———. 1973. Bears of the south far-eastern USSR. New Delhi: Indian Nat. Scient. Documentation Centre.

Brouin, G. 1950. Notes sur les ongulés du Cercle d'Agadez et leur chasse. *In* Contribution a l'étude de L'Air, ed. G. Brouin, pp. 425–54.

Mem. l'institut français d'Afrique noire, no. 10. Paris: Librairie Larose.

Buechner, H. 1960. The bighorn sheep in the United States, its past, present and future. Wildl. Monogr. No. 4. 174 pp.

Bunch, T., and Valdez, R. In press. Comparative morphology, karyotypes and transferrins of Iranian and North American desert wild sheep. Trans. Desert Bighorn Council.

Burrard, G. 1925. Big game hunting in the Himalayas and Tibet. London: H. Jenkins.

Burrard, S., and Hayden, H. 1907. A sketch of the geography and geology of the Himalaya Mountains and Tibet. Calcutta: Superint. Governm. Print., India.

Burt, W., and Grossenheider, R. 1952. A field guide to the mammals. Boston: Houghton Mifflin.

Burton, R. 1915. Weights and measurements of game animals. J. Bombay Nat. Hist. Soc. 24:186.

———. 1926. Three months up the valley of the Sutlej River. J. Bombay Nat. Hist. Soc. 30:352–67.

Calvin, L. 1969. A brief note on the birth of snow leopards *Panthera uncia* at Dallas Zoo. *In* Int. Zoo Yearbook, vol. 9, ed. J. Lucas, p. 96. London: Zool. Soc. London.

Carruthers, D. 1913. Unknown Mongolia. 2 vols. London: Hutchinson.

Caughley, G. 1963. Dispersal rates of several ungulates introduced into New Zealand. Nature. 200:280–81.

———. 1965. Horn rings and tooth eruption as criteria of age in the Himalayan thar, *Hemitragus jemlahicus*. N. Zealand J. Science. 8:333–51.

———. 1966. Mortality patterns in mammals. Ecology. 47:333–51.

———. 1967. Growth, stabilization and decline in New Zealand populations of the Himalayan thar (*Hemitragus jemlahicus*). Ph.D. thesis, Univ. Canterbury, New Zealand.

———. 1969. Wildlife and recreation in the Trisuli Watershed and other areas in Nepal. FAO, UN Dev. Progr. Project Rep. No 6.

———. 1970a. *Cervus elaphus* in southern Tibet. J. Mammal. 51:611–14.

———. 1970b. Habitat of the Himalayan tahr *Hemitragus jemlahicus* (H. Smith). J. Bombay Nat. Hist. Soc. 67:103–5.

———. 1970c. Eruption of ungulate populations with emphasis on Himalayan thar in New Zealand. Ecology. 51:53–72.

———. 1970d. Fat reserves of Himalayan thar in New Zealand by season, sex, area and age. N. Zealand J. Science. 13:209–19.

———. 1971. The season of births for northern-hemisphere ungulates in New Zealand. Mammalia. 25:204–19.

Champion, H., Seth, S., and Khattak, G. N.d. Forest types of Pakistan. Peshawar: Pakistan Forest Institute.

Charles, R. 1957. Morphologie dentaire du thar et du bouquetin espèces actuelles et sub fossiles des gisements préhistoriques. Mammalia. 21:136–41.

Chitty, D. 1960. Population processes in the vole and their relevance to general theory. Can. J. Zool. 38:99–113.

Christie, A. 1964. A note on the chamois in New Zealand. Proc. N. Z. Ecol. Soc. 11:32–36.

———, and Andrews, J. 1964. Introduced ungulates in New Zealand. (a) Himalayan Tahr. Tuatara. 12:69–77.

Church, P. 1901. Chinese Turkestan with caravan and rifle. London: Rivingtons.

Clark, J. 1964. The great arc of the wild sheep. Norman: Univ. Oklahoma Press.

Cobbold, R. 1900. Innermost Asia. London: W. Heinemann.

Coblentz, B. 1976. Wild goats of Santa Catalina. Nat. Hist. 85:71–77.

Collias, N. 1956. Analysis of socialization in sheep and goats. Ecology. 37:228–39.

Cooper, H. 1923. The Mishmi takin (Budorcas taxicolor). J. Bombay Nat. Hist. Soc. 29:550–51.

Corbet, G. 1966. The terrestrial mammals of western Europe. London: G. Foulis.

Couturier, M. 1960. Ecologie et protection du Bouquetin (Capra aegagrus ibex ibex L.) et du chamois (Rupicapra rupicapra rupicapra L.) dans les Alpes. In Ecology and management of wild grazing animals in temperate zones, ed. F. Bourlière, pp. 54–73. Morges: IUCN.

———. 1962. Le bouquetin des Alpes. Grenoble: Privately printed.

Cowan, I. 1940. Distribution and variation in the native sheep of North America. Am. Midl. Nat. 24:505–80.

———, and McCrory, W. 1970. Variation in the mountain goat, Oreamnos americanus (Blainville). J. Mammal. 51:60–73.

Cracraft, J. 1973. Continental drift, paleoclimatology, and the evolution and biogeography of birds. J. Zool., London. 169:455–545.

Crandall, L. 1964. The management of wild mammals in captivity. Chicago: Univ. Chicago Press.

Crook, I. 1969. Feral goats of North Wales. Animals. 12:13–15.

Crook, J. 1965. The adaptive significance of avian social organization. Symp. Zool. Soc. London. 14:181–218.

———. 1970. Social organization and the environment: aspects of contemporary social ethology. Anim. Beh. 18:197–209.

———, and Gartlan, J. 1966. Evolution of primate societies. Nature. 210:1200–1203.

Cumberland, C. 1895. Sport on the Pamirs and Turkistan steppes. Edinburgh: Blackwood and Sons.

Cunningham, A. 1854, reprinted 1970. Ladak. New Delhi: Sagar Publ.

Dalimier, P. 1954. Quelques observations au sujet du mouflon à manchettes. Mammalia. 18:331–33.

Danford, C. 1875. Note on the wild goat (Capra aegagrus Gm). Proc. Zool. Soc. London, pp. 458–68.

Dang, H. 1962. Our less known species—the serow. The Cheetal. 5:18–19.

————. 1964. A natural sanctuary in the Himalaya: Nanda Devi and the Rishiganga Basin. The Cheetal. 7:34–40.

————. 1967. The snow leopard and its prey. The Cheetal. 10:72–84.

————. 1968a. Govind Pashu Vihar. The Cheetal. 11:65–83.

————. 1968b. The goral of Benog. The Cheetal. 11:47–58.

Daniel, J. 1971. The Nilgiri tahr, *Hemitragus hylocrius* Ogilby, in the High Range, Kerala and the southern hills of the Western Ghats. J. Bombay Nat. Hist. Soc. 67:535–42.

Darrah, H. 1898. Sport in the highlands of Kashmir. London: Rowland Ward.

Das, S. 1902. Journey to Lhasa and Central Tibet. London: John Murray.

Dasmann, R., and Mossman, A. 1962. Population studies of impala in Southern Rhodesia. J. Mammal. 43:375–95.

Davidar, E. 1975. The Nilgiri tahr. Oryx. 13:205–11.

De Beaux, O. 1956. Posizione sistematica degli stambecchi e capre selvatiche viventi (*Capra* Linneo 1758) e loro distribuzione geografica. Atti dela Accademia Ligure. 12:123–228.

Decker, E., and Kowalski, G. 1972. The behavior and ecology of urial sheep. Dept. Fish and Wildlife Biol., Col. State Univ., Ft. Collins, Col. 152 pp.

Demidoff, E. 1900. After wild sheep in the Altai and Mongolia. London: Rowland Ward.

Dolan, J. 1963. Beitrag zur systematischen Gliederung des Tribus Rupicaprini Simpson. Z. Zool. Sys. und Evolutionsforschung. 1:311–407.

Dollman, J., and Burlace, J. 1922. Rowland Ward's records of big game. London: Rowland Ward.

————. 1935. Rowland Ward's records of big game. London: Rowland Ward.

Douglas, M. 1971. Behaviour responses of red deer and chamois to cessation of hunting. N. Z. J. Science. 14:507–18.

Dubost, G. 1970. L'organisation spatiale et sociale de *Muntiacus reevesi* Ogilby 1839 en semi-liberté. Mammalia. 34:331–55.

Dunmore, Earl of. 1893. The Pamirs. 2 vols. London: John Murray.

Eaton, O. 1952. Weight and length measurements of fetuses of Karakul sheep and goats. Growth. 16:175–87.

Eckholm, E. 1975. The deterioration of mountain environments. Science. 189:764–70.

Eichler, R. 1973. Photographing the Spanish ibex. Animals. 15:352–56.

Eisenberg, J. 1966. The social organization of mammals. Handbuch der Zoologie. 8:1–92.

————, and Lockhart, M. 1972. An ecological reconnaissance of Wilpattu National Park, Ceylon. Smith. Contr. to Zool., No. 101., Smith. Inst., Washington. 118 pp.

————, Muckenhirn, N., and Rudran, R. 1972. The relationship between ecology and social structure in primates. Science. 176:863–74.

Ellerman, J., and Morrison-Scott, T. 1951. Checklist of Palaearctic and Indian mammals 1758 to 1946. London: British Museum.

Elliott, J. 1973. Field sports in India 1800–1947. London: Gentry Books.

Elliott, P. 1975. Longevity and the evolution of polygamy. Am. Nat. 109:281–87.

Engelmann, C. 1938. Über die Grosssäuger Szetschwans, Sikongs und Osttibets. Z. Säugetierk. 13:1–76.

Epstein, H. 1972. The origin of the domestic animals of Africa, vol. 2. New York: Africana Publ. Corp.

Estes, R. 1966. Behaviour and life history of the wildebeest (*Connochaetes taurinus* Burchell). Nature. 212:999–1000.

————. 1972. The role of the vomeronasal organ in mammalian reproduction. Mammalia. 36:315–41.

————. 1974. Social organization of the African Bovidae. *In* The behaviour of ungulates and its relation to management, ed. V. Geist and F. Walther, pp. 166–205. IUCN Publ. No. 24. Morges: IUCN.

Etherton, P. 1911. Across the roof of the world. New York: Frederick A. Stokes.

Evans, P. 1967. The aoudad sheep, an exotic introduced in the Palo Duro Canyon of Texas, pp. 183–88. Proc. 21st Annual Conf. S. E. Ass. Game and Fish Comm.

Fairley, J. 1975. The lion river: the Indus. New York: John Day.

Falconer, D. 1953. Selection for large and small size in mice. J. Genet. 51:470–501.

Fletcher, F. 1911. Sport on the Nilgiris and in Wynaad. London: Macmillan.

Flint, V., Chugonov, U., and Smirin, V. 1965. Mammals of the USSR. Moscow: Meesl Publ.

Flower, W. 1932. Notes on the recent mammals of Egypt, with a list of the species recorded from that kingdom. Proc. Zool. Soc. London, pp. 369–450.

Fraser, A. 1968. Reproductive behaviour in ungulates. London: Academic Press.

Freeland, W., and Janzen, D. 1974. Strategies in herbivory by mammals: the role of plant secondary compounds. Am. Nat. 108:269–86.

Freeman, H. 1975. A preliminary study of the behaviour of captive snow leopard, *Panthera uncia*. *In* Int. zoo yearbook, vol. 15, ed. N. Duplaix-Hall, pp. 217–22. London: Zool. Soc. London.

————, and Hutchins, M. In press. Captive management of snow leopard cubs: an overview. Zool. Garten.

Frenzel, B. 1968. The Pleistocene vegetation of Northern Eurasia. Science. 161:637–49.

Frueh, R. 1968. A note on breeding snow leopards at St. Louis zoo. *In* Int. zoo yearbook, vol. 8, ed. C. Jarvis, pp. 74–76. London: Zool. Soc. London.

Gansser, A. 1964. Geology of the Himalayas. London: John Wiley.

Gee, E. 1967a. Occurrence of the Nayan or great Tibetan sheep, *Ovis ammon hodgsoni* Blyth in Bhutan. J. Bombay Nat. Hist. Soc. 64:553.

————. 1967b. Occurrence of brown bear, *Ursus arctos* Linnaeus, in Bhutan. J. Bombay Nat. Hist. Soc. 64:551–52.

Geist, V. 1965. On the rutting behavior of the mountain goat. J. Mammal. 45:551–68.

————. 1966. The evolution of horn-like organs. Behaviour. 27:177–214.

————. 1968. On the inter-relation of external appearance, social behaviour and social structure of mountain sheep. Z. Tierpsych. 25:199–215.

————. 1969. *Ovis canadensis* (Bovidae); social behaviour of males. Ency. Cinematographia, Göttingen.

————. 1971a. Mountain sheep. Chicago: Univ. Chicago Press.

————. 1971b. The relation of social evolution and dispersal in ungulates during the Pleistocene, with emphasis on the Old World deer and the genus *Bison*. Quat. Research. 1:285–315.

————. 1971c. A behavioural approach to the management of wild ungulates. *In* The scientific management of animal and plant communities for conservation, ed. E. Duffey and A. Watt, pp. 413–24. Oxford: Blackwell Scientific Publ.

————. 1974a. On the relationship of social evolution and ecology in ungulates. Am. Zool. 14:205–20.

————. 1974b. On the relationship of ecology and behaviour in the evolution of ungulates: theoretical considerations. *In* The behaviour of ungulates and its relation to management, eds. V. Geist and F. Walther, pp. 235–46. IUCN Publ. No. 24. Morges: IUCN.

Gentry, A. 1968. Historical zoogeography of antelopes. Nature. 217:874–75.

Gillan, G. 1935. The distribution of the Great Pamir sheep (*Ovis ammon poli*, Blyth). J. Bombay Nat. Hist. Soc. 37:216–17.

Goethe, F., and Goethe, E. 1939. Aus dem Jugendleben des Muffelwildes. Zool. Gart. 11:1–22.

Goodwin, H., and Holloway, C. 1972. IUCN Red Data Book, Mammalia. Lausanne: Heliographia.

Gordon, T. 1876. The roof of the world. Edinburgh: Edmonston and Douglas.

Gray, A. 1954. Mammalian hybrids. Commonwealth Agr. Bureaux, Farnham Royal, England. 144 pp.

Gray, D. 1973. Social organization and behavior of muskoxen (*Ovibos moschatus*) on Bathurst Island, N.W.T. Ph.D. Thesis, Univ. Alberta, Edmonton. 212 pp.

————. 1974. The defence formation of the musk-ox. The Musk-ox (Inst. North. Studies, Univ. Sask.) 14:25–29.

Grubb, P. 1974a. Social organization of Soay sheep and the behaviour of ewes and lambs. *In* Island survivors, ed. P. Jewell, C. Milner, and J. Boyd, pp. 131–59. London: Athlone Press.

————. 1974b. The rut and behaviour of Soay rams. *In* Island survivors, ed. P. Jewell, C. Milner, and J. Boyd, pp. 195–223. London: Athlone Press.

—— 1974c. Population dynamics of the Soay sheep. *In* Island survivors, ed. P. Jewell, C. Milner, and J. Boyd, pp. 242–72. London: Athlone Press.

——. 1974d. Mating activity and the social significance of rams in a feral sheep community. *In* The behaviour of ungulates and its relation to management, ed. V. Geist and F. Walther, pp. 457–76. IUCN Publ. No. 24. Morges: IUCN.

——, and Jewell, P. 1974. Movement, daily activity and home range of Soay sheep. *In* Island survivors, ed. P. Jewell, C. Milner, and J. Boyd, pp. 160–94. London: Athlone Press.

Gruber, U. 1969. Tiergeographische, Ökologische und Bionomische Untersuchungen an kleinen Säugetieren in Ost-Nepal. Khumbu Himal. 3:197–312.

Guggisberg, C. 1975. Wild cats of the world. New York: Taplinger.

Gulisashvili, V. 1973. On the refuges of arboreal flora in the Caucasus during the glacial period. Bull. Moscow Soc. Naturalists, Biol. Ser. 78:82–88.

Guthrie, R. 1968. Paleoecology of the large-mammal community in interior Alaska during the late Pleistocene. Am. Mid. Nat. 79:346–63.

——. 1971. A new theory of mammalian rump patch evolution. Behaviour. 38:132–45.

——, and Petocz, R. 1970. Weapon automimicry among mammals. Am. Nat. 104:585–88.

Haas, G. 1958. Untersuchungen über angeborene Verhaltensweisen bei Mähnenspringern (*Ammotragus lervia* Pallas). Z. Tierpsych. 16:218–42.

Hafez, E., and Scott, J. 1962. The behaviour of sheep and goats. *In* The behaviour of domestic animals, ed. E. Hafez, pp. 297–333. Baltimore: Williams and Wilkins.

Hagen, T. 1963. The evolution of the highest mountains in the world. *In* Mount Everest, ed. T. Hagen, G. Dyhrenfurth, C. v Führer-Haimendorf, and E. Schneider, pp. 1–96. London: Oxford Univ. Press.

——. 1970. Nepal. Chicago: Rand McNally.

Haltenorth, T. 1961. Fruchtbare Rückkreuzung eines weiblichen Mähnenschaf-Hausziegenbastards mit einem Steinbock. Säugtierkundl. Mitt. 9:105–9.

——. 1963. Klassifikation der Säugetiere: Artiodactyla. Handbuch der Säugetiere. 8:1–167.

——, and Trense, W. 1956. Das Grosswild der Erde und seine Trophäen. Bonn: Bay. Landwirtschaftsverlag.

Harper, F. 1945. Extinct and vanishing mammals of the Old World. Am. Comm. Internatl. Wildlife Prot. Special Publ. No. 12. New York: New York Zool. Park.

Harper, L. 1970. Ontogenetic and phylogenetic functions of the parent-offspring relationship in mammals. *In* Advances in the study of animal behavior, ed. D. Lehrman, R. Hinde, and E. Shaw, pp. 75–117. New York: Academic Press.

Harrer, H. 1953. Seven years in Tibet. London: Hart-Davis.

Harris, A., and Mundel, P. 1974. Size reduction in bighorn sheep (*Ovis canadensis*) at the close of the Pleistocene. J. Mammal. 55:678–80.

Harrison, D. 1968a. The mammals of Arabia, vol. 2. London: Ernest Benn.

—————. 1968b. On three mammals new to the fauna of Oman, Arabia, with the description of a new subspecies of bat. Mammalia. 32:317–25.

Haughton, H. 1913. Sport and folklore in the Himalaya. London: Edward Arnold.

Hayden, H., and Cosson, C. 1927. Sport and travel in the highlands of Tibet. London: R. Cobden-Sanderson.

Hedin, S. 1898. Through Asia. 2 vols. London: Methuen.

—————. 1903. Central Asia and Tibet, vol. 2. London: Hurst and Blackett.

Hemmer, H. 1966. Untersuchungen zur Stammesgeschichte der Pantherkatzen (Pantherinae). Veröff. Zool. Staatssamml. München. 11:1–121.

Heptner, V., Nasimovic, A., and Bannikov, A. 1966. Die Säugetiere der Sowjetunion. Vol. 1, Paarhufer und Unpaarhufer. Jena: Gustav Fischer Verlag.

Herre, W. and Kesper, K. 1953. Zur Verbreitung von *Ovis ammon* in Europa. Zool. Anzeiger. 151:204–9.

—————, and Röhrs, M. 1955. Über die Formenmannigfaltigkeit des Gehörns der *Caprini* Simpson 1945. Zool. Gart. 22:85–110.

Herscher, L., Richmond, J., and Moore, A. 1963. Maternal behavior in sheep and goats. *In* Maternal behavior in mammals, ed. H. Rheingold, pp. 203–32. New York: John Wiley.

Hingston, R. 1925. Animal life at high altitudes. Geogr. J. 65:185–98.

Hjeljord, O. 1973. Mountain goat forage and habitat preference in Alaska. J. Wildl. Manage. 37:353–62.

Hoefs, M. 1974. Food selection by Dall's sheep (*Ovis dalli dalli* Nelson). *In* The behaviour of ungulates and its relation to management, ed. V. Geist and F. Walther, pp. 758–86. IUCN Publ. No. 24. Morges: IUCN.

Holdsworth, R. 1932. The flowers of the Kamet and Badrinath Ranges. *In* Kamet conquered, F. Smythe, pp. 348–71. London: Victor Gollancz.

Holloway, C., and Jungius, H. 1973. Reintroduction of certain mammals and bird species into the Gran Paradiso National Park. Zool. Anz. 191:1–44.

Holroyd, J. 1967. Observations of Rocky Mountain goats on Mount Wardle, Kootenay National Park, British Columbia. Can. Field Nat. 81:1–24.

Hoogstraal, H., Kaiser, M., and Mitchell, R. 1970. *Anomalohimalaya lama*, new genus and new species (Ixodoidea: Ixodidae), a tick parasitizing rodents, shrews, and hares in the Tibetan highlands of Nepal. Ann. Ent. Soc. Am. 63:1576–85.

Hooker, J. 1854. Himalayan Journals. 2 vols. London: John Murray.

Hornocker, M. 1969. Winter territoriality in mountain lions. J. Wildl. Manage. 33:457–64.

Hsü, K. 1972. When the Mediterranean dried up. Scient. Am. 227:27–36.

Hudson, W. 1892. The naturalist in La Plata. London: J. Dent.

Hutton, A. 1947. The Nilgiri tahr (*Hemitragus hylocrius*). J. Bombay Nat. Hist. Soc. 47:374–76.

Jarman, P. 1974. The social organization of antelope in relation to their ecology. Behaviour. 48:215–67.

Jerdon, T. 1874. The mammals of India. London: John Weldon.

Jewell, P., and Grubb, P. 1974. The breeding cycle, the onset of oestrus and conception in Soay sheep. *In* Island survivors, ed. P. Jewell, C. Milner, and J. Boyd, pp. 224–41. London: Athlone Press.

Johnson, B. 1969. South Asia. London: Heinemann.

Joleaud, L. 1928. Le mouflon à Manchettes. Mem. de la Soc. de Biogéographie, Paris. 2:35–37.

Jonkel, C., Gray, D., and Hubert, B. 1975. Immobilizing and marking wild muskoxen in Arctic Canada. J. Wildl. Manage. 39:112–17.

Katz, I. 1949. Behavioral interactions in a herd of Barbary sheep (*Ammotragus lervia*). Zoologica. 34:9–18.

Kelsall, J. 1968. The caribou. Can. Wildl. Serv. Monogr. No. 3. Ottawa: Queen's Printer.

———. 1969. Structural adaptations of moose and deer for snow. J. Mammal. 50:302–10.

Kennion, R. 1910. Sport and life in the Further Himalaya. Edinburgh: Blackwood.

Kinloch, A. 1892. Large game shooting in Thibet, the Himalayas, Northern and Central India. Bombay: Thacker, Spink and Co.

Kitchener, S., Meritt, D., and Rosenthal, M. 1975. Observations on the breeding and husbandry of snow leopards, *Panthera uncia. In* Int. Zoo Yearbook, vol. 15, ed. N. Duplaix-Hall, pp. 212–17. London: Zool. Soc. London.

Klein, D. 1965. Ecology of deer range in Alaska. Ecol. Monogr. 35:259–84.

———. 1968. The introduction, increase and crash of reindeer on St. Matthew Island. J. Wildl. Manage. 32:350–67.

———. 1970. Food selection by North American deer and their response to over-utilization of preferred plant species. *In* Animal populations in relation to their food resources, ed. A. Watson, pp. 25–46. Oxford: Blackwell Sci. Publ.

Klingel, H. 1967. Soziale Organisation und Verhalten freilebender Steppenzebras. Z. Tierpsych. 24:580–624.

Klopfer, P., and Klopfer, M. 1968. Maternal "imprinting" in goats: the fostering of alien young. Z. Tierpsych. 23:588–93.

Knaus, W. and Schröder, W. 1975. Das Gamswild. Paul Parey, Hamburg.

Koby, P. 1956. Une représentation de Tahr (*Hemitragus*) à Cougnac? Bull. Soc. Prehisto. Fr. 53:103–7.

Koch, W. 1931. Das Gehörn der Schraubenziege (*Capra falconeri* Wagn.)

Zool. Anzeiger 93:275–78.

Komori, A. N.d. Survey on the breeding of Japanese serows, *Capricornis crispus*, in captivity. Mimeographed, seen 1976. 6 pp.

Korobitsyna, K., Nadler, C., Vorontsov, N., and Hoffmann, R. 1974. Chromosomes of the Siberian snow sheep, *Ovis nivicola*, and implications concerning the origin of Amphiberingian wild sheep (Subgenus *Pachyceros*). Quat. Res. 4:235–45.

Krämer, A. 1969. Soziale Organisation und Sozialverhalten einer Gemspopulation (*Rupicapra rupicapra* L.) der Alpen Z. Tierpsych. 26:889–964.

———, and Aeschbacher, A. 1971. Zum Fluchtverhalten des Steinwildes (*Capra ibex*) im Oberengadin, Schweiz. Säugetierk. Mitt. 19:164–71.

Kukla, G. 1976. Around the ice age world. Nat. Hist. 85:56–61.

Kumerloeve, H. 1967. Zur Verbreitung kleinasiatischer Raub—und Huftiere sowie einiger Grossnager. Säugetierk. Mitt. 15:337–409.

Kurtén, B. 1968. Pleistocene mammals of Europe. London: Weidenfeld and Nicolson.

———. 1972. The age of mammals. New York: Columbia Univ. Press.

Kurup, G. 1966. Mammals of Assam and adjoining areas. Proc. Zool. Soc., Calcutta. 19:1–21.

Kusnetzov, G., and Matjushkin, E. 1962. The snow leopard is hunting. Nature. 12:65–67. (In Russian.)

Lambden, J. 1966. A note on the breeding and hand-rearing of a Marco Polo sheep *Ovis ammon poli* at London zoo. *In* Int. Zoo Yearbook, vol. 6, ed. C. Jarvis, pp. 90–93. London: Zool. Soc. London.

Lay, D. 1967. A study of the mammals of Iran. Fieldiana. 54:1–282.

———, Nadler, C., and Hassinger, J. 1971. The transferrins and hemoglobins of wild Iranian sheep (*Ovis* Linnaeus). Comp. Biochem. Physiol. 40:521–29.

Lent, P. 1971. Muskox management controversies in North America. Biol. Cons. 3:255–63.

———. 1974. Mother-infant relationships in ungulates. *In* The behaviour of ungulates and its relation to management, ed. V. Geist and F. Walther, pp. 14–55. IUCN Publ. No. 24. Morges: IUCN.

———, and Knutson, D. 1971. Muskox and snow cover on Nunivak Island, Alaska. *In* Snow and ice in relation to wildlife and recreation, pp. 50–62. Iowa Coop. Wildl. Res. Unit, Iowa State Univ., Ames.

Leuthold, W. 1967. Beobachtungen zum Jugendverhalten der Kob-Antilopen. Z. Säugetierk. 32:59–62.

———. 1974. Observations on home range and social organization of lesser kudu *Tragelaphus imberbis*. *In* The behaviour of ungulates and its relation to management, ed. V. Geist and F. Walther, pp. 206–34. IUCN Publ. No. 24. Morges: IUCN.

Lydekker, R. 1898. Wild oxen, sheep, and goats of all lands. London: Rowland Ward.

———. 1902. The wild sheep of the Upper Ili and Yana Valleys. Proc. Zool. Soc. London, pp. 80–85.

———. 1913. Catalogue of the ungulate mammals in the British

Museum, vol. 1. London: British Museum.

―――. 1924. The game animals of India, Burma, Malaya and Tibet. London: Rowland Ward.

MacArthur, R. 1965. Patterns of species diversity. Biol. Rev. 40: 510–33.

―――, and Wilson, E. 1963. An equilibrium theory in island biogeography. Evolution. 17:373–87.

Macintyre, D. 1891. Hindu-Koh. Edinburgh: William Blackwood and Sons.

Mani, M. 1962. Introduction to high altitude entomology. London: Methuen.

―――. 1968. Ecology and biogeography of high altitude insects. The Hague: W. Junk Publ.

―――, ed. 1974. Ecology and biogeography in India. The Hague: W. Junk Publ.

Markham, F. 1854. Shooting in the Himalayas. London: Richard Bentley.

Marma, B., and Yunchis, V. 1968. Observations on the breeding, management and physiology of snow leopards *Panthera u. uncia* at Kaunas Zoo from 1962 to 1967. *In* Int. Zoo Yearbook, vol. 8, ed. C. Jarvis, pp. 65–74. London: Zool. Soc. London.

Mason, K. 1955. Abode of snow. New York: E. P. Dutton.

Mayr, E. 1969. Species, speciation and chromosomes. *In* Comparative mammalian cytogenetics, ed. K. Benirschke, pp. 1–7. New York: Springer Verlag.

McDougall, P. 1975. The feral goats of Kielderhead Moor. J. Zool. Lond. 176:215–46.

McElroy, C. 1971. Sky-high hunt for a giant ram. Sports Afield. 165:50–51, 138–48.

Meinertzhagen, R. 1927. Ladakh, with special reference to its natural history. Geogr. J. 70:129–63.

Mellon, J. 1972. The thirteenth ram. Outdoor Life. 149:42–45, 156–59.

Miller, J. 1913. Sport on the plateaux of Mongolia. *In* Unknown Mongolia, vol. 2, D. Carruthers, pp. 319–50. London: Hutchinson.

Moody, P. 1958. Serological evidence on the relationships of the musk-ox. J. Mammal. 39:554–59.

Mountfort, G. 1969. The vanishing jungle. London: Collins.

Müller-Schwarze, D. 1974. Social functions of various scent glands in certain ungulates and the problems encountered in experimental studies of scent communication. *In* The behaviour of ungulates and its relation to management. ed. V. Geist and F. Walther, pp. 107–13. IUCN Publ. No. 24. Morges: IUCN.

Nadler, C., Hoffmann, R., and Woolf, A. 1973. G-band patterns as chromosomal markers and the interpretation of chromosomal evolution in wild sheep (*Ovis*). Experientia. 29: 117–19.

―――. 1974. G-band patterns, chromosomal homologies, and evolutionary relationships among wild sheep, goats, and aoudads (Mammalia, Artiodactyla). Experientia. 30:744–46.

Nadler, C., Korobitsina, K., Hoffmann, R., and Vorontsov, N. 1973. Cytogenetic differentiation, geographic distribution, and domestication in Palearctic sheep (*Ovis*). Z. Säugetierk. 38:109–25.

————, and Lay, D. 1975. Chromosomes of some Asian wild sheep (*Ovis*) and goats (*Capra*). Chrom. Inf. Service. 18:28–31.

————, Lay, D., and Hassinger, J. 1971. Cytogenetic analyses of wild sheep populations in northern Iran. Cytogenetics. 10:137–52.

Nasimovich, A. 1955. The role of the regime of snow cover in the life of ungulates in the USSR. Moskva: Akad. Nauk. (translated from the Russian by Canadian Wildlife Service, Ottawa).

Newsome, A. 1965. Reproduction in natural populations of the red kangaroo *Megaleia rufa* in central Australia. Aust. J. Zool. 13:735–59.

Nievergelt, B. 1966a. Der Alpensteinbock (*Capra ibex* L.) in seinem Lebensraum. Mammalia depicta. Hamburg: Paul Parey.

————. 1966b. Unterschiede in der Setzzeit beim Alpensteinbock (*Capra ibex* L.). Revue Suisse de Zool. 73:446–54.

————. 1967. Die Zusammensetzung der Gruppen beim Alpensteinbock. Z. Säugetierk. 32:129–44.

————. 1970. Simien: Ethiopia's threatened mountain area. *In* The mountain world, ed. M. Barnes, pp. 132–37. London: George Allen and Unwin.

————. 1974. A comparison of rutting behaviour and grouping in the Ethiopian and Alpine ibex. *In* The behaviour of ungulates and its relation to management, ed. V. Geist and F. Walther, pp. 324–40. IUCN Pub. No. 24. Morges: IUCN.

Novikov, G. 1962. Fauna of USSR., No. 62., Carnivorous mammals. Jerusalem: Israel Program Scient. Trans.

Numata, M. 1966. Vegetation and conservation in Eastern Nepal. J. College Arts and Sciences, Chiba Univ. 4:559–69.

Ognev, S. 1962. Mammals of USSR and adjacent countries, vol. 3. Jerusalem: Israel Program Scient. Trans.

Ogren, H. 1965. Barbary sheep. Department of Game and Fish Bull. No. 13., Santa Fe. 117 pp.

Okada, Y., and Kakuta, T. 1970. Studies on the Japanese serow (*Capricornis crispus* (Temminck). *In* Soc. Pres. Japanese Serow, Suzuka Mountain (10th anniv. vol.), pp. 1–15.

Orians, G. 1969. On the evolution of mating systems in birds and mammals. Amer. Nat. 103:589–603.

Owen-Smith, N. 1974. The social system of the white rhinoceros. *In* The behaviour of ungulates and its relation to management, ed. V. Geist and F. Walther, pp. 341–51. IUCN Publ No. 24. Morges: IUCN.

Papageorgiou, N. 1974. Population energy relationships of the agrimi (*Capra aegagrus cretica*) on Theodorou Island, Greece. Ph.D. thesis, Michigan State Univ.

Parker, G. 1974. Courtship persistence and female-guarding as male time investment strategies. Behaviour. 48:157–84.

Payne, S. 1968. The origins of domestic sheep and goats: a reconsidera-

tion in the light of fossil evidence. Proc. Prehist. Soc. 34:368–84.
Peissel, M. 1967. Mustang, the forbidden kingdom. New York: E. P. Dutton.
Petocz, R. 1973a. Conservation of wild ungulates. *In* World Wildlife Yearbook, 1972–73, ed. P. Jackson, pp. 117–20. Morges: World Wildlife Fund.
———. 1973b. The effect of snow cover on the social behavior of bighorn rams and mountain goats. Can. J. Zool. 51:987–93.
Petzsch, H. 1957. Reflexionen zur Phylogenie der *Capridae* im allgemeinen und der Hausziege im besonderen. Wiss. Zeit. der Martin-Luther-Univ., Halle-Wittenberg. 6:995–1019.
———, and Witstruk, K. 1958. Beobachtungen an daghestanischen Turen (*Capra caucasica cylindricornis* Blyth) im Berg-Zoo Halle. Zool. Garten. 25:6–29.
Pfeffer, P. 1967. Le mouflon de Corse (*Ovis ammon musimon* Schreber, 1782); position systématique, écologie, et ethologie comparées. Mammalia. 31 (supplement):1–262.
———, and Settimo, R. 1973. Déplacements saisonniers et compétition vitale entre mouflons, chamois et bouquetins dans la réserve du Mercantour (Alpes Maritimes). Mammalia. 37:203–19.
Philip, Duke of Edinburgh, and Fisher, J. 1970. Wildlife Crisis. London: Hamish Hamilton.
Pilgrim, G. 1925. The migration of Indian mammals. Proc. 12th Indian Sci. Congress, pp. 200–218.
———. 1939. The fossil Bovidae of India. Pal. Ind. N. S. 26:1–356.
———. 1947. The evolution of the buffaloes, oxen, sheep, and goats. Linn. Soc. London; J. Zool. 41:272–86.
Pleticha, P. 1972. Die Setzzeiten des Alpensteinbocks (*Capra ibex ibex*) in Gefangenschaft. Säugetierk. Mitt. 20:354–59.
Pocock, R. 1910. On the specialized cutaneous glands of ruminants. Proc. Zool. Soc. London, pp. 840–986.
———. 1918. On some external characters of ruminant Artiodactyla-Part 3; The Antilopinae, Rupicaprinae, and Caprinae, with a note on the penis of the Cephalophinae and Neotraginae. Ann. Mag. Nat. Hist. 9:125–44.
———. 1939. The fauna of British India. Mammalia, vol. 1. London: Taylor and Francis.
Pohle, C. 1974. Haltung und Zucht der Saiga-Antilope (*Saiga tatarica*) im Tierpark Berlin. Zool. Garten. 44:387–409.
Pohle, H. 1944. *Hemitragus jemlahicus schaeferi* sp. n., die östliche Form des Thars. Zool. Anzeiger. 144:184–91.
Polo, M. 1958. The travels. Harmondsworth: Penguin Books.
Powell-Cotton, P. 1902. A sporting trip through Abyssinia. London: Rowland Ward.
Prakash, I. 1963. Zoogeography and evolution of the mammalian fauna of Rajasthan Desert, India. Mammalia. 27:342–51.
Prater, S. 1965. The book of Indian animals. Bombay: Bombay Nat. Hist. Soc.

Przewalski, N. 1884. Reisen in Tibet am oberen Lauf des Gelben Flusses in den Jahren 1879–1880. Jena.

Raibes, R., and Dyson, R. 1961. The prehistoric climate of Baluchistan and the Indus Valley. Am. Anth. 63:265–81.

Ralls, K. 1975. Agonistic behavior in Maxwell's duiker, *Cephalophus maxwelli*. Mammalia. 39:241–49.

Rau, M. 1974. Vegetation and phytogeography of the Himalaya. *In* Ecology and biogeography in India, ed. M. Mani, pp. 247–80. The Hague: W. Junk Publ.

Rawling, C. 1905. The great plateau. London: Edward Arnold.

Reed, C., and Palmer, H. 1964. A late Quaternary goat (*Capra*) in North America? Z. Säugetierk. 29:372–78.

———, and Schaffer, W. 1972. How to tell the sheep from the goats. Field Mus. Bull. 43:1–7.

Roberts, G. 1971. Chamois and thar in New Zealand. Animals. 13:772–75.

Roberts, T. 1967. A note on *Capra hircus Blythi* Hume, 1875. J. Bombay Nat. Hist. Soc. 64:358–65.

———. 1969. A note on *Capra falconeri* (Wagner, 1839). Z. Säugetierk. 34:238–49.

———. 1973. Conservation problems in Baluchistan with particular reference to wildlife preservation. Pak. J. Forestry. 23:117–27.

Rodd, F. 1926. People of the veil. London: Macmillan.

Roosevelt, T., and Roosevelt, K. 1926. East of the sun and west of the moon. New York: Charles Scribner's Sons.

Rothschild, Lord. 1921. Ungulate mammals collected by Captain Angus Buchanan. Novitates Zoologicae. 28:75.

Rudge, M. 1969. Reproduction of feral goats *Capra hircus* L. near Wellington, New Zealand. N. Z. J. Science. 12:817–27.

———. 1970. Mother and kid behaviour in feral goats (*Capra hircus* L.). Z. Tierpsych. 27:687–92.

Ryder, M. 1958. Follicle arrangement in skin from wild sheep, primitive domestic sheep and in parchment. Nature. 182:781–83.

Sadleir, R. 1969. The ecology of reproduction in wild and domestic mammals. London: Methuen.

Sambraus, H. 1973. Das Sexualverhalten der domestizierten einheimischen Wiederkäuer. Fortschritte d. Verhaltenforschung. No. 12. Berlin: Paul Parey.

Schäfer, E. 1933. Berge, Buddhas und Bären. Berlin: Paul Parey.

———. 1937. Über das Zwergblauschaf (*Pseudois* spec. nov) und das Grossblauschaf (*Pseudois nahoor* Hdgs.) in Tibet. Zool. Gart. 9:263–78.

———. 1950. Über den Schapi (*Hemitragus jemlahicus Schaeferi*) Zool. Anz. 145:247–60.

Schaffer, W., and Reed, C. 1972. The co-evolution of social behavior and cranial morphology in sheep and goats (Bovidae Caprini). Fieldiana, Zoology. 61:1–88.

Schaller, G. 1967. The deer and the tiger. Chicago: Univ. Chicago Press.

———. 1969a. Observations on the hangul or Kashmir stag (*Cervus elaphus hanglu* Wagner). J. Bombay Nat. Hist. Soc. 66:1–7.

———. 1969b. Food habits of the black bear (*Selenarctos thibetanus*) in the Dachigam Sanctuary, Kashmir. J. Bombay Nat. Hist. Soc. 66:156–59.

———. 1971. Observations on the Nilgiri tahr (*Hemitragus hylocrius* Ogilby, 1838). J. Bombay Nat. Hist. Soc. 67:365–89.

———. 1972a. On meeting a snow leopard. Animal Kingdom. 75:7–13.

———. 1972b. The Serengeti lion. Chicago: Univ. of Chicago Press.

———. 1973a. Observations on Himalayan tahr (*Hemitragus jemlahicus*). J. Bombay Nat. Hist. Soc. 70:1–24.

———. 1973b. On the behaviour of blue sheep (*Pseudois nayaur*). J. Bombay Nat. Hist. Soc. 69:523–37.

———. 1973c. Chitral Gol reserve. *In* World Wildlife Yearbook, 1972–73, ed. P. Jackson, pp. 136–38. Morges: World Wildlife Fund.

———. 1975. A walk in the Hindu Kush. Animal Kingdom. 78:8–19.

———. 1976a. Pakistan: Status of wildlife and research needs. *In* Ecological guidelines for the use of natural resources in the Middle East and South West Asia, pp. 133–43. IUCN Publ. No. 34. Morges: IUCN.

———. 1976b. Mountain mammals in Pakistan. Oryx. 13:351–56.

———. In press. A note on a population of *Gazella gazella bennetti*. J. Bombay Nat. Hist. Soc.

———, and Hamer, A. In press. Rutting behavior of Père David's deer. Zool. Garten.

———, and Khan, S. 1975. The status and distribution of markhor (*Capra falconeri*). Biol. Cons. 7:185–98.

———, and Laurie, A. 1974. Courtship behaviour of the wild goat. Z. Säugetierk. 39:115–27.

———, and Mirza, Z. 1971. On the behaviour of Kashmir markhor (*Capra falconeri cashmiriensis*). Mammalia. 35:548–66.

———, and Mirza, Z. 1974. On the behaviour of Punjab urial (*Ovis orientalis punjabiensis*). *In* The behaviour of ungulates and its relation to management, ed. V. Geist and F. Walther, pp. 306–23. IUCN Publ. No. 24. Morges: IUCN.

Schmitt, J. 1963. *Ammotragus lervia* Pallas. Mähnenschaf oder Mähnenziege? Z. Säugetierk. 28:7–12.

Scholander, P., Hock, R., Walters, V., and Irving, L. 1950. Adaptations to cold in Arctic and tropical mammals and birds in relation to body temperature, insulation and basal metabolic rate. Biol. Bull. 99:259–71.

Schultze-Westrum, T. 1963. Die Wildziegen der agäischen Inseln. Säugetierk. Mitt. 11:145–82.

Schweinfurth, U. 1957. Die horizontale und vertikale Verbreitung der Vegetation im Himalaya. Bonner Geograph. Abhandlungen. 20:1–373.

Shackleton, D. 1973. Population quality and bighorn sheep (*Ovis canadensis canadensis* Shaw). Ph.D. thesis, University of Calgary.

———, and Hutton, D. 1971. An analysis of the mechanisms of brooming of mountain sheep horns. Z. Säugetierk. 6:342–50.

Shank, C. 1972. Some aspects of social behaviour in a population of feral goats (*Capra hircus* L.). Z. Tierpsychol. 30:488–528.

Sheldon, W. 1975. The wilderness home of the giant panda. Amherst: Univ. Mass. Press.

Shelton, M. 1960. Influence of the presence of a male goat on the initiation of estrous cycling and ovulation of Angora does. J. Anim. Sci. 19:368–75.

Shipton, E. 1936. Nanda Devi. London: Hodder and Stoughton.

———. 1938. Blank on the map. London: Hodder and Stoughton.

———. 1951. Mountains of Tartary. London: Hodder and Stoughton.

Simpson, G. 1945. Principles of classification and a classification of mammals. Bull. Am. Mus. Nat. Hist. 85:1–350.

Sinclair, A. 1974. The natural regulation of buffalo populations in East Africa. E. Afr. Wild. J. 12:169–83.

———. 1975. The resource limitation of trophic levels in tropical grassland ecosystems. J. Anim. Ecol. 44:497–520.

Smith, D. 1954. The bighorn sheep in Idaho. Dept. Fish and Game, Wildl. Bull. No. 1, Boise, Idaho. 154 pp.

Smith, F., van Toller, C., and Boyes, T. 1966. The "critical period" in the attachment of lambs and ewes. Anim. Behav. 14:120–25.

Smith, I. 1965. The influence of level of nutrition during winter and spring upon oestrous activity in the ewe. World Rev. Animal Prod. 4:95–102.

Spencer, D., and Lensink, C. 1970. The muskox of Nunivak Island, Alaska. J. Wildl. Manage. 34:1–15.

Squires, V. 1975. Ecology and behaviour of domestic sheep (*Ovis aries*): a review. Mammal Review. 5:35–57.

Stainton, J. 1972. Forests of Nepal. London: John Murray.

Stebbins, E. 1912. Stalks in the Himalaya. London: John Lane.

Stein, A. 1912. Ruins of desert Cathay. 2 vols. London: Macmillan.

Steinhauf, D. 1958. Beobachtungen zum Brunftverhalten des Steinwildes (*Capra ibex*). Säugetierk. Mitt. 7:5–10.

Sterndale, R. 1884. Natural history of the mammalia of India and Ceylon. Calcutta: Thacker, Spink.

Stock, A., and Stokes, W. 1969. A re-evaluation of Pleistocene bighorn sheep from the Great Basin and their relationship to living members of the genus *Ovis*. J. Mammal. 50:805–07.

Stockley, C. 1928. Big game shooting in the Indian Empire. London: Constable and Co.

———. 1936. Stalking in the Himalayas and Northern India. London: Herbert Jenkins.

Stroganov, S. 1969. Carnivorous mammals of Siberia. Jerusalem: Israel Program Scient. Trans.

Struhsaker, T. 1967. Behavior of elk (*Cervus canadensis*) during the rut. Z. Tierpsych. 24:80–114.

Sullivan, A., and Shaffer, M. 1975. Biogeography of the megazoo. Science. 189:13–17.

Sushkin, P. 1925. Wild sheep of the world and their distribution. J. Mammal. 6:145–57.

Swan, L. 1961. The ecology of the high Himalayas. Scient. Am. 205:68–78.

————, and Leviton, A. 1962. The herpetology of Nepal: a history, check list, and zoogeographical analysis of the herpetofauna. Proc. Calif. Acad. Sci. 32:103–47.

Taylor, C. 1966. The vascularity and possible thermoregulatory function of the horns in goats. Physiol. Zool. 39:127–39.

————. 1969. The eland and the oryx. Scient. Am. 220:89–95.

Teer, J., Thomas, J. and Walker, E. 1965. Ecology and management of white-tailed deer in the Llano Basin of Texas. Wildl. Monogr. No. 15. 62 pp.

Tener, J. 1965. Muskoxen in Canada. Can. Wildl. Serv. Ottawa: Queen's Printer.

Terborgh, J. 1974. Preservation of natural diversity: the problem of extinction prone species. Bioscience. 24:715–22.

Thenius, E., and Hofer, H. 1960. Stammesgeschichte der Säugetiere. Berlin: Springer Verlag.

Thesiger, W. 1959. Arabian sands. New York: E. P. Dutton.

Thomson, A., and Thomson, W. 1949. Lambing in relation to the diet of the pregnant ewe. Br. J. Nutr. 2:290–305.

Tilman, H. 1935. Nanda Devi and the sources of the Ganges. Him. J. 7:1–26.

————. 1949. Two mountains and a river. Cambridge: Cambridge Univ. Press.

Trinkler, E. 1930. The ice-age on the Tibetan plateau and in the adjacent regions. Geogr. J. 75:225–32.

Trivers, R., and Willard, D. 1973. Natural selection of parental ability to vary the sex ratio of offspring. Science. 179:90–92.

Troll, C. 1969. Die Klimatische und Vegetationsgeographische Gliederung des Himalaya Systems. Khumbu Himal. 1:353–88.

Trouessart, E-L. 1904–05. Catalogues mammalium. Berolini: R. Friedländer.

Tschanz, B. 1962. Über die Beziehung zwischen Muttertier und Jungen beim Moufflon (Ovis aries musimon Pall.). Experientia. 18:1–8.

Ulmer, F. 1941. On the weights of Dall's sheep. J. Mammal. 22:448–49.

————. 1966. Voices of the Felidae. In Int. Zoo Yearbook, vol. 6, ed, C. Jarvis, pp. 259–62. London: Zool. Soc. London.

Uloth, W. 1966. Die Taxonomie der recenten Wildschafe im Blickpunkt der Verkreuzung. Säugetierkundl. Mitt. 14:273–78.

Valdez, R. 1976. Fecundity of wild sheep (Ovis orientalis) in Iran. J. Mammal. 57:762–63.

Van den Brink, F. 1967. Guide des Mammifères D'Europe. Neuchâtel: Delachaux et Niestlé.

Van Zeist, W. 1969. Reflections on prehistoric environments in the Near East. In The domestication and exploitation of plants and animals, ed. P. Ucko and G. Dimbleby, pp. 35–46. London: Gerald Duckworth.

Vaurie, C. 1972. Tibet and its birds. London: Witherby.

Vereshchagin, N. 1967. The mammals of the Caucasus. Jerusalem: Israel Program Scient. Trans.

Verme, L. 1965. Reproduction studies on penned white-tailed deer. J. Wildl. Manage. 29:74–79.

von Wissmann, H. 1961. Stufen und Gürtel der Vegetation und des Klimas in Hochasien und seinen Randgebieten. Erdkunde, Arch. f. Wiss. Geogr. 15:19–44.

Vorontsov, N., Korobitsina, K., Nadler, C., Hoffman, R., Sapozhnikov, G., and Gorelov, Y. 1972. Cytogenetic differentiation and species borders in true Palearctic sheep. Zoologichesky Zhurnal. 51:1109–21.

Wadia, D. 1966. Geology of India. London: Macmillan.

Wahby, A. 1931. Vie et moeurs des *Capra aegagrus* (Pallas) des Mts. Taurus (région d'Alaya). Archiv. Zool. Italiano. 16:545–49.

Walker, E. 1968. Mammals of the world. Vol. 2. Baltimore: Johns Hopkins Univ. Press.

Wallace, H. 1913. The big game of central and western China. London: John Murray.

Walther, F. 1961. Einige Verhaltensbeobachtungen am Bergwild des Georg von Opel-Freigeheges. Jahrbuch G. v. Opel-Freigehege. 3:53–89.

———. 1968. Verhalten der Gazellen. Die Neue Brehm Bücheri No. 373. Wittenberg-Lutherstadt: A. Ziemsen Verlag.

———. 1974. Some reflections on expressive behaviour in combats and courtship of certain horned ungulates. *In* The behaviour of ungulates and its relation to management, ed. V. Geist and F. Walther, pp. 56–106. IUCN Publ. No. 24. Morges: IUCN.

Ward, A. 1923. Game animals of Kashmir and adjacent hill provinces. J. Bombay Nat. Hist. Soc. 28:334–44.

———. 1924. Game animals of Kashmir and adjacent hill provinces. J. Bombay Nat. Hist. Soc. 29:879–87.

Ward, F. 1921. Some observations on the birds and mammals of Imaw Bum. J. Bombay Nat. Hist. Soc. 27:754–58.

———. 1934. A plant hunter in Tibet. London: Jonathan Cape.

———. 1936. A sketch of the vegetation and geography of Tibet. Proc. Linn. Soc. London. 148:133–60.

Ward, M. 1966. Some geographical and medical observations in North Bhutan. Geogr. J. 132:491–505.

Wellby, M. 1898. Through unknown Tibet. London: T. Fisher Unwin.

Welles, R., and Welles, F. 1961. The bighorn of Death Valley. Fauna of the Natl. Parks of the U.S.; Fauna Series No. 6. Washington: U.S. Govt. Printing Office.

West, E. 1926. Takin shooting in the spring. J. Bombay Nat. Hist. Soc. 31:273–75.

Wheeler, M. 1968. Early India and Pakistan. London: Thames and Hudson.

White, J. 1910. Journeys in Bhutan. Geogr. J. 35:18–42.

Wickler, W. 1967. Socio-sexual signals and their intra-specific imitation among primates. *In* Primate ethology, ed. D. Morris, pp. 69–147. Chicago: Aldine Publ. Co.

Wilkinson, P., Shank, C., and Penner, D. 1976. Muskox-caribou summer range relations on Banks Island, N.W.T. J. Wildl. Manage. 40:151–62.

Williams, G., and Rudge, M. 1969. A population study of feral goats (*Capra hircus* L.) from Macauley Island, New Zealand. Proc. N.Z. Ecol. Soc. 16:17–28.

Wood, J. 1910. Travels and sport in Turkestan. London: Chapman and Hall.

Wurster, D., and Benirschke, K. 1968. Chromosome studies in the Superfamily Bovoidea. Chromosoma. 25:152–71.

Yin, U. 1967. Wild animals of Burma. Rangoon: Rangoon Gazette.

Younghusband, F. 1896. The heart of a continent. London: John Murray.

Zeuner, F. 1963. A history of domesticated animals. New York: Harper and Row.

Zuckerman, S. 1953. The breeding season of mammals in captivity. Proc. Zool. Soc. London. 122:827–950.

Zugmayer, E. 1908. Eine Reise durch Zentralasien im Jahre 1906. Berlin.

INDEX

Adams, A., 79–80, 132
Aeschbacher, A., 185, 186, 188, 237, 265, 341
Afzal, M., 340
Akasaka, T., 167, 168, 201, 204, 221, 242, 266, 290, 292, 341
Alexander, G., 118, 300
Ali, Hamid, 340
Ali, S., 14
Ali, Syed Asad, 340
Ali, Syed Babar, 339
Allen, G., 22, 40, 46, 48, 62, 64, 68, 87, 91, 201, 205
Altemus, R., 341
Altmann, D., 282
American Museum of Natural History, 341
Ammotragus. See Aoudad
Anderson, J., 84, 186, 209, 271–72
Andrews, J., 182
Antelope, 288, 289, 293, 296, 317, 318, 327
Antelope, African, 76, 206, 285, 286, 287, 288, 309
Antelope, Tragelaphini, 293
Aoudad: aggressive behavior among, 220–21, 226, 228, 238, 248, 319–21; breeding among, 121, 126; callus on front knees of, 87; courtship behavior of, 266; cranial shape of, 307; drinking habits of, 176; ecological distribution of, 76, 80; evolution of, 327, 330–31; feeding habits of, 166, 287; geographical distribution of, 46, 63, 76; grooming among, 184; herding habits of, 197–98; horns of, 25, 93–100, 306, 309; mother-young relations among, 281; pelage of, 91, 92, 93; reproductive potential of, 106; resting behavior of, 177, sexual dimor-

phism among, 85, 305; social behavior of, 310; stockiness of, 86; taxonomic affinities of, with bharal, 320–21; taxonomy of, 22, 27, 43
Ape, 143
Appleton, H., 29
Argali: aggressive behavior of, 322; body build of, 86, 87; body size of, 84; coexistence of, with other species, 288; cranial shape of, 307; ecological distribution of, 65, 74–75, 77, 79, 80; evolution of, 330–33; fecundity of, 125, 127; geographical distribution of, 60–63; horns of, 95–96, 101, 306–7; life span of, 303; mortality of, 131; pelage of, 91, 93; sexual segregation of, 203, 295; in snow, 175; tails of, 92; taxonomy of, 37, 29–41; wolf as predator of, 140
Argali, Altai, 84, 94, 96, 259, 308, 331
Argali, Severtzov's, 95
Argali, Tibetan: ecological distribution of, 75; fecundity of, 126; feeding of, with other species, 79, 80; geographical distribution of, 62–63; golden eagle as predator on, 133; horns of, 96, 97, 298, 306–7; rump patches on, 250; tail of, 92; taxonomy of, 39–40
Argali, Tien Shan, 39, 40, 60, 84, 88, 99, 91, 92, 95, 99, 126
Asad-ur-Rehman, Shazadas, 145, 340
Ashraf, M., 340
Ass, wild, 288
Autenrieth, R., 277

Backhaus, D., 238
Bailey, F., 46, 62, 183
Baldwin, J., 133, 149, 152

413